Proceedings of the 2011 Villa de Leyva Summer School

Geometric, Algebraic and Topological Methods for Quantum Field Theory

Proceedings of the 2011 Villa de Leyva Summer School

Geometric, Algebraic and Topological Methods for Quantum Field Theory

Villa de Leyva, Colombia 4 – 22 July 2011

Editors

Alexander Cardona
Universidad de los Andes, Colombia

Carolina Neira-Jiménez
Universität Regensburg, Germany

Hernán Ocampo
Universidad del Valle, Colombia

Sylvie Paycha
Universität Potsdam, Germany

Andrés F Reyes-Lega
Universidad de los Andes, Colombia

World Scientific

NEW JERSEY • LONDON • SINGAPORE • BEIJING • SHANGHAI • HONG KONG • TAIPEI • CHENNAI

Published by

World Scientific Publishing Co. Pte. Ltd.

5 Toh Tuck Link, Singapore 596224

USA office: 27 Warren Street, Suite 401-402, Hackensack, NJ 07601

UK office: 57 Shelton Street, Covent Garden, London WC2H 9HE

Library of Congress Cataloging-in-Publication Data

Villa de Leyva Summer School (7th : 2011 : Leiva, Boyacá, Colombia)

Geometric, algebraic and topological methods for quantum field theory : proceedings of the 2011 Villa de Leyva Summer School, Villa de Leyva, Colombia, 4–22 July 2011 / edited by Alexander Cardona (Universidad de los Andes, Colombia), Carolina Neira-Jiménez (Universität Regensburg, Germany), Hernán Ocampo (Universidad del Valle, Colombia), Sylvie Paycha (Universität Potsdam, Germany) & Andrés F Reyes-Lega (Universidad de los Andes, Colombia).

pages cm

Includes bibliographical references and index.

ISBN 978-981-4460-04-0 (hardcover : alk. paper)

1. Geometric quantization--Congresses. 2. Quantum field theory--Mathematics--Congresses. 3. Topology--Congresses. I. Cardona, Alexander, editor of compilation. II. Title.

QC174.17.G46G455 2011

530.14'30151--dc23

2013037184

British Library Cataloguing-in-Publication Data

A catalogue record for this book is available from the British Library.

Printed in Singapore by World Scientific Printers.

INTRODUCTION

This volume offers an introduction to some recent developments in several active topics at the interface between geometry, topology, analysis and quantum field theory:

- Spectral geometry and index theory,
- Noncommutative models and the standard model,
- Algebraic geometry and Feynman periods,
- Generalized geometries and string theory,
- Differential geometry and gravity,
- Conformal field theory and integrability.

It is based on lectures and short communications delivered during a summer school on *"Geometric, Algebraic and Topological Methods for Quantum Field Theory"* held in Villa de Leyva, Colombia, in July 2011. This school was the seventh of a series of summer schools which have taken place every other year since July 1999. The invited lectures, aimed at graduate students in physics or mathematics, start with introductory material before presenting more advanced results. Each lecture is self-contained and can be read independently of the rest.

The volume begins with the lectures on Spectral Geometry by Bruno Iochum in which, strongly motivated by physics, the author presents the fundamental concepts and results in the spectral approach to noncommutative geometry. After the analytic preliminaries necessary to define the spectral action for noncommutative spectral triples, computations are carried out in some particular cases of interest in mathematical physics, such as for Riemannian manifolds (Einstein-Hilbert action) and for the noncommutative torus.

The Atiyah-Singer index theorem and some of its generalizations are the subject of the contribution by Maxim Braverman and Leonardo Cano, based on the lectures given by the first author during the summer school. The lectures start with a review of the mathematical tools required to understand the significance of the classical Atiyah-Singer index theorem, then provides a presentation of its equivariant counterpart and goes on to a discussion of the recent developments aimed at generalizations of the index theorem for transversally elliptic operators acting on non-compact manifolds.

The third lecture, by Paolo Aluffi, is motivated by an algebro-geometric interpretation of Feynman amplitudes in terms of periods. It starts with an informal presentation of background in algebraic geometry on the Grothendieck group of varieties and on characteristic classes, both viewed as generalizations of the ordinary topological Euler characteristic. The last part of the lecture reviews some recent work on Grothendieck classes and Chern classes of graph hypersurfaces.

The fourth lecture byJorge Zanelli is an elementary introduction to Chern-Simons forms aimed at understanding their role in physics and the reason for their usefulness in gravity. The emphasis in this presentation is on the construction of the action principles and on the geometric features that make Chern-Simons forms particularly well suited for the physics of geometric systems. The gauge invariance is reviewed, putting in the forefront the local Lorentz symmetry of space-time; the abundance of local Lorentz-invariant gravitational theories is reduced by choosing the arbitrary coefficients in the Lagrangian in such a way that, in odd dimensions, the gauge symmetry becomes enlarged.

In the fifth lecture, Matilde Marcolli covers recent applications to particle physics of models inspired by noncommutative geometry. Complementing the formal mathematical aspects of the spectral action covered by Bruno Iochum's lectures, the author discusses how an action functional for gravity coupled with matter on an ordinary space-time can be seen as pure gravity on a noncommutative product geometry. Also, it is shown how the type of noncommutative geometry involved determines the physical content of the theory, i.e. the type of fermionic and bosonic fields present in the theory. The lecture ends with an analysis of physical consequences of such particle models which include the Higgs mass prediction.

The sixth lecture by A.P. Balachandran begins with a review of the motivation for noncommutative spacetimes which arises from quantum gravity. After a physically motivated introduction to Hopf algebras, the construction of covariant quantum fields on such spacetimes is discussed. This formalism is then applied to cosmic mircowave background and to the prediction of apparent non-Pauli transitions. The lecture ends with a discussion of limits on the scale of noncommutativity, derived from experimental data.

Integrability, a way to test the so-called Maldacena conjecture, is the theme of Matthias Staudacher's Lecture. It first gives a presentation of the

Yang Baxter equations presented as an algebraic integrability condition by means of the one dimensional spin chain and the Bethe Ansatz. The author then discusses the AdS/CFT duality and its relation to integrability in a $\mathcal{N} = 4$ Supersymmetric Yang-Mills Theory, after which he describes some predictions resulting from the theory.

Mariana Graña and Hagen Triendl's contribution, based on the first author's series of lectures in Villa de Leyva, provides an introduction to pure spinors, generalized complex geometry and their role in flux compactifications in the context of string theory. After a short review of some useful mathematical and physical tools in string theory and geometry, including the twisted integrability of the algebraic structure to which the pure spinor is attached, the authors give an interpretation of the manifolds arising in flux compactification in string theory from the point of view of generalized complex geometry.

The invited lectures are followed by four short communications on a wide spectrum of topics on the borderline of mathematics and physics. In their paper, Iván Contreras and Alberto Cattaneo give an overview of a novel approach to symplectic groupoids in the context of Poisson sigma models with boundary. In his contribution, Andrés Angel reviews some classical results on string topology of manifolds, and their recent extensions to the theory of orbifolds. Pedro Morales-Almazán's contribution, which discusses the Grothendieck ring class of banana and flower graphs, can be viewed as an illustration of the lectures by P. Aluffi. The last communication in this volume, by Rodrigo Vargas Le-Bert, studies the geometry underlying the non-integrable and cyclic representations of a real Lie algebra.

We hope that these self-contained and introductory but yet scientifically very challenging lectures and short communications will serve as an incentive for a broad reading audience of both mathematicians and physicists. We hope that they will motivate, as much as the school itself seems to have done, young students to pursue what might be their first acquaintance with some of the problems on the edge of mathematics and physics presented here. We further hope that the more advanced reader will find some pleasure in reading about different outlooks on related topics and seeing how the well-known mathematical tools prove to be very useful in some areas of quantum field theory.

We are indebted to various organizations for their financial support for this school. Let us first of all thank the Universidad de los Andes, which was our main source of financial support in Colombia, and the Centre International de Mathématiques Pures et Appliquées - CIMPA, for their significant contribution to the success of this school. Other organizations such as CLAF in Brazil, the Abdus Salam Internacional Centre for Theoretical Physics and the International Mathematical Union, through the CDE program, also contributed in a substantial way to the financial support needed for this event.

Special thanks to Mónica Castillo and Ana Catalina Salazar, who did a great job for the practical organization of the school, the quality of which was very much appreciated by participants and lecturers. We are also very indebted to Manuel Triana, Luz Malely Gutiérrez and Mauricio Morales for their help in various essential tasks needed for the successful development of the school.

We would also like to thank the administrative staff at the Universidad de los Andes, particularly Carl Langebaek, Vice-rector of Research, Carlos Montenegro, Dean of the Faculty of Sciences, Alf Onshuus, Head of the Mathematics Department, and Carlos Ávila, Head of the Physics Department, for their constant encouragement and support.

Without the people named here, all of whom helped in the organization in one way or another, before, during and after the school, this scientific event would not have left such vivid memories in the lecturers' and participants' minds. Last but not least, thanks to all the participants who gave us all, lecturers and editors, the impulse to prepare this volume through the enthusiasm they showed during the school, and thank you to all the contributors and referees for their participation in the realization of these proceedings.

The editors,

Alexander Cardona, Carolina Neira, Hernán Ocampo, Sylvie Paycha and Andrés Reyes-Lega.*

*On leave from University Blaise Pascal, Clermont-Ferrand.

CONTENTS

PART A
LECTURES

SPECTRAL GEOMETRY

Bruno Iochum

Aix-Marseille Université,
CNRS UMR 7332, CPT, 13288 Marseille France,
E-mail: iochum@cpt.univ-mrs.fr

The goal of these lectures is to present the few fundamentals of noncommutative geometry looking around its spectral approach. Strongly motivated by physics, in particular by relativity and quantum mechanics, Chamseddine and Connes have defined an action based on spectral considerations, the so-called spectral action.

The idea here is to review the necessary tools which are behind this spectral action to be able to compute it first in the case of Riemannian manifolds (Einstein–Hilbert action). Then, all primary objects defined for manifolds will be generalized to reach the level of noncommutative geometry via spectral triples, with the concrete analysis of the noncommutative torus which is a deformation of the ordinary one.

The basics of different ingredients will be presented and studied like, Dirac operators, heat equation asymptotics, zeta functions and then, how to get within the framework of operators on Hilbert spaces, the notion of noncommutative residue, Dixmier trace, pseudodifferential operators etc. These notions are appropriate in noncommutative geometry to tackle the case where the space is swapped with an algebra like for instance the noncommutative torus.

Keywords: Spectral geometry; spectral triples; noncommutative manifolds.

1. Motivations

Let us first expose a few motivations from physics to study noncommutative geometry which is by essence a spectral geometry. Of course, precise mathematical definitions and results will be given in other sections.

The notion of spectrum is quite important in physics, for instance in classical mechanics, the Fourier spectrum is essential to understand vibrations or the light spectrum in electromagnetism. The notion of spectral theory is also important in functional analysis, where the spectral theorem tells us that any selfadjoint operator A can be seen as an integral over its spectral measure $A = \int_{a\in \mathrm{Sp}(a)} a\, dP_a$ if $\mathrm{Sp}(A)$ is the spectrum of A. This is of course essential in the axiomatic formulation of quantum mechanics, especially in the Heisenberg picture where the tools are the observables namely are selfadjoint operators.

But this notion is also useful in geometry. In special relativity, we consider fields $\psi(\vec{x})$ for $\vec{x} \in \mathbb{R}^4$ and the electric and magnetic fields E, $B \in$ Function$(M = \mathbb{R}^4, \mathbb{R}^3)$. Einstein introduced in 1915 the gravitational field and the equation of motion of matter. But a problem appeared: what is the physical meaning of coordinates x^μ and equations of fields? Assume the general covariance of field equations. If $g_{\mu\nu}(x)$ or the tetradfield $e^I_\mu(x)$ is a solution (where I is a local inertial reference frame), then, for any diffeomorphism ϕ of M which is active or passive (i.e. change of coordinates), $e'^I_\nu(x) = \frac{\partial x^\mu}{\partial \phi(x)^\nu}\, e^I_\mu(x)$ is also a solution. As a consequence, when relativity became general, the points disappeared and only fields over fields remained, in the sense that there are no fields on a given space-time. But how to practice geometry without space, given usually by a manifold M? In this latter case, the spectral approach, namely the control of eigenvalues of the scalar (or spinorial) Laplacian yield important information on M and one can address the question whether they are sufficient: can one hear the shape of M?

There are two natural points of view on the notion of space: one is based on points (of a manifold), this is the traditional geometrical one. The other is based on algebra and this is the spectral one. So the idea is to use algebra of the dual spectral quantities.

This is of course more in the spirit of quantum mechanics but it remains to know what is a quantum geometry with bosons satisfying the Klein-Gordon equation $(\Box + m^2)\psi(\vec{x}) = s_b(\vec{x})$ and fermions satisfying $(i\partial\!\!\!/ - m)\psi(\vec{x}) = s_f(\vec{x})$ for sources s_b, s_f. Here $\partial\!\!\!/$ can be seen as a square root of $\Box$ and the Dirac operator will play a key role in noncommutative geometry.

In some sense, quantum forces and general relativity drive us to a spectral approach of physics, especially of space-time.

Noncommutative geometry, mainly pioneered by A. Connes (see[24,30]), is based on a spectral triple $(\mathcal{A}, \mathcal{H}, \mathcal{D})$ where the $*$-algebra $\mathcal{A}$ generalizes smooth functions on space-time M (or the coordinates) with pointwise product, $\mathcal{H}$ generalizes the Hilbert space of above quoted spinors ψ and $\mathcal{D}$ is a selfadjoint operator on $\mathcal{H}$ which generalizes $\partial\!\!\!/$ via a connection on a vector bundle over M. The algebra $\mathcal{A}$ also acts, via a representation of $*$-algebra, on $\mathcal{H}$.

Noncommutative geometry treats space-time as quantum physics does for the phase-space since it gives an uncertainty principle: under a certain scale, phase-space points are indistinguishable. Below the scale Λ^{-1}, a certain renormalization is necessary. Given a geometry, the notion of action plays an essential role in physics, for instance, the Einstein–Hilbert action in gravity or the Yang–Mills–Higgs action in particle physics. So here, given the data $(\mathcal{A}, \mathcal{H}, \mathcal{D})$, the appropriate notion of action was introduced by Chamseddine and Connes[10] and defined as $S(\mathcal{D}, \Lambda, f) = \mathrm{Tr}\big(f(\mathcal{D}/\Lambda)\big)$ where $\Lambda \in \mathbb{R}^+$ plays the role of a cut-off and f is a positive even function. The asymptotic series in $\Lambda \to \infty$ yields to an effective theory. For instance, this action applied to a noncommutative model of space-time $M \times F$ with a fine structure for fermions encoded in a finite geometry F gives rise from pure gravity to the standard model coupled with gravity.[11,20,30]

The purpose of these notes is mainly to compute this spectral action on a few examples such as manifolds and the noncommutative torus.

In section 2, we present standard material on pseudodifferential operators (ΨDO's) over a compact Riemannian manifold. A description of the behavior of the kernel of a ΨDO near the diagonal is given with the important example of elliptic operators. Then follows the notion of Wodzicki residue and its computation, the main point being to understand why it is a residue.

In section 3, the link with the Dixmier trace is shown. Different subspaces of compact operators are described in particular, the ideal $\mathcal{L}^{1,\infty}(\mathcal{H})$. Its definition is relevant in view of renormalization theory, where one needs to control the logarithmic divergency of the series $\sum_{n=1}^{\infty} n^{-1}$. We will see that this convergence "defect" of the Riemann zeta function (in the sense that this generates a lot of complications of convergence in physics) actually turns out to be an "asset" because it is precisely the Dixmier trace and more generally the Wodzicki residue which are the right tools to mimic this zeta function: firstly, they controls the spectral aspects of a manifold and secondly they can be generalized to any spectral triple.

In section 4, we recall the basic definition of a Dirac (or Dirac-like) operator on a compact Riemannian manifold $(M,\, g)$ endowed with a vector bundle E. An example is the (Clifford) bundle $E = \mathcal{C}\ell^*\, M$ where $\mathcal{C}\ell\, T^*_x M$ is the Clifford algebra for $x \in M$. This leads to the notion of spin structure, spin connection ∇^S and Dirac operator $D\!\!\!\!/ = -ic \circ \nabla^S$ where c is the Clifford multiplication. A special focus is put on the change of metrics g under conformal transformations.

In section 5 are presented the fundamentals of heat kernel theory, namely the Green function of the heat operator $e^{t\Delta}$, $t \in \mathbb{R}^+$. In particular, its expansion as $t \to 0^+$ in terms of coefficients of the elliptic operator

Δ, with a method to compute the coefficients of this expansion is explained, the idea being to replace the Laplacian Δ by $\mathcal{D}^2$ later on.

In section 6, a noncommutative integration theory is developed around the notion of spectral triple. This requires understanding the notion of differential (or pseudodifferential) operators in this context. Within differential calculus, the link between the one-form and the fluctuations of the given $\mathcal{D}$ is outlined.

Section 7 concerns a few actions in physics, such as the Einstein–Hilbert or Yang–Mills actions. The spectral action $\mathrm{Tr}\big(f(\mathcal{D}/\Lambda)\big)$ is justified and the link between its asymptotic expansion in Λ and the heat kernel coefficients is given via noncommutative integrals of powers of $|\mathcal{D}|$.

For each section, we suggest references since this review is by no means original.

Notations:

$\mathbb{N} = \{1, 2, \dots\}$ is the set of positive integers and $\mathbb{N}_0 = \mathbb{N} \cup \{0\}$ the set of non negative integers.

On $\mathbb{R}^d$, the volume form is $dx = dx^1 \wedge \cdots \wedge dx^d$.

$\mathbb{S}^d$ is the sphere of radius 1 in dimension d. The induced metric $d\xi = |\sum_{j=1}^{d} (-1)^{j-1} \xi_j \, d\xi_1 \wedge \cdots \wedge \widehat{d\xi_j} \wedge \cdots \wedge d\xi_d|$ restricts to the volume form on $\mathbb{S}^{d-1}$.

M is a d-dimensional manifold with metric g.

U, V are open set either in M or in $\mathbb{R}^d$.

We denote by $dvol_g$, for all positively oriented g-orthonormal basis $\{\xi_1, \cdots, \xi_d\}$ of T_xM, for $x \in M$, the unique volume element such that $dvol_g(\xi_1, \cdots, \xi_d) = 1$. Thus, in a local chart, $\sqrt{\det g_x}\, |dx| = |dvol_g|$.

When $\alpha \in \mathbb{N}^d$ is a multi-index, $\partial_x^\alpha = \partial_{x_1}^{\alpha_1} \partial_{x_2}^{\alpha_2} \cdots \partial_{x_d}^{\alpha_d}$, $\quad |\alpha| = \sum_{i=1}^{d} \alpha_i$ and $\alpha! = \alpha_1 a_2 \cdots \alpha_d$. For $\xi \in \mathbb{R}^d$, $|\xi| = \big(\sum_{k=1}^{d} |\xi_k|^2\big)^{1/2}$ is the Euclidean metric.

$\mathcal{H}$ is a separable Hilbert space and $\mathcal{B}(\mathcal{H}), \mathcal{K}(\mathcal{H}), \mathcal{L}^p(\mathcal{H})$ denote respectively the set of bounded, compact and p-Schatten-class operators, so $\mathcal{L}^1(\mathcal{H})$ are trace-class operators.

2. Wodzicki residue and kernel near the diagonal

The aim of this section is to show that the Wodzicki residue *WRes* is a trace on the set $\Psi DO(M)$ of pseudodifferential operators on a compact manifold M of dimension d.

References for this section: Classical books are.[99,102] For an orientation more in the spirit of noncommutative geometry here we follow[86,87] based on,[3,33] see also the excellent books.[49,82,83,104,105]

2.1. *A quick overview on pseudodifferential operators*

Definition 2.1. In the following, $m \in \mathbb{C}$.
A symbol $\sigma(x,\xi)$ of order m is a C^∞ function: $(x,\xi) \in U \times \mathbb{R}^d \to \mathbb{C}$ satisfying

(i) $|\partial_x^\alpha \partial_\xi^\beta \sigma(x,\xi)| \leq C_{\alpha\beta}(x)\,(1+|\xi|)^{\Re(m)-|\beta|}$, $C_{\alpha\beta}$ bounded on U.

(ii) We suppose that $\sigma(x,\xi) \simeq \sum_{j\geq 0} \sigma_{m-j}(x,\xi)$ where σ_k is homogeneous of degree k in ξ where $\simeq$ means a controlled asymptotic behavior

$$|\partial_x^\alpha \partial_\xi^\beta (\sigma - \sum_{j<N} \sigma_{m-j})(x,\xi)| \leq C_{N\alpha\beta}(x)\,|\xi|^{\Re(m)-N-|\beta|}$$

for $|\xi| \geq 1$, with $C_{N\alpha\beta}$ bounded on U. The set of symbols of order m is denoted by $S^m(U\times\mathbb{R}^d)$.

A function $a \in C^\infty(U\times U\times\mathbb{R}^d)$ is an amplitude of order m, if for any compact $K \subset U$ and any $\alpha,\beta,\gamma \in \mathbb{N}^d$ there exists a constant $C_{K\alpha\beta\gamma}$ such that

$$|\partial_x^\alpha \partial_y^\gamma \partial_\xi^\beta\, a(x,y,\xi)| \leq C_{K\alpha\beta\gamma}\,(1+|\xi|)^{\Re(m)-|\beta|}, \quad \forall\, x,y \in K, \xi \in \mathbb{R}^d.$$

The set of amplitudes is written $A^m(U)$.

For $\sigma \in S^m(U\times\mathbb{R}^d)$, we get a continuous operator $\sigma(\cdot,D) : u \in C_c^\infty(U) \to C^\infty(U)$ given by

$$\sigma(\cdot,D)(u)(x) = \sigma(x,D)(u) = \tfrac{1}{(2\pi)^d} \int_{\mathbb{R}^d} \sigma(x,\xi)\,\widehat{u}(\xi)\,e^{ix\cdot\xi}\,d\xi \qquad (1)$$

where $\widehat{\ }$ means the Fourier transform. This operator $\sigma(\cdot,D)$ will be also denoted by $Op(\sigma)$. For instance, if $\sigma(x,\xi) = \sum_\alpha a_\alpha(x)\,\xi^\alpha$, then $\sigma(x,D) = \sum_\alpha a_\alpha(x)\,D_x^\alpha$ with $D_x = -i\partial_x$. Remark that, by transposition, there is a natural extension of $\sigma(\cdot,D)$ from the set $\mathcal{D}'_c(U)$ of distributions with compact support in U to the set of distributions $\mathcal{D}'(U)$.

By definition, the leading term for $|\alpha| = m$ is the *principal symbol* and the *Schwartz kernel of* $\sigma(x,D)$ is defined by

$$k^{\sigma(x,D)}(x,y) = \tfrac{1}{(2\pi)^d} \int_{\mathbb{R}^d} \sigma(x,\xi)\,e^{i(x-y)\cdot\xi}\,d\xi = \check{\sigma}_{\xi\to y}(x,x-y)$$

where $\check{\ }$ is the Fourier inverse in variable ξ. Similarly, if the kernel of the operator $Op(a)$ associated to the amplitude a is

$$k^a(x,y) = \tfrac{1}{(2\pi)^d} \int_{\mathbb{R}^d} a(x,y,\xi)\,e^{i(x-y)\cdot\xi}\,d\xi. \qquad (2)$$

Definition 2.2. $P : C_c^\infty(U) \to C^\infty(U)$ (or $\mathcal{D}'(U)$) is said to be smoothing if its kernel is in $C^\infty(U\times U)$ and $\Psi DO^{-\infty}(U)$ will denote the set of smoothing operators.

For $m \in \mathbb{C}$, the set $\Psi DO^m(U)$ of pseudodifferential operators of order m is the set of P such that $P : C_c^\infty(U) \to C^\infty(U)$, $Pu(x) = \big(\sigma(x,D) + R\big)(u)$ where $\sigma \in S^m(U \times \mathbb{R}^d)$, $R \in \Psi DO^{-\infty}$. We call σ the symbol of P.

It is important to note that a smoothing operator is a pseudodifferential operator whose amplitude is in $A^m(U)$ for all $m \in \mathbb{R}$: by (2), $a(x,y,\xi) = e^{-i(x-y)\cdot\xi} k(x,y)\,\phi(\xi)$ where the function $\phi \in C_c^\infty(\mathbb{R}^d)$ satisfies $\int_{\mathbb{R}^d} \phi(\xi)\,d\xi = (2\pi)^d$. The main obstruction to smoothness is on the diagonal since $k^{\sigma(x,D)}$ is C^∞ outside the diagonal.

Also, one should have in mind the duality between symbols and pseudodifferential operators:

$$\begin{aligned}\sigma(x,\xi) \in S^m(U \times \mathbb{R}^d) &\longleftrightarrow k_\sigma(x,y) \in C_c^\infty(U \times U \times \mathbb{R}^d)\\ &\longleftrightarrow A = Op(\sigma) \in \Psi DO^m\end{aligned}$$

where we used the following definition $\sigma^A(x,\xi) = e^{-ix\cdot\xi} A(x \to e^{ix\cdot\xi})$. Moreover,

$$\sigma^A \simeq \sum_\alpha \tfrac{(-i)^\alpha}{\alpha!}\, \partial_\xi^\alpha \partial_y^\alpha\, k_\sigma^A(x,y,\xi)_{|y=x}$$

and

$$k_\sigma^A(x,y) = \tfrac{1}{(2\pi)^d} \int_{\mathbb{R}^d} e^{i(x-y)\cdot\xi}\, k^A(x,y,\xi)\,d\xi\,,$$

where $k^A(x,y,\xi)$ is the amplitude of $k_\sigma^A(x,y)$. Actually, $\sigma^A(x,\xi) = e^{iD_\xi D_y}\, k^A(x,y,\xi)_{|y=x}$ and $e^{iD_\xi D_y} = 1 + iD_\xi D_y - \frac{1}{2}(D_\xi D_y)^2 + \cdots$. Thus $A = Op(\sigma^A) + R$ where R is a regularizing operator on U.

There are two fundamental points concerning pseudodifferential operators: they form an algebra and this notion is stable by diffeomorphism justifying its extension to manifolds and then to bundles:

Theorem 2.1. *(i) If $P_1 \in \Psi DO^{m_1}$ and $P_2 \in \Psi DO^{m_2}$, then $P_1P_2 \in \Psi DO^{m_1+m_2}$ with symbol*

$$\sigma^{P_1P_2}(x,\xi) \simeq \sum_{\alpha\in\mathbb{N}^d} \tfrac{(-i)^\alpha}{\alpha!}\, \partial_\xi^\alpha \sigma^{P_1}(x,\xi)\, \partial_x^\alpha \sigma^{P_2}(x,\xi).$$

The principal symbol of P_1P_2 is $\sigma^{P_1P_2}_{m_1+m_2}(x,\xi) = \sigma^{P_1}_{m_1}(x,\xi)\,\sigma^{P_2}_{m_2}(x,\xi)$.

(ii) Let $P \in \Psi DO^m(U)$ and $\phi \in Diff(U,V)$ where V is another open set of $\mathbb{R}^d$. The operator $\phi_*P : f \in C^\infty(V) \to P(f\circ\phi)\circ\phi^{-1}$ satisfies $\phi_*P \in \Psi DO^m(V)$ and its symbol is

$$\sigma^{\phi_*P}(x,\xi) = \sigma_m^P(\phi^{-1}(x),(d\phi)^t\xi) + \sum_{|\alpha|>0} \tfrac{(-i)^\alpha}{\alpha!}\, \phi_\alpha(x,\xi)\, \partial_\xi^\alpha \sigma^P\big(\phi^{-1}(x),(d\phi)^t\xi\big)$$

where ϕ_α is a polynomial of degree α in ξ. Moreover, its principal symbol is

$$\sigma_m^{\phi_* P}(x,\xi) = \sigma_m^P\big(\phi^{-1}(x), (d\phi)^t \xi\big).$$

In other terms, the principal symbol is covariant by diffeomorphism: $\sigma^{\phi_* P}{}_m = \phi_* \sigma_m^P$.

While the proof of formal expressions is a direct computation, the asymptotic behavior requires some care, see.[99,102]

An interesting remark is in order: $\sigma^P(x,\xi) = e^{-ix\cdot\xi}\, P(x \to e^{ix\cdot\xi})$, thus the dilation $\xi \to t\xi$ with $t > 0$ gives $t^{-m}\, e^{-itx\cdot\xi} P\, e^{itx\cdot\xi} = t^{-m}\sigma^P(x,t\xi) \simeq t^{-m}\sum_{j\geq 0}\sigma_{m-j}^P(x,t\xi) = \sigma_m^P(x,\xi) + o(t^{-1})$. Thus, if $P \in \Psi DO^m(U)$ with $m \geq 0$,

$$\sigma_m^P(x,\xi) = \lim_{t\to\infty} t^{-m}\, e^{-ith(x)}\, P\, e^{ith(x)},$$

where $h \in C^\infty(U)$ is (almost) defined by $dh(x) = \xi$.

2.2. *Case of manifolds*

Let M be a (compact) Riemannian manifold of dimension d. Thanks to Theorem 2.1, the following makes sense:

Definition 2.3. $\Psi DO^m(M)$ is defined as the set of operators $P : C_c^\infty(M) \to C^\infty(M)$ such that

(i) the kernel $k^P \in C^\infty(M \times M)$ off the diagonal,

(ii) The map : $f \in C_c^\infty\big(\phi(U)\big) \to P(f \circ \phi) \circ \phi^{-1} \in C^\infty\big(\phi(U)\big)$ is in $\Psi DO^m\big(\phi(U)\big)$ for every coordinate chart $(U, \phi : U \to \mathbb{R}^d)$.

Of course, this can be generalized: Given a vector bundle E over M, a linear map $P : \Gamma_c^\infty(M,E) \to \Gamma^\infty(M,E)$ is in $\Psi DO^m(M,E)$ when k^P is smooth off the diagonal, and local expressions are ΨDO's with matrix-valued symbols.

The covariance formula implies that σ_m^P is independent of the chosen local chart so is globally defined on the bundle $T^*M \to M$ and σ_m^P is defined for every $P \in \Psi DO^m$ using overlapping charts and patching with partition of unity.

An important class of pseudodifferential operators are those which are invertible modulo regularizing ones:

Definition 2.4. $P \in \Psi DO^m(M,E)$ is elliptic if $\sigma_m^P(x,\xi)$ is invertible for all $0 \neq \xi \in TM_x^*$.

This means that $|\sigma^P(x,\xi)| \geq c_1(x)|\xi|^m$ for $|\xi| \geq c_2(x)$, $x \in U$ where c_1, c_2 are strictly positive continuous functions on U. This also means that there exists *a parametrix:*

Lemma 2.1. *The following are equivalent:*
(i) $Op(\sigma) \in \Psi DO^m(U)$ *is elliptic.*
(ii) There exist $\sigma' \in S^{-m}(U \times \mathbb{R}^d)$ *such that* $\sigma \circ \sigma' = 1$ *or* $\sigma' \circ \sigma = 1$.
(iii) $Op(\sigma)\, Op(\sigma') = Op(\sigma')\, Op(\sigma) = 1$ *modulo* $\Psi DO^{-\infty}(U)$.
Thus $Op(\sigma') \in \Psi DO^{-m}(U)$ *is also elliptic.*

Remark that any $P \in \Psi DO(M,E)$ can be extended to a bounded operator on $L^2(M,E)$ when $\Re(m) \leq 0$ but this needs an existing scalar product for given metrics on M and E.

Theorem 2.2. *(see*[49]*) When* $P \in \Psi DO^{-m}(M,E)$ *is elliptic with* $\Re(m) > 0$, *its spectrum is discrete when* M *is compact.*

We rephrase a previous remark (see [4, Proposition 2.1]): Let E be a vector bundle of rank r over M. If $P \in \Psi DO^{-m}(M,E)$, then for any couple of sections $s \in \Gamma^\infty(M,E)$, $t^* \in \Gamma^\infty(M,E^*)$, the operator $f \in C^\infty(M) \to \langle t^*, P(fs)\rangle \in C^\infty(M)$ is in $\Psi DO^m(M)$. This means that in a local chart (U,ϕ), these operators are $r \times r$ matrices of pseudodifferential operators of order $-m$. The total symbol is in $C^\infty(T^*U) \otimes End(E)$ with $End(E) \simeq M_r(\mathbb{C})$. The principal symbol can be globally defined: $\sigma^P_{-m}(x,\xi) : E_x \to E_x$ for $x \in M$ and $\xi \in T^*_x M$, can be seen as a smooth homomorphism homogeneous of degree $-m$ on all fibers of T^*M. We get the simple formula which could be seen as a definition of the principal symbol

$$\sigma^P_{-m}(x,\xi) = \lim_{t\to\infty} t^{-m}\left(e^{-ith} \cdot P \cdot e^{ith}\right)(x) \;\text{ for } x \in M,\, \xi \in T^*_x M \qquad (3)$$

where $h \in C^\infty(M)$ is such that $d_x h = \xi$.

2.3. *Singularities of the kernel near the diagonal*

The question to be solved is to define a homogeneous distribution which is an extension on $\mathbb{R}^d$ of a given homogeneous symbol on $\mathbb{R}^d \backslash \{0\}$. Such extension is a regularization used for instance by Epstein–Glaser in quantum field theory.

The Schwartz space on $\mathbb{R}^d$ is denoted by $\mathcal{S}$ and the space of tempered distributions by $\mathcal{S}'$.

Definition 2.5. For $f_\lambda(\xi) = f(\lambda\xi)$, $\lambda \in \mathbb{R}^*_+$, define $\tau \in \mathcal{S}' \to \tau_\lambda$ by $\langle \tau_\lambda, f\rangle = \lambda^{-d}\langle \tau, f_{\lambda^{-1}}\rangle$ for all $f \in \mathcal{S}$. A distribution $\tau \in \mathcal{S}'$ is homogeneous of order $m \in \mathbb{C}$ when $\tau_\lambda = \lambda^m\, \tau$.

Prop 2.1. Let $\sigma \in C^\infty(\mathbb{R}^d \backslash \{0\})$ be a homogeneous symbol of order $k \in \mathbb{Z}$.
(i) If $k > -d$, then σ defines a homogeneous distribution.
(ii) If $k = -d$, there is a unique obstruction to an extension of σ given by $c_\sigma = \int_{\mathbb{S}^{d-1}} \sigma(\xi)\, d\xi$, namely, one can at best extend σ in $\tau \in \mathcal{S}'$ such that $\tau_\lambda = \lambda^{-d}\big(\tau + c_\sigma \log(\lambda)\, \delta_0\big)$.

In the following result, we are interested by the behavior near the diagonal of the kernel k^P for $P \in \Psi DO$. For any $\tau \in \mathcal{S}'$, we choose the decomposition as $\tau = \phi \circ \tau + (1-\phi) \circ \tau$ where $\phi \in C_c^\infty(\mathbb{R}^d)$ and $\phi = 1$ near 0. We can look at the infrared behavior of τ near the origin and its ultraviolet behavior near infinity. Remark first that, since $\phi \circ \tau$ has a compact support, $(\phi \circ \tau)\check{} \in \mathcal{S}'$, so the regularity of $\check{\tau}$ depends only of its ultraviolet part $\big((1-\phi)\circ\tau\big)\check{}$.

Prop 2.2. Let $P \in \Psi DO^m(U)$, $m \in \mathbb{Z}$. Then, in local form near the diagonal,

$$k^P(x,y) = \sum_{-(m+d) \le j \le 0} a_j(x, x-y) - c_P(x) \log|x-y| + \mathcal{O}(1)$$

where $a_j(x,y) \in C^\infty\big(U \times U \backslash \{x\}\big)$ is homogeneous of order j in y and $c_P(x) \in C^\infty(U)$ with

$$c_P(x) = \tfrac{1}{(2\pi)^d} \int_{\mathbb{S}^{d-1}} \sigma^P_{-d}(x,\xi)\, d\xi. \tag{4}$$

We deduce readily the trace behavior of the amplitude of P:

Theorem 2.3. *Let $P \in \Psi DO^m(M,E)$, $m \in \mathbb{Z}$. Then, for any trivializing local coordinates*

$$tr\big(k^P(x,y)\big) = \sum_{j=-(m+d)}^{0} a_j(x, x-y) - c_P(x) \log|x-y| + \mathcal{O}(1),$$

where a_j is homogeneous of degree j in y, c_P is intrinsically locally defined by

$$c_P(x) = \tfrac{1}{(2\pi)^d} \int_{\mathbb{S}^{d-1}} tr\big(\sigma^P_{-d}(x,\xi)\big)\, d\xi. \tag{5}$$

Moreover, $c_P(x)|dx|$ is a 1-density over M which is functorial for the diffeomorphisms ϕ:

$$c_{\phi_* P}(x) = \phi_*\big(c_p(x)\big). \tag{6}$$

The main difficulty is of course to prove that c_P is well defined.

2.4. *Wodzicki residue*

The claim is that $\int_M c_P(x)|dx|$ is a residue. For this, we embed everything in $\mathbb{C}$. In the same spirit as in Proposition 2.1, one obtains the following

Lemma 2.2. *Every $\sigma \in C^\infty\big(\mathbb{R}^d \backslash \{0\}\big)$ which is homogeneous of degree $m \in \mathbb{C}\backslash\mathbb{Z}$ can be uniquely extended to a homogeneous distribution.*

Definition 2.6. Let U be an open set in $\mathbb{R}^d$ and Ω be a domain in $\mathbb{C}$. A map $\sigma : \Omega \to S^m(U \times \mathbb{R}^d)$ is said to be holomorphic when

the map $z \in \Omega \to \sigma(z)(x, \xi)$ is analytic for all $x \in U$, $\xi \in \mathbb{R}^d$,

the order $m(z)$ of $\sigma(z)$ is analytic on Ω,

the two bounds of Definition 2.1 (i) and (ii) of the asymptotics $\sigma(z) \simeq \sum_j \sigma_{m(z)-j}(z)$ are locally uniform in z.

This hypothesis is sufficient to get that the map $z \to \sigma_{m(z)-j}(z)$ is holomorphic from Ω to $C^\infty\big(U \times \mathbb{R}^d \backslash \{0\}\big)$ and the map $\partial_z\sigma(z)(x,\xi)$ is a classical symbol on $U \times \mathbb{R}^d$ such that $\partial_z\sigma(z)(x,\xi) \simeq \sum_{j\geq 0} \partial_z\sigma_{m(z)-j}(z)(x,\xi)$.

Definition 2.7. The map $P : \Omega \subset \mathbb{C} \to \Psi DO(U)$ is said to be holomorphic if it has the decomposition $P(z) = \sigma(z)(\cdot, D) + R(z)$ (see definition (1)) where $\sigma : \Omega \to S(U \times \mathbb{R}^d)$ and $R : \Omega \to C^\infty(U \times U)$ are holomorphic.

As a consequence, there exists a holomorphic map from Ω into $\Psi DO(M, E)$ with a holomorphic product (when M is compact).

Elliptic operators: Recall that $P \in \Psi DO^m(U)$, $m \in \mathbb{C}$, is elliptic essentially means that P is invertible modulo smoothing operators. More generally, $P \in \Psi DO^m(M, E)$ is elliptic if its local expression in each coordinate chart is elliptic.

Let $Q \in \Psi DO^m(M, E)$ with $\Re(m) > 0$. We assume that M is compact and Q is elliptic. Thus Q has a discrete spectrum. We assume $\text{Spectrum}(Q) \cap \mathbb{R}^- = \emptyset$ and the existence of a curve Γ whose interior contains the spectrum and avoid branch points of λ^z at $z = 0$:

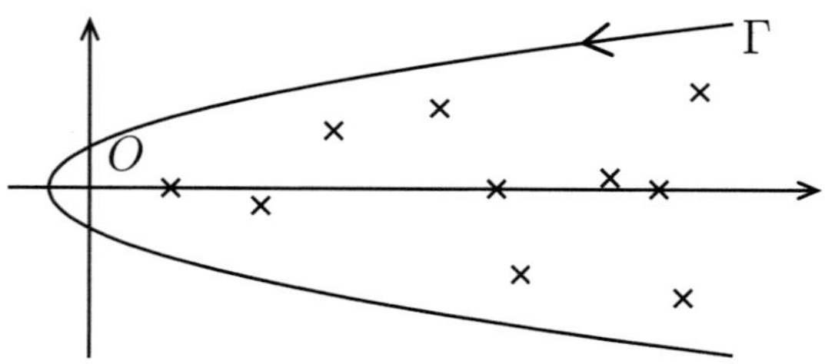

When $\Re(s) < 0$, $Q^s = \frac{1}{i2\pi}\int_\Gamma \lambda^s\,(\lambda - Q)^{-1}\,d\lambda$ makes sense as operator on $L^2(M,E)$.

The map $s \to Q^s$ is a one-parameter group containing Q^0 and Q^1 which is holomorphic on $\Re(s) \leq 0$. We want to integrate symbols, so we will need the set S_{int} of integrable symbols. Using same type of arguments as in Proposition 2.1 and Lemma 2.2, one proves

Prop 2.3. Let $L : \sigma \in S^{\mathbb{Z}}_{int}(\mathbb{R}^d) \to L(\sigma) = \check{\sigma}\,(0) = \frac{1}{(2\pi)^d}\int_{\mathbb{R}^d}\sigma(\xi)\,d\xi$. Then L has a unique holomorphic extension $\widetilde{L}$ on $S^{\mathbb{C}\backslash\mathbb{Z}}(\mathbb{R}^d)$. Moreover, when $\sigma(\xi) \simeq \sum_j \sigma_{m-j}(\xi)$, $m \in \mathbb{C}\backslash\mathbb{Z}$, $\widetilde{L}(\sigma) = \big(\sigma - \sum_{j\leq N}\tau_{m-j}\big)\check{}\,(0) = \frac{1}{(2\pi)^d}\int_{\mathbb{R}^d}\big(\sigma - \sum_{j\leq N}\tau_{m-j}\big)(\xi)\,d\xi$ where m is the order of σ, N is an integer with $N > \Re(m) + d$ and τ_{m-j} is the extension of σ_{m-j} of Lemma 2.2.

Corollary 2.1. *If $\sigma : \mathbb{C} \to S(\mathbb{R}^d)$ is holomorphic and $order\big(\sigma(s)\big) = s$, then $\widetilde{L}\big(\sigma(s)\big)$ is meromorphic with simple poles on $\mathbb{Z}$ and for $p \in \mathbb{Z}$, $\operatorname*{Res}_{s=p}\widetilde{L}\big(\sigma(s)\big) = -\frac{1}{(2\pi)^d}\int_{\mathbb{S}^{d-1}}\sigma_{-d}(p)(\xi)\,d\xi$.*

We are now ready to get the main result of this section which is due to Wodzicki.[109,110]

Definition 2.8. Let $\mathcal{D} \in \Psi DO(M,E)$ be an elliptic pseudodifferential operator of order 1 on a boundary-less compact manifold M endowed with a vector bundle E.

Let $\Psi DO_{int}(M,E) = \{\,Q \in \Psi DO^{\mathbb{C}}(M,E) \mid \Re\big(order(Q)\big) < -d\,\}$ be the class of pseudodifferential operators whose symbols are in S_{int}, i.e. integrable in the ξ-variable.

In particular, if $P \in \Psi DO_{int}(M,E)$, then its kernel $k^P(x,x)$ is a smooth density on the diagonal of $M \times M$ with values in $End(E)$.

For $P \in \Psi DO^{\mathbb{Z}}(M,E)$, define

$$WResP = \operatorname*{Res}_{s=0}\operatorname{Tr}\big(P|\mathcal{D}|^{-s}\big). \tag{7}$$

This makes sense because:

Theorem 2.4. *(i) The map $P \in \Psi DO_{int}(M,E) \to \operatorname{Tr}(P) \in \mathbb{C}$ has a unique analytic extension on $\Psi DO^{\mathbb{C}\backslash\mathbb{Z}}(M,E)$.*

(ii) If $P \in \Psi DO^{\mathbb{Z}}(M,E)$, the map: $s \in \mathbb{C} \to \operatorname{Tr}\big(P|\mathcal{D}|^{-s}\big)$ has at most simple poles on $\mathbb{Z}$ and

$$WRes\ P = -\int_M c_P(x)\,|dx| \tag{8}$$

is independent of $\mathcal{D}$. Recall (see Theorem 2.3) that $c_P(x) = \frac{1}{(2\pi)^d}\int_{\mathbb{S}^{d-1}} tr\big(\sigma^P_{-d}(x,\xi)\big)\, d\xi$.

(iii) WRes is a trace on the algebra $\Psi DO^{\mathbb{Z}}(M,E)$.

Proof. (i) The map $s \to \mathrm{Tr}\big(P|\mathcal{D}|^{-s}\big)$ is holomorphic on $\mathbb{C}$ and connect $P \in \Psi DO^{\mathbb{C}\backslash\mathbb{Z}}(M,E)$ to the set $\Psi DO_{int}(M,E)$ within $\Psi DO^{\mathbb{C}\backslash\mathbb{Z}}(M,E)$, so a analytic extension is necessarily unique.

(ii) one apply the above machinery:

(1) Notice that Tr is holomorphic on smoothing operator, so, using a partition of unity, we can reduce to a local study of scalar ΨDO's.

(2) First, fix $s = 0$. We are interested in the function $L_\phi(\sigma) = \mathrm{Tr}\big(\phi\,\sigma(x,D)\big)$ with $\sigma \in S_{int}(U \times \mathbb{R}^d)$ and $\phi \in C^\infty(U)$. For instance, if $P = \sigma(\cdot, D)$,

$$\begin{aligned}\mathrm{Tr}(\phi\, P) &= \int_U \phi(x)\, k^P(x,x)\, |dx| = \tfrac{1}{(2\pi)^d}\int_U \phi(x)\,\sigma(x,\xi)\, d\xi\, |dx| \\ &= \int_U \phi(x)\, L(\sigma(x,\cdot))\, |dx|,\end{aligned}$$

so one extends L_ϕ to $S^{\mathbb{C}\backslash\mathbb{Z}}(U \times \mathbb{R}^d)$ with Proposition 2.3 via $\widetilde{L}_\phi(\sigma) = \int_U \phi(x)\, \widetilde{L}_\phi\big(\sigma(x,\cdot)\big)\, |dx|$.

(3) If now $\sigma(x,\xi) = \sigma(s)(x,\xi)$ depends holomorphically on s, we get uniform bounds in x, thus we get, via Lemma 2.2 applied to $\widetilde{L}_\phi\big(\sigma(s)(x,\cdot)\big)$ uniformly in x, yielding a natural extension to $\widetilde{L}_\phi\big(\sigma(s)\big)$ which is holomorphic on $\mathbb{C}\backslash\mathbb{Z}$.
When $\mathrm{order}(\sigma(s)) = s$, the map $\widetilde{L}_\phi\big(\sigma(s)\big)$ has at most simple poles on $\mathbb{Z}$ and for each $p \in \mathbb{Z}$, $\underset{s=p}{\mathrm{Res}}\, \widetilde{L}_\phi\big(\sigma(s)\big) = -\frac{1}{(2\pi)^d}\int_U\int_{\mathbb{S}^{d-1}} \phi(x)\,\sigma_{-d}(p)(x,\xi)\, d\xi\, |dx| = -\int_U \phi(x)\, c_{P_p}(x)\, |dx|$ where we used (5) with $P = Op\big(\sigma_p(x,\xi)\big)$.

(4) In the general case, we get a unique meromorphic extension of the usual trace Tr on $\Psi DO^{\mathbb{Z}}(M,E)$ that we still denoted by Tr.

When $P : \mathbb{C} \to \Psi DO^{\mathbb{Z}}(M,E)$ is meromorphic with $\mathrm{order}\big(P(s)\big) = s$, then $\mathrm{Tr}\big(P(s)\big)$ has at most poles on $\mathbb{Z}$ and $\underset{s=p}{\mathrm{Res}}\, \mathrm{Tr}\big(P(s)\big) = -\int_M c_{P(p)}(x)\, |dx|$ for $p \in \mathbb{Z}$. So we get the claim for the family $P(s) = P|\mathcal{D}|^{-s}$.

(iii) Let $P_1, P_2 \in \Psi DO^{\mathbb{Z}}(M,E)$. Since Tr is a trace on $\Psi DO^{\mathbb{C}\backslash\mathbb{Z}}(M,E)$, we get by (i), $\mathrm{Tr}\big(P_1P_2|\mathcal{D}|^{-s}\big) = \mathrm{Tr}\big(P_2|\mathcal{D}|^{-s}P_1\big)$. Moreover

$$WRes(P_1P_2) = \underset{s=0}{\mathrm{Res}}\, \mathrm{Tr}\big(P_2|\mathcal{D}|^{-s}P_1\big) = \underset{s=0}{\mathrm{Res}}\, \mathrm{Tr}\big(P_2P_1|\mathcal{D}|^{-s}\big) = WRes(P_2P_1)$$

where for the second equality we used (8) so the residue depends only of the value of $P(s)$ at $s = 0$. □

Note that $WRes$ is invariant by diffeomorphism:

$$\text{if } \phi \in \text{Diff}(M),\ WRes(P) = WRes(\phi_* P) \tag{9}$$

which follows from (6). The next result is due to Guillemin and Wodzicki.

Corollary 2.2. *The Wodzicki residue WRes is the only trace (up to multiplication by a constant) on the algebra $\Psi DO^{-m}(M,E)$, $m \in \mathbb{N}$, when M is connected and $d \geq 2$.*

Proof. The restriction to $d \geq 2$ is used only in the part *3)* below. When $d = 1$, T^*M is disconnected and they are two residues.

1) On symbols, derivatives are commutators:

$$[x^j, \sigma] = i\partial_{\xi_j}\sigma, \quad [\xi_j, \sigma] = -i\partial_{x^j}\sigma.$$

2) If $\sigma_{-d}^P = 0$, then $\sigma^P(x,\xi)$ is a finite sum of commutators of symbols: If $\sigma^P \simeq \sum_j \sigma_{m-j}^P$ with $m = order(P)$, by Euler's theorem, $\sum_{k=1}^d \xi_k\,\partial_{\xi_k}\sigma_{m-j}^P = (m-j)\,\sigma_{m-j}^P$ (this is false for $m = j$!) and $\sum_{k=1}^d [x^k,\, \xi_k\,\sigma_{m-j}^P] = i\sum_{k=1}^d \partial_{\xi_k}\xi_k\,\sigma_{m-j}^P = i(m-j+d)\,\sigma_{m-j}^P$. So $\sigma^P = \sum_{k=1}^d [\xi_k\,\tau, x^k]$.

Let T be another trace on $\Psi DO^{\mathbb{Z}}(M,E)$. Then $T(P)$ depends only on σ_{-d}^P because $T([\cdot,\cdot]) = 0$.

3) We have $\int_{\mathbb{S}^{d-1}} \sigma_{-d}^P(x,\xi)\, d|\xi| = 0$ if and only if σ_{-d}^P is sum of derivatives:
The if part is direct. Only if part: σ_{-d}^P is orthogonal to constant functions on the sphere $\mathbb{S}^{d-1}$ and these are kernels of the Laplacian: $\Delta_{\mathbb{S}} f = 0 \Longleftrightarrow df = 0 \Longleftrightarrow f = cst$. Thus $\Delta_{\mathbb{S}^{d-1}} h = \sigma_{-d}^P{}_{\restriction \mathbb{S}^{d-1}}$ has a solution h on $\mathbb{S}^{d-1}$. If $\tilde h(\xi) = |\xi|^{-d+2}\, h\big(\frac{\xi}{|\xi|}\big)$ is its extension to $\mathbb{R}^d \backslash \{0\}$, then we get $\Delta_{\mathbb{R}^d}\tilde h(\xi) = |\xi|\sigma_{-d}^P\big(\frac{\xi}{|\xi|}\big) = \sigma_{-d}^P(\xi)$ because $\Delta_{\mathbb{R}^d} = r^{1-d}\,\partial_r\big(r^{d-1}\,\partial_r\big) + r^{-2}\,\Delta_{\mathbb{S}^{d-1}}$. This means that $\tilde h$ is a symbol of order $d-2$ and $\partial_\xi \tilde h$ is a symbol of order $d-1$. As a consequence,

$$\sigma_{-d}^P = \sum_{k=1}^d \partial_{\xi_k}^2 \tilde h = -i\sum_{k=1}^d [\partial_{\xi_k}\tilde h,\, x^k]$$

is a sum of commutators.

4) End of proof: the symbol $\sigma_{-d}^P(x,\xi) - \frac{|\xi|^{-d}}{\text{Vol}(\mathbb{S}^{d-1})}\, c_P(x)$ is of order $-d$ with zero integral, thus is a sum of commutators by *3)* and $T(P) = T\big(Op(|\xi|^{-d}\, c_p(x))\big)$, $\forall T \in \Psi DO^{\mathbb{Z}}(M,E)$. So the map $\mu : f \in C_c^\infty(U) \to T\big(Op(f|\xi|^{-d})\big)$ is linear, continuous and satisfies $\mu(\partial_{x^k} f) = 0$ because

$\partial_{x^k}(f)\,|\xi|^{-d}$ is a commutator if f has a compact support and U is homeomorphic to $\mathbb{R}^d$. As a consequence, μ is a multiple of the Lebesgue integral

$$T(P) = \mu\big(c_P(x)\big) = c \int_M c_P(x)\,|dx| = c\, WRes(P). \qquad \square$$

Example 2.1. Laplacian on a manifold: Let M be a compact Riemannian manifold of dimension d and Δ be the scalar Laplacian, which is a differential operator of order 2. Then

$$\mathrm{WRes}\big((1+\Delta)^{-d/2}\big) = \mathrm{Vol}\big(\mathbb{S}^{\mathrm{d}-1}\big) = \tfrac{2\pi^{\mathrm{d}/2}}{\Gamma(\mathrm{d}/2)}\,.$$

Proof. $(1+\Delta)^{-d/2} \in \Psi DO(M)$ has order $-d$ and its principal symbol $\sigma_{-d}^{(1+\Delta)^{-d/2}}$ satisfies $\sigma_{-d}^{(1+\Delta)^{-d/2}}(x,\xi) = -\big(g_x^{ij}\,\xi_i\xi_j\big)^{-d/2} = -||\xi||_x^{-d}$. So (8) gives

$$\begin{aligned} WRes\big((1+\Delta)^{-d/2}\big) &= \int_M |dx| \int_{\mathbb{S}^{d-1}} ||\xi||_x^{-d}\, d\xi = \int_M |dx| \sqrt{det\, g_x}\, \mathrm{Vol}\big(\mathbb{S}^{\mathrm{d}-1}\big) \\ &= \mathrm{Vol}\big(\mathbb{S}^{\mathrm{d}-1}\big) \int_{\mathrm{M}} |\mathrm{dvol_g}| = \mathrm{Vol}\big(\mathbb{S}^{\mathrm{d}-1}\big). \qquad \square \end{aligned}$$

3. Dixmier trace

References for this section:[33,49,68,87,104,105]

The trace Tr on the operators on a Hilbert space $\mathcal{H}$ has an interesting property, it is *normal.* Recall that Tr acting on $\mathcal{B}(\mathcal{H})$ is a particular case of a weight ω acting on a von Neumann algebra $\mathcal{M}$: it is a homogeneous additive map from $\mathcal{M}^+ = \{\, aa^* \mid a \in \mathcal{M}\,\}$ to $[0,\infty]$.
A state is a weight $\omega \in \mathcal{M}^*$ (so $\omega(a) < \infty$, $\forall a \in \mathcal{M}$) such that $\omega(1) = 1$.
A trace is a weight such that $\omega(aa^*) = \omega(a^*a)$ for all $a \in \mathcal{M}$.
A weight ω is normal if $\omega(\sup_\alpha a_\alpha) = \sup_\alpha \omega(a_\alpha)$ whenever $(a_\alpha) \subset \mathcal{M}^+$ is an increasing bounded net. This is equivalent to say that ω is lower semicontinuous with respect to the σ-weak topology.

In particular, *the usual trace* Tr *is normal on* $\mathcal{B}(\mathcal{H})$. Remark that the net $(a_\alpha)_\alpha$ converges in $\mathcal{B}(\mathcal{H})$ and this property looks innocent since a trace preserves positivity.
Nevertheless it is natural to address the question: are all traces (in particular on an arbitrary von Neumann algebra) normal? In 1966, Dixmier answered by the negative[34] by exhibiting non-normal, say singular, traces. Actually, his motivation was to answer the following related question: is any trace ω on $\mathcal{B}(\mathcal{H})$ proportional to the usual trace on the set where ω is finite?

The aim of this section is first to define this Dixmier trace, which essentially means $\text{Tr}_{Dix}(T)$ " = " $\lim_{N\to\infty} \frac{1}{\log N} \sum_{n=0}^{N} \mu_n(T)$, where the $\mu_n(T)$ are the singular values of T ordered in decreasing order and then to relate this to the Wodzicki trace. It is a non-normal trace on some set that we have to identify. Naturally, the reader can feel the link with the Wodzicki trace via Proposition 2.2. We will see that if $P \in \Psi DO^{-d}(M)$ where M is a compact Riemannian manifold of dimension d, then,

$$\text{Tr}_{Dix}(P) = \tfrac{1}{d}\, WRes(P) = \tfrac{1}{d} \int_M \int_{S^*M} \sigma_{-d}^P(x,\xi)\, d\xi |dx|$$

where S^*M is the cosphere bundle on M.

The physical motivation is quite essential: we know how $\sum_{n\in\mathbb{N}^*} \frac{1}{n}$ diverges and this is related to the fact the electromagnetic or Newton gravitational potentials are in $\frac{1}{r}$ which has the same singularity (in one-dimension as previous series). Actually, this (logarithmic-type) divergence appears everywhere in physics and explains the widely use of the Riemann zeta function $\zeta : s \in \mathbb{C} \to \sum_{n\in\mathbb{N}^*} \frac{1}{n^s}$. This is also why we have already seen a logarithmic obstruction in Theorem 2.3 and define a zeta function associated to a pseudodifferential operator P by $\zeta_P(s) = \text{Tr}\big(P|\mathcal{D}|^{-s}\big)$ in (7).

We now have a quick review on the main properties of singular values of an operator.

3.1. *Singular values of compact operators*

In noncommutative geometry, infinitesimals correspond to compact operators: for $T \in \mathcal{K}(\mathcal{H})$ (compact operators), let $\mu_n(T) = \inf\{\, \|T_{\restriction E^\perp}\| \mid E$ subspace of $\mathcal{H}$ with $\dim(E) = n\,\}$, $n \in \mathbb{N}$. This could looks strange but actually, by mini-max principle, $\mu_n(T)$ is nothing else than the $(n+1)$th of eigenvalues of $|T|$ sorted in decreasing order. Since $\lim_{n\to\infty} \mu_n(T) = 0$, for any $\epsilon > 0$, there exists a finite-dimensional subspace E_ϵ such that $\|T_{\restriction E_\epsilon^\perp}\| < \epsilon$ and this property being equivalent to T compact, T deserves the name of infinitesimal.

Moreover, we have following properties:

$\mu_n(T) = \mu_n(T^*) = \mu_n(|T|)$.
$T \in \mathcal{L}^1(\mathcal{H})$ (meaning $\|T\|_1 = \text{Tr}(|T|) < \infty$) $\Longleftrightarrow \sum_{n\in\mathbb{N}} \mu_n(T) < \infty$.
$\mu_n(A\,T\,B) \leq \|A\|\,\mu_n(T)\,\|B\|$ when $A, B \in \mathcal{B}(\mathcal{H})$.
$\mu_N(U\,T\,U^*) = \mu_N(T)$ when U is a unitary.

Definition 3.1. For $T \in \mathcal{K}(\mathcal{H})$, the partial trace of order $N \in \mathbb{N}$ is $\sigma_N(T) = \sum_{n=0}^{N} \mu_n(T)$.

Remark that $\|T\| \leq \sigma_N(T) \leq N\|T\|$ which implies $\sigma_n \simeq \|\cdot\|$ on $\mathcal{K}(\mathcal{H})$. Then

$$\sigma_N(T_1 + T_2) \leq \sigma_N(T_1) + \sigma_N(T_2),$$
$$\sigma_{N_1}(T_1) + \sigma_{N_2}(T_2) \leq \sigma_{N_1+N_2}(T_1 + T_2) \ \text{ when } T_1, T_2 \geq 0.$$

This norm σ_N splits:

$$\sigma_N(T) = \inf \{ \, \|x\|_1 + N \, \|y\| \mid T = x + y \text{ with } x \in \mathcal{L}^1(\mathcal{H}),\, y \in \mathcal{K}(\mathcal{H}) \} .$$

This justifies a continuous approach with the

Definition 3.2. The partial trace of T of order $\lambda \in \mathbb{R}^+$ is

$$\sigma_\lambda(T) = \inf \{ \, \|x\|_1 + \lambda \|y\| \mid T = x + y \text{ with } x \in \mathcal{L}^1(\mathcal{H}),\, y \in \mathcal{K}(\mathcal{H}) \} .$$

As before, $\sigma_{\lambda_1}(T_1) + \sigma_{\lambda_2}(T_2) = \sigma_{\lambda_1+\lambda_2}(T_1 + T_2)$, for $\lambda_1, \lambda_2 \in \mathbb{R}^+$, $0 \leq T_1, T_2 \in \mathcal{K}(\mathcal{H})$. We define a real interpolate space between $\mathcal{L}^1(\mathcal{H})$ and $\mathcal{K}(\mathcal{H})$ by

$$\mathcal{L}^{1,\infty} = \{ T \in \mathcal{K}(\mathcal{H}) \mid \|T\|_{1,\infty} = \sup_{\lambda \geq e} \tfrac{\sigma_\lambda(T)}{\log \lambda} < \infty \} .$$

If $\mathcal{L}^p(\mathcal{H})$ is the ideal of operators T such that $\mathrm{Tr}\big(|T|^p\big) < \infty$, so $\sigma_\lambda(T) = \mathcal{O}(\lambda^{1-1/p})$, we have

$$\mathcal{L}^1(\mathcal{H}) \subset \mathcal{L}^{1,\infty} \subset \mathcal{L}^p(\mathcal{H}) \quad \text{for } p > 1, \qquad \|T\| \leq \|T\|_{1,\infty} \leq \|T\|_1 . \tag{10}$$

Lemma 3.1. *$\mathcal{L}^{1,\infty}$ is a C^*-ideal of $\mathcal{B}(\mathcal{H})$ for the norm $\|\cdot\|_{1,\infty}$. Moreover, it is equal to the Macaev ideal*

$$\mathcal{L}^{1,+} = \{ T \in \mathcal{K}(\mathcal{H}) \mid \|T\|_{1,+} = \sup_{N \geq 2} \tfrac{\sigma_N(T)}{\log(N)} < \infty \} .$$

3.2. *Dixmier trace*

We begin with a Cesàro mean of $\frac{\sigma_\rho(T)}{\log \rho}$ with respect of the Haar measure of the group $\mathbb{R}^*_+$:

Definition 3.3. For $\lambda \geq e$ and $T \in \mathcal{K}(\mathcal{H})$, let $\tau_\lambda(T) = \frac{1}{\log \lambda} \int_e^\lambda \frac{\sigma_\rho(T)}{\log \rho} \frac{d\rho}{\rho}$.

Clearly, $\sigma_\rho(T) \leq \log \rho \, \|T\|_{1,\infty}$ and $\tau_\lambda(T) \leq \|T\|_{1,\infty}$, thus the map: $\lambda \to \tau_\lambda(T)$ is in $C_b([e, \infty])$. It is not additive on $\mathcal{L}^{1,\infty}$ but this defect is under control:

$$\tau_\lambda(T_1 + T_2) - \tau_\lambda(T_1) - \tau_\lambda(T_2) \underset{\lambda \to \infty}{\simeq} \mathcal{O}\big(\tfrac{\log (\log \lambda)}{\log \lambda}\big), \ \text{ when } 0 \leq T_1, T_2 \in \mathcal{L}^{1,\infty}.$$

Lemma 3.2. *In fact, we have*

$$|\, \tau_\lambda(T_1 + T_2) - \tau_\lambda(T_1) - \tau_\lambda(T_2) \,| \leq \big(\tfrac{\log 2(2 + \log \log \lambda)}{\log \lambda}\big) \, \|T_1 + T_2\|_{1,\infty},$$

when $T_1, T_2 \in \mathcal{L}^{1,\infty}_+$.

The Dixmier's idea was to force additivity: since the map $\lambda \to \tau_\lambda(T)$ is in $C_b([e,\infty])$ and $\lambda \to \big(\frac{\log 2(2+\log\log\lambda)}{\log\lambda}\big)$ is in $C_0([e,\infty[)$, consider the C^*-algebra $\mathcal{A} = C_b([e,\infty])/C_0([e,\infty[)$.
If $[\tau(T)]$ is the class of the map $\lambda \to \tau_\lambda(T)$ in $\mathcal{A}$, previous lemma shows that $[\tau] : T \to [\tau(T)]$ is additive and positive homogeneous from $\mathcal{L}^{1,\infty}_+$ into $\mathcal{A}$ satisfying $[\tau(UTU^*)] = [\tau(T)]$ for any unitary U.

Now let ω be a state on $\mathcal{A}$, namely a positive linear form on $\mathcal{A}$ with $\omega(1) = 1$. Then, $\omega \circ [\tau(\cdot)]$ is a tracial weight on $\mathcal{L}^{1,\infty}_+$. Since $\mathcal{L}^{1,\infty}$ is a C^*-ideal of $\mathcal{B}(\mathcal{H})$, each of its element is generated by (at most) four positive elements, and this map can be extended to a map $\omega \circ [\tau(\cdot)] : T \in \mathcal{L}^{1,\infty} \to \omega([\tau(T)]) \in \mathbb{C}$ such that $\omega([\tau(T_1T_2)]) = \omega([\tau(T_2T_1)])$ for $T_1, T_2 \in \mathcal{L}^{1,\infty}$. This leads to the following definition and result:

Definition 3.4. The Dixmier trace Tr_ω associated to a state ω on $\mathcal{A}$ is $\mathrm{Tr}_\omega(\cdot) = \omega \circ [\tau(\cdot)]$.

Theorem 3.1. Tr_ω *is a trace on* $\mathcal{L}^{1,\infty}$ *which depends only on the locally convex topology of* $\mathcal{H}$, *not of its scalar product.*

1) Note that $\mathrm{Tr}_\omega(T) = 0$ if $T \in \mathcal{L}^1(\mathcal{H})$ and all Dixmier traces vanish on the closure for the norm $\|.\|_{1,\infty}$ of the ideal of finite rank operators. So Dixmier traces are not normal.

2) The C^*-algebra $\mathcal{A}$ is not separable, so it is impossible to exhibit any state ω! Despite (10) and the fact that the $\mathcal{L}^p(\mathcal{H})$ are separable ideals for $p \geq 1$, $\mathcal{L}^{1,\infty}$ is not a separable.
Moreover, as for Lebesgue integral, there are sets which are not measurable. For instance, a function $f \in C_b([e,\infty])$ has a limit $\ell = \lim_{\lambda\to\infty} f(\lambda)$ if and only if $\ell = \omega(f)$ for all state ω.

Definition 3.5. The operator $T \in \mathcal{L}^{1,\infty}$ is said to be measurable if $\mathrm{Tr}_\omega(T)$ is independent of ω. In this case, Tr_ω is denoted Tr_{Dix}.

Lemma 3.3. *The operator* $T \in \mathcal{L}^{1,\infty}$ *is measurable and* $\mathrm{Tr}_\omega(T) = \ell$ *if and only if the map* $\lambda \in \mathbb{R}^+ \to \tau_\lambda(T) \in \mathcal{A}$ *converges at infinity to* ℓ.

After Dixmier, singular (i.e. non normal) traces have been deeply investigated, see for instance,[72,74,75] but we quote only the following characterization of measurability:

If $T \in \mathcal{K}_+(\mathcal{H})$, *then* T *is measurable if and only if* $\lim_{N\to\infty} \frac{1}{\log N} \sum_{n=0}^{N} \mu_n(T)$ *exists.*

Example 3.1. Computation of the Dixmier trace of the inverse Laplacian on the torus:

Let $\mathbb{T}^d = \mathbb{R}^d/2\pi\mathbb{Z}^d$ be the d-dimensional torus and $\Delta = -\sum_{i=1}^d \partial^2_{x^i}$ be the scalar Laplacian seen as unbounded operator on $\mathcal{H} = L^2(\mathbb{T}^d)$. We want to compute $\mathrm{Tr}_\omega\big((1+\Delta)^{-p}\big)$ for $\frac{d}{2} \leq p \in \mathbb{N}^*$. We use $1+\Delta$ to avoid the kernel problem with the inverse. As the following proof shows, 1 can be replaced by any $\epsilon > 0$ and the result does not depends on ϵ.

Notice that the functions $e_k(x) = \frac{1}{2\pi}\, e^{ik\cdot x}$ with $x \in \mathbb{T}^d$, $k \in \mathbb{Z}^d = (\mathbb{T}^d)^*$ form a basis of $\mathcal{H}$ of eigenvectors: $\Delta\, e_k = |k|^2\, e_k$. For $t \in \mathbb{R}^*_+$, $e^t\mathrm{Tr}\big(e^{-t(1+\Delta)}\big) = \sum_{k\in\mathbb{Z}^d} e^{-t|k|^2} = \big(\sum_{k\in\mathbb{Z}} e^{-tk^2}\big)^d$. We know that $|\int_{-\infty}^\infty e^{-tx^2}\, dx - \sum_{k\in\mathbb{Z}} e^{-tk^2}| \leq 1$, and since the first integral is $\sqrt{\frac{\pi}{t}}$, we get $e^t\mathrm{Tr}\big(e^{-t(1+\Delta)}\big) \underset{t\downarrow 0^+}{\simeq} \big(\frac{\pi}{t}\big)^{d/2} = \alpha\, t^{-d/2}$.

We will use a Tauberian theorem: $\mu_n\big((1+\Delta)^{-d/2})\big) \underset{n\to\infty}{\simeq} \big(\alpha\, \frac{1}{\Gamma(d/2+1)}\big)\, \frac{1}{n}$, see.[49,54] Thus

$$\lim_{N\to\infty} \tfrac{1}{\log N} \textstyle\sum_{n=0}^N \mu_n\big((1+\Delta)^{-d/2}\big) = \frac{\alpha}{\Gamma(d/2+1)} = \frac{\pi^{d/2}}{\Gamma(d/2+1)}.$$

So $(1+\Delta)^{-d/2}$ is measurable and $\mathrm{Tr}_{Dix}\big((1+\Delta)^{-d/2})\big) = \mathrm{Tr}_\omega\big((1+\Delta)^{-d/2})\big) = \frac{\pi^{d/2}}{\Gamma(d/2+1)}$. Since $(1+\Delta)^{-p}$ is traceable for $p > \frac{d}{2}$, $\mathrm{Tr}_{Dix}\big((1+\Delta)^{-p}\big) = 0$.

This result has been generalized in Connes' trace theorem:[23] since *WRes* and Tr_{Dix} are traces on $\Psi DO^{-m}(M,E)$, $m \in \mathbb{N}$, we get the following

Theorem 3.2. *Let M be a d-dimensional compact Riemannian manifold, E a vector bundle over M and $P \in \Psi DO^{-d}(M,E)$. Then, $P \in \mathcal{L}^{1,\infty}$, is measurable and* $\mathrm{Tr}_{Dix}(P) = \frac{1}{d}$ *WRes*(P).

4. Dirac operator

There are several ways to define a Dirac-like operator. The best one is to define Clifford algebras, their representations, the notion of Clifford modules, spin^c structures on orientable manifolds M defined by Morita equivalence between the C^*-algebras $C(M)$ and $\Gamma(\mathcal{C}\ell\, M)$. Then the notion of spin structure and finally, with the notion of spin and Clifford connection, we reach the definition of a (generalized) Dirac operator.

Here we try to bypass this approach to save time.

References: a classical book is,[70] but I suggest.[46] Here we follow,[86] see also.[49]

4.1. *Definition and main properties*

Let $(M,\, g)$ be a compact Riemannian manifold with metric g, of dimension d and E be a vector bundle over M. An example is the (Clifford) bundle

$E = C\ell T^*M$ where the fiber $C\ell T^*_x M$ is the Clifford algebra of the real vector space $T^*_x M$ for $x \in M$ endowed with the nondegenerate quadratic form g.

Given a connection ∇ on E, a differential operator P of order $m \in \mathbb{N}$ on E is an element of $\mathrm{Diff}^m(M,E) = \Gamma\big(M, End(E)\big) \cdot Vect\{\nabla_{X_1} \cdots \nabla_{X_j} \,|\, X_j \in \Gamma(M,TM),\, j \leq m\}$.

In particular, $\mathrm{Diff}^m(M,E)$ is a subalgebra of $End\big(\Gamma(M,E)\big)$ and the operator P has a principal symbol σ^P_m in $\Gamma\big(T^*M, \pi^* End(E)\big)$ where $\pi : T^*M \to M$ is the canonical submersion and $\sigma^P_m(x,\xi)$ is given by (3).

Example: Let $E = \bigwedge T^*M$. The exterior product and the contraction given on $\omega, \omega_j \in E$ by

$$\epsilon(\omega_1)\,\omega_2 = \omega_1 \wedge \omega_2,$$

$$\iota(\omega)\,(\omega_1 \wedge \cdots \wedge \omega_m) = \sum_{j=1}^{m} (-1)^{j-1} g(\omega, \omega_j)\, \omega_1 \wedge \cdots \wedge \widehat{\omega_j} \wedge \cdots \wedge \omega_m$$

suggest the definition $c(\omega) = \epsilon(\omega) + \iota(\omega)$, and one checks that

$$c(\omega_1)\, c(\omega_2) + c(\omega_2) c(\omega_1) = 2g(\omega_1, \omega_2)\, \mathrm{id}_E. \tag{11}$$

E has a natural scalar product: if $e_1, \cdots, e_d$ is an orthonormal basis of $T^*_x M$, then the scalar product is chosen such that $e_{i_1} \wedge \cdots \wedge e_{i_p}$ for $i_1 < \cdots < i_p$ is an orthonormal basis.

If $d \in \mathrm{Diff}^1$ is the exterior derivative and d^* is its adjoint for the deduced scalar product on $\Gamma(M,E)$, then their principal symbols are

$$\sigma^d_1(\omega) = i\epsilon(\omega), \qquad \sigma^{d^*}_1(\omega) = -i\iota(\omega). \tag{12}$$

This follows from

$$\sigma^d_1(x,\xi) = \lim_{t\to\infty} \tfrac{1}{t}\big(e^{-ith(x)} d e^{ith(x)}\big)(x) = \lim_{t\to\infty} \tfrac{1}{t}\, it\, d_x h = i\, d_x h = i\,\xi,$$

where h is such that $d_x h = \xi$, so $\sigma^d_1(x,\xi) = i\,\xi$, and similarly for $\sigma^{d^*}_1$.

More generally, if $P \in \mathrm{Diff}^m(M)$, then $\sigma^P_m(dh) = \frac{1}{i^m m!}\,(ad\,h)^m(P)$ with $ad\,h = [\cdot, h]$ and $\sigma^{P^*}_m(\omega) = \sigma^P_m(\omega)^*$ where the adjoint P^* is for the scalar product on $\Gamma(M,E)$ associated to an hermitean metric on E: $\langle \psi, \psi' \rangle = \int_M \langle \psi(x), \psi'(x) \rangle_x \,|dx|$ is a scalar product on $\Gamma(M,E)$.

Definition 4.1. The operator $P \in \mathrm{Diff}^2(M,E)$ is called a generalized Laplacian when its symbol satisfies $\sigma^P_2(x,\xi) = |\xi|^2_x \, id_{E_x}$ for $x \in M$, $\xi \in T^*_x M$.

This is equivalent to say that, in local coordinates, $P = -\sum_{i,j} g^{ij}(x)\partial_{x^i}\partial_{x^j} + b^j(x)\partial_{x^j} + c(x)$, where the b^j are smooth and c is in $\Gamma\big(M, End(E)\big)$.

Definition 4.2. Assume that $E = E^+ \oplus E^-$ is a $\mathbb{Z}_2$-graded vector bundle. When $D \in \text{Diff}^{\,1}(M,E)$ and $D = \begin{pmatrix} 0 & D^+ \\ D^- & 0 \end{pmatrix}$ (i.e. D is odd) where $D^\pm : \Gamma(M, E^\mp) \to \Gamma(M, E^\pm)$, D is called a Dirac operator if $D^2 = \begin{pmatrix} D^-D^+ & 0 \\ 0 & D^+D^- \end{pmatrix}$ is a generalized Laplacian.

A good example is given by $E = \bigwedge T^*M = \bigwedge^{even} T^*M \oplus \bigwedge^{odd} T^*M$ and the de Rham operator $D = d + d^*$. It is a Dirac operator since $D^2 = dd^* + d^*d$ is a generalized Laplacian according to (12). D^2 is also called the Laplace–Beltrami operator.

Definition 4.3. Define $\mathcal{C}\ell\, M$ as the vector bundle over M whose fiber in $x \in M$ is the Clifford algebra $\mathcal{C}\ell\, T^*_xM$ (or $\mathcal{C}\ell\, T_xM$ using the musical isomorphism $X \in TM \leftrightarrow X^\flat \in T^*M$).

A bundle E is called a Clifford bundle over M when there exists a $\mathbb{Z}_2$-graduate action $c : \Gamma(M, \mathcal{C}\ell\, M) \to End\big(\Gamma(M,E)\big)$.

The main idea which drives this definition is that Clifford actions correspond to principal symbols of Dirac operators:

Prop 4.1. If E is a Clifford module, every odd $D \in \text{Diff}^{\,1}$ such that $[D,\, f] = i\, c(df)$ for $f \in C^\infty(M)$ is a Dirac operator.
Conversely, if D is a Dirac operator, there exists a Clifford action c with $c(df) = -i\,[D,\, f]$.

Consider previous example: $E = \bigwedge T^*M = \bigwedge^{even} T^*M \oplus \bigwedge^{odd} T^*M$ is a Clifford module for $c = i(\epsilon + \iota)$ coming from the Dirac operator $D = d + d^*$: by (12)

$$i[D, f] = i[d + d^*,\, f] = i\big(i\sigma_1^d(df) - i\sigma_1^{d^*}(df)\big) = -i(\epsilon + \iota)(df).$$

Definition 4.4. Let E be a Clifford module over M. A connection ∇ on E is a Clifford connection if for $a \in \Gamma(M, \mathcal{C}\ell\, M)$ and $X \in \Gamma(M, TM)$, $[\nabla_X, c(a)] = c(\nabla^{LC}_X\, a)$ where ∇^{LC}_X is the Levi-Civita connection after its extension to the bundle $\mathcal{C}\ell\, M$ (here $\mathcal{C}\ell\, M$ is the bundle with fiber $\mathcal{C}\ell\, T_xM$).
A Dirac operator D_∇ is associated to a Clifford connection ∇:

$$D_\nabla = -i\, c \circ \nabla, \qquad \Gamma(M,E) \xrightarrow{\nabla} \Gamma(M, T^*M \otimes E) \xrightarrow{c\otimes 1} \Gamma(M,E),$$

where we use c for $c \otimes 1$.

Thus if in local coordinates, $\nabla = \sum_{j=1}^{d} dx^j \otimes \nabla_{\partial_j}$, the associated Dirac operator is given by $D_\nabla = -i\sum_j c(dx^j)\, \nabla_{\partial_i}$. In particular, for $f \in C^\infty(M)$, $[D_\nabla, f\, id_E] = -i\sum_{i=1}^{d} c(dx^i)\, [\nabla_{\partial_j}, f] = \sum_{j=1}^{d} -i\, c(dx^j)\, \partial_j f = -ic(df)$. By Proposition 4.1, D_∇ deserves the name of Dirac operator!

Examples:

1) For the previous example $E = \bigwedge T^*M$, the Levi-Civita connection is indeed a Clifford connection whose associated Dirac operator coincides with the de Rham operator $D = d + d^*$.

2) *The spinor bundle*: Recall that the spin group Spin_d is the non-trivial two-fold covering of SO_d, so we have $0 \longrightarrow \mathbb{Z}_2 \longrightarrow \mathrm{Spin}_d \xrightarrow{\xi} SO_d \longrightarrow 1$. Let $\mathrm{SO}(TM) \to M$ be the SO_d-principal bundle of positively oriented orthonormal frames on TM of an oriented Riemannian manifold M of dimension d.

A *spin structure* on an oriented d-dimensional Riemannian manifold (M, g) is a Spin_d-principal bundle $\mathrm{Spin}(TM) \xrightarrow{\pi} M$ with a two-fold covering map $\mathrm{Spin}(TM) \xrightarrow{\eta} \mathrm{SO}(TM)$ such that the following diagram commutes:

$$\begin{array}{ccccc} \mathrm{Spin}(TM) \times \mathrm{Spin}_d & \longrightarrow & \mathrm{Spin}(TM) & \xrightarrow{\pi} & \\ \Big\downarrow{\scriptstyle \eta\times\xi} & & \Big\downarrow{\scriptstyle \eta} & & M \\ \mathrm{SO}(TM) \times \mathrm{SO}_d & \longrightarrow & \mathrm{SO}(TM) & \xrightarrow{\pi} & \end{array}$$

where the horizontal maps are the actions of Spin_d and SO_d on the principal fiber bundles $\mathrm{Spin}(TM)$ and $\mathrm{SO}(TM)$.

A *spin manifold* is an oriented Riemannian manifold admitting a spin structure.
In above definition, one can replace Spin_d by the group Spin_d^c which is a central extension of SO_d by $\mathbb{T}$: $0 \longrightarrow \mathbb{T} \longrightarrow \mathrm{Spin}_d^c \xrightarrow{\xi} SO_d \longrightarrow 1$.

An oriented Riemannian manifold (M, g) is spin if and only if the second Stiefel–Whitney class of its tangent bundle vanishes. Thus a manifold is spin if and only both its first and second Stiefel–Whitney classes vanish. In this case, the set of spin structures on (M, g) stands in one-to-one correspondence with $H^1(M, \mathbb{Z}_2)$. In particular the existence of a spin structure does not depend on the metric or the orientation of a given manifold.

Let ρ be an irreducible representation of $\mathcal{C}\ell\,\mathbb{C}^d \to End_{\mathbb{C}}(\Sigma_d)$ with $\Sigma_d \simeq \mathbb{C}^{2^{\lfloor d/2 \rfloor}}$ as set of complex spinors. Of course, $\mathcal{C}\ell\,\mathbb{C}^d$ is endowed with its canonical complex bilinear form.

The *spinor bundle* S of M is the complex vector bundle associated to the principal bundle $\mathrm{Spin}(TM)$ with the spinor representation, namely $S = \mathrm{Spin}(TM) \times_{\rho_d} \Sigma_d$. Here ρ_d is a representation of Spin_d on $\mathrm{Aut}(\Sigma_d)$ which is the restriction of ρ.

More precisely, if $d = 2m$ is even, $\rho_d = \rho^+ + \rho^-$ where $\rho^\pm$ are two nonequivalent irreducible complex representations of Spin_{2m} and $\Sigma_{2m} = \Sigma_{2m}^+ \oplus \Sigma_{2m}^-$, while for $d = 2m+1$ odd, the spinor representation ρ_d is irreducible.

In practice, M is a spin manifold means that there exists a Clifford bundle $S = S^+ \oplus S^-$ such that $S \simeq \bigwedge T^*M$. Due to the dimension of M, the Clifford bundle has fiber

$$\mathcal{C}\ell_x M = \begin{cases} M_{2^m}(\mathbb{C}) \text{ when } d = 2m \text{ is even,} \\ M_{2^m}(\mathbb{C}) \oplus M_{2^m}(\mathbb{C}) \text{ when } d = 2m+1. \end{cases}$$

Locally, the spinor bundle satisfies $S \simeq M \times \mathbb{C}^{d/2}$.

A spin connection $\nabla^S : \Gamma^\infty(M,S) \to \Gamma^\infty(M,S) \otimes \Gamma^\infty(M, T^*M)$ is any connection which is compatible with Clifford action: $[\nabla^S, c(\cdot)] = c(\nabla^{LC} \cdot)$. It is uniquely determined by the choice of a spin structure on M (once an orientation of M is chosen).

Definition 4.5. The Dirac (called Atiyah–Singer) operator given by the spin structure is

$$\not{D} = -i\, c \circ \nabla^S. \tag{13}$$

In coordinates, $\not{D} = -ic(dx^j)\big(\partial_j - \omega_j(x)\big)$ where ω_j is the spin connection part which can be computed in the coordinate basis $\omega_j = \frac{1}{4}\big(\Gamma^k_{ji}\, g_{kl} - \partial_i(h^\alpha_j)\delta_{\alpha\beta} h^\beta_l\big) c(dx^i)\, c(dx^l)$ and the matrix $H = [h^\alpha_j]$ is such that $H^t H = [g_{ij}]$ (we use Latin letters for coordinate basis indices and Greek letters for orthonormal basis indices).

This gives $\sigma_1^D(x,\xi) = c(\xi) + ic(dx^j)\,\omega_j(x)$. Thus in *normal coordinates around* x_0,

$$c(dx^j)(x_0) = \gamma^j, \qquad \sigma_1^D(x_0,\xi) = c(\xi) = \gamma^j \xi_j$$

where the γ's are constant hermitean matrices.

The Hilbert space of spinors is

$$\mathcal{H} = L^2\big((M,g),S)\big) = \{\,\psi \in \Gamma^\infty(M,S) \mid \int_M \langle \psi,\psi\rangle_x \, dvol_g(x) < \infty\,\} \tag{14}$$

where we have a scalar product which is $C^\infty(M)$-valued. On its domain $\Gamma^\infty(M,S)$, $\not{D}$ is symmetric: $\langle \psi, \not{D}\phi\rangle = \langle \not{D}\psi, \phi\rangle$. Moreover, it has a selfadjoint closure (which is $\not{D}^{**}$):

Theorem 4.1. *(see[49,70,104,105]) Let (M,g) be an oriented compact Riemannian spin manifold without boundary. By extension to $\mathcal{H}$, $D\!\!\!/$ is essentially selfadjoint on its original domain $\Gamma^\infty(M,S)$. It is a differential (unbounded) operator of order one which is elliptic.*

There is a nice formula which relates the Dirac operator $D\!\!\!/$ to the spinor Laplacian

$$\Delta^S = -\mathrm{Tr}_g(\nabla^S \circ \nabla^S) : \Gamma^\infty(M,S) \to \Gamma^\infty(M,S).$$

Before giving it, we need to fix a few notations: let $R \in \Gamma^\infty\big(M, \bigwedge^2 T^*M \otimes End(TM)\big)$ be the *Riemann curvature tensor* with components $R_{ijkl} = g(\partial_i, R(\partial_k, \partial_l)\partial_j)$, the *Ricci tensor* components are $R_{jl} = g^{ik} T_{ijkl}$ and the *scalar curvature* is $s = g^{jl} R_{jl}$.

Prop 4.2. Schrödinger–Lichnerowicz formula: if s is the scalar curvature of M, $D\!\!\!/^2 = \Delta^S + \frac{1}{4}s$.

The proof is just a lengthy computation (see for instance[49]).
We already know via Theorems 2.2 and 4.1 that $D\!\!\!/^{-1}$ is compact so has a discrete spectrum.
For $T \in \mathcal{K}_+(\mathcal{H})$, we denote by $\{\lambda_n(T)\}_{n\in\mathbb{N}}$ its spectrum sorted in decreasing order including multiplicity (and in increasing order for an unbounded positive operator T such that T^{-1} is compact) and by $N_T(\lambda) = \#\{\,\lambda_n(T) \mid \lambda_n \leq \lambda\,\}$ its counting function.

Theorem 4.2. *With same hypothesis, the asymptotics of the Dirac operator counting function is $N_{|D\!\!\!/|}(\lambda) \underset{\lambda\to\infty}{\sim} \frac{2^d\, Vol(\mathbb{S}^{d-1})}{d(2\pi)^d}\, Vol(M)\, \lambda^d$ where* $\mathrm{Vol(M)} = \int_{\mathrm{M}} \mathrm{dvol}$.

We already encounter such computation in Example 3.1.

4.2. *Dirac operators and change of metrics*

Recall that the spinor bundle S_g and square integrable spinors $\mathcal{H}_g$ defined in (14) depends on the chosen metric g, so we note M_g instead of M and $\mathcal{H}_g := L^2(M_g, S_g)$ and a natural question is: what happens to a Dirac operator when the metric changes?

Let g' be another Riemannian metric on M. Since the space of d-forms is one-dimensional, there exists a positive function $f_{g,g'} : M \to \mathbb{R}^+$ such that $dvol_{g'} = f_{g',g}\, dvol_g$.

Let $I_{g,g'}(x) : S_g \to S_{g'}$ the natural injection on the spinors spaces above point $x \in M$ which is a pointwise linear isometry: $|I_{g,g'}(x)\,\psi(x)|_{g'} =$

$|\psi(x)|_g$. Let us first see its construction: there always exists a g-symmetric automorphism $H_{g,g'}$ of the $2^{\lfloor d/2 \rfloor}$- dimensional vector space TM such that $g'(X,Y) = g(H_{g,g'}X,Y)$ for $X,Y \in TM$ so define $\iota_{g,g'}\,X = H_{g,g'}^{-1/2}\,X$. Note that $\iota_{g,g'}$ commutes with right action of the orthogonal group O_d and can be lifted up to a diffeomorphism Pin_d-equivariant on the spin structures associated to g and g' and this lift is denoted by $I_{g,g'}$ (see[6]). This isometry is extended as operator on the Hilbert spaces $I_{g,g'} : \mathcal{H}_g \to \mathcal{H}_{g'}$ with $(I_{g,g'}\,\psi)(x) := I_{g,g'}(x)\,\psi(x)$.

Now define

$$U_{g',g} := \sqrt{f_{g,g'}}\,I_{g',g} : \mathcal{H}_{g'} \to \mathcal{H}_g\,.$$

Then by construction, $U_{g',g}$ is a unitary operator from $\mathcal{H}_{g'}$ onto $\mathcal{H}_g$: for $\psi' \in \mathcal{H}_{g'}$,

$$\begin{aligned}\langle U_{g',g}\psi', U_{g',g}\psi'\rangle_{\mathcal{H}_g} &= \int_M |U_{g,g'}\psi'|_g^2\,dvol_g = \int_M |I_{g',g}\psi'|_g^2\,f_{g',g}\,dvol_g \\ &= \int_M |\psi'|_{g'}^2\,dvol_{g'} = \langle \psi_{g'}, \psi_{g'}\rangle_{\mathcal{H}_{g'}}.\end{aligned}$$

So we can realize $\not{D}_{g'}$ as an operator $D_{g'}$ acting on $\mathcal{H}_g$ with

$$D_{g'} : \mathcal{H}_g \to \mathcal{H}_g, \quad D_{g'} := U_{g,g'}^{-1}\,\not{D}_{g'}\,U_{g,g'}. \tag{15}$$

This is an unbounded operator on $\mathcal{H}_g$ which has the same eigenvalues as $\not{D}_{g'}$.

In the same vein, the k-th Sobolev space $H^k(M_g,S_g)$ (which is the completion of the space $\Gamma^\infty(M_g,S_g)$ under the norm $\|\psi\|_k^2 = \sum_{j=0}^k \int_M |\nabla^j\psi(x)|^2\,dx$; be careful, ∇ applied to $\nabla\psi$ is the tensor product connection on $T^*M_g \otimes S_g$ etc, see Theorem 2.2) can be transported: the map $U_{g,g'} : H^k(M_g,S_g) \to H^k(M_{g'},S_{g'})$ is an isomorphism, see.[96] In particular, (after the transport map U), the domain of $D_{g'}$ and $\not{D}_{g'}$ are the same.

A nice example of this situation is when g' is in the conformal class of g where we can compute explicitly $\not{D}_{g'}$ and $D_{g'}$.[2,6,46,56]

Theorem 4.3. *Let $g' = e^{2h}g$ with $h \in C^\infty(M,\mathbb{R})$. Then there exists an isometry $I_{g,g'}$ between the spinor bundle S_g and $S_{g'}$ such that for $\psi \in \Gamma^\infty(M,S_g)$,*

$$\begin{aligned}&\not{D}_{g'}\,I_{g,g'}\,\psi = e^{-h}\,I_{g,g'}\,\big(\not{D}_g\,\psi - i\tfrac{d-1}{2}\,c_g(grad\ h)\,\psi\big),\\ &\not{D}_{g'} = e^{-\frac{d+1}{2}h}\,I_{g,g'}\,\not{D}_g\,I_{g,g'}^{-1}\,e^{\frac{d-1}{2}h},\\ &D_{g'} = e^{-h/2}\,\not{D}_g\,e^{-h/2}.\end{aligned}$$

Note that $D_{g'}$ is not a Dirac operator as defined in (13) since its principal symbol has an x-dependence: $\sigma^{D_{g'}}(x,\xi) = e^{-h(x)}\, c_g(\xi)$. The principal symbols of $\not{D}_{g'}$ and $\not{D}_g$ are related by

$$\sigma_d^{\not{D}_{g'}}(x,\xi) = e^{-h(x)/2}\, U_{g',g}^{-1}(x)\, \sigma_d^{\not{D}_g}(x,\xi)\, U_{g',g}(x)\, e^{-h(x)/2}, \quad \xi \in T_x^* M.$$

Thus $c_{g'}(\xi) = e^{-h(x)}\, U_{g',g}^{-1}(x)\, c_g(\xi)\, U_{g',g}(x)$, $\xi \in T_x^*M$. This formula gives a verification of $g'(\xi,\eta) = e^{-2h} g(\xi,\eta)$ using $c_g(\xi)c_g(\eta) + c_g(\eta)c_g(\xi) = 2g(\xi,\eta)\,\mathrm{id}_{S_g}$.

It is also natural to look at the changes on a Dirac operator when the metric g is modified by a diffeomorphism α which preserves the spin structure. The diffeomorphism α can be lifted to a diffeomorphism O_d-equivariant on the O_d-principal bundle of g-orthonormal frames with $\tilde{\alpha} = H_{\alpha^* g,g}^{-1/2}\, T\alpha$, and this lift also exists on S_g when α preserves both the orientation and the spin structure. However, the last lift is defined up to a $\mathbb{Z}_2$-action which disappears if α is connected to the identity.

The pull-back $g' = \alpha^* g$ of the metric g is defined by $(\alpha^* g)_x(\xi,\eta) = g_{\alpha(x)}(\alpha_*(\xi), \alpha_*\eta)$, $x \in M$, where α_* is the push-forward map : $T_xM \to T_{\alpha(x)}M$. Of course, the metric g' and g are different but the geodesic distances are the same and one checks that $d_{g'} = \alpha^* d_g$.

The principal symbol of a Dirac operator D is $\sigma_d^D(x,\xi) = c_g(\xi)$ so gives the metric g by (11). This information will be used in the definition of a spectral triple. A commutative spectral triple associated to a manifold generates the so-called Connes' distance which is nothing else but the metric distance; see the remark after (24). The link between $d_{\alpha^* g}$ and d_g is explained by (15), since the unitary induces an automorphism of the C^*-algebra $C^\infty(M)$.

5. Heat kernel expansion

References for this section:[4,43,44] and especially.[107]

The heat kernel is a Green function of the heat operator $e^{t\Delta}$ (recall that $-\Delta$ is a positive operator) which measures the temperature evolution in a domain whose boundary has a given temperature. For instance, the heat kernel of the Euclidean space $\mathbb{R}^d$ is

$$k_t(x,y) = \tfrac{1}{(4\pi t)^{d/2}} e^{-|x-y|^2/4t} \text{ for } x \neq y \tag{16}$$

and it solves the heat equation

$$\partial_t k_t(x,y) = \Delta_x k_t(x,y), \quad \forall t > 0,\, x, y \in \mathbb{R}^d,$$

with initial condition $\lim_{t\downarrow 0} k_t(x,y) = \delta(x-y)$. Actually, $k_t(x,y) = \frac{1}{2\pi}\int_{-\infty}^{\infty} e^{-ts^2}\, e^{is(x-y)}\, ds$ when $d = 1$.

Note that for $f \in \mathcal{D}(\mathbb{R}^d)$, we have $\lim_{t\downarrow 0} \int_{\mathbf{R}^d} k_t(x,y)\, f(y)\, dy = f(x)$.
For a connected domain (or manifold with boundary with vector bundle V) U, let λ_n be the eigenvalues for the Dirichlet problem of minus the Laplacian

$$-\Delta\phi = \lambda\psi \text{ in } U, \qquad \psi = 0 \text{ on } \partial U.$$

If $\psi_n \in L^2(U)$ are the normalized eigenfunctions, the inverse Dirichlet Laplacian Δ^{-1} is a selfadjoint compact operator, $0 \leq \lambda_1 \leq \lambda_2 \leq \lambda_3 \leq \cdots, \lambda_n \to \infty$.

The interest for the heat kernel is that, if $f(x) = \int_0^\infty dt\, e^{-tx}\, \phi(x)$ is the Laplace transform of ϕ, then $\mathrm{Tr}\big(f(-\Delta)\big) = \int_0^\infty dt\, \phi(t)\, \mathrm{Tr}\big(e^{t\,\Delta}\big)$ (if everything makes sense) is controlled by $\mathrm{Tr}\big(e^{t\Delta}\big) = \int_M dvol(x)\, \mathrm{tr}_{V_x} k_t(x,x)$ since $\mathrm{Tr}\big(e^{t\,\Delta}\big) = \sum_{n=1}^\infty e^{t\,\lambda_n}$ and

$$\begin{aligned} k_t(x,y) = \langle x, e^{t\,\Delta}\, y\rangle &= \sum_{n,m=1}^{\infty} \langle x, \psi_m\rangle\, \langle \psi_m, e^{t\,\Delta}\psi_n\rangle\, \langle \psi_n, y\rangle \\ &= \sum_{n,m=1}^{\infty} \overline{\psi_n(x)}\, \psi_n(y)\, e^{t\,\lambda_n}. \end{aligned}$$

So it is useful to know the asymptotics of the heat kernel k_t on the diagonal of $M \times M$ especially near $t = 0$.

5.1. *The asymptotics of heat kernel*

Let now M be a smooth compact Riemannian manifold without boundary, V be a vector bundle over M and $P \in \Psi DO^m(M,V)$ be a positive elliptic operator of order $m > 0$. If $k_t(x,y)$ is the kernel of the heat operator e^{-tP}, then the following asymptotics exits on the diagonal:

$$k_t(x,x) \underset{t\downarrow 0^+}{\sim} \sum_{k=0}^{\infty} a_k(x)\, t^{(-d+k)/m}$$

which means that $\big|k_t(x,x) - \sum_{k\leq k(n)} a_k(x)\, t^{(-d+k)/m}\big|_{\infty,n} < c_n\, t^n$ for $0 < t < 1$ where we used $|f|_{\infty,n} := \sup_{x\in M} \sum_{|\alpha|\leq n} |\partial_x^\alpha f|$ (since P is elliptic, $k_t(x,y)$ is a smooth function of (t,x,y) for $t > 0$, see [43, section 1.6, 1.7]).

More generally, we will use $k(t,f,P) = \mathrm{Tr}\big(f\, e^{-tP}\big)$ where f is a smooth function. We have similarly

$$k(t,f,P) \underset{t\downarrow 0^+}{\sim} \sum_{k=0}^{\infty} a_k(f,P)\, t^{(-d+k)/m}. \tag{17}$$

The utility of function f will appear later for the computation of coefficients a_k.

The following points are of importance:

1) The existence of this asymptotics is non-trivial.[43,44]

2) The coefficients $a_{2k}(f,P)$ can be computed locally as integral of local invariants: Recall that a locally computable quantity is the integral on the manifold of a local frame-independent smooth function of one variable, depending only on a finite number of derivatives of a finite number of terms in the asymptotic expansion of the total symbol of P.

In noncommutative geometry, local generally means that it is concentrated at infinity in momentum space.

3) The odd coefficients are zero: $a_{2k+1}(f,P) = 0$.

For instance, let us assume from now on that P is a Laplace type operator of the form

$$P = -(g^{\mu\nu}\partial_\mu\partial_\nu + \mathbb{A}^\mu\partial_\mu + \mathbb{B}) \tag{18}$$

where $(g^{\mu\nu})_{1\leq\mu,\nu\leq d}$ is the inverse matrix associated to the metric g on M, and $\mathbb{A}^\mu$ and $\mathbb{B}$ are smooth $L(V)$-sections on M (endomorphisms) (see also Definition 4.1). Then (see [44, Lemma 1.2.1]) there is a unique connection ∇ on V and a unique endomorphism E such that

$$P = -(\mathrm{Tr}_g\nabla^2 + E), \quad \nabla^2(X,Y) := [\nabla_X, \nabla_Y] - \nabla_{\nabla^{LC}_X Y}\,,$$

X, Y are vector fields on M and ∇^{LC} is the Levi-Civita connection on M. Locally

$$\mathrm{Tr}_g\nabla^2 := g^{\mu\nu}(\nabla_\mu\nabla_\nu - \Gamma^\rho_{\mu\nu}\nabla_\rho)$$

where $\Gamma^\rho_{\mu\nu}$ are the Christoffel coefficients of ∇^{LC}. Moreover (with local frames of T^*M and V), $\nabla = dx^\mu \otimes (\partial_\mu + \omega_\mu)$ and E are related to $g^{\mu\nu}$, $\mathbb{A}^\mu$ and $\mathbb{B}$ through

$$\begin{aligned} \omega_\nu &= \tfrac{1}{2} g_{\nu\mu}(\mathbb{A}^\mu + g^{\sigma\varepsilon}\Gamma^\mu_{\sigma\varepsilon}\ \mathrm{id}_V)\,, \\ E &= \mathbb{B} - g^{\nu\mu}(\partial_\nu\omega_\mu + \omega_\nu\omega_\mu - \omega_\sigma\Gamma^\sigma_{\nu\mu})\,. \end{aligned}$$

In this case, the coefficients $a_k(f,P) = \int_M dvol_g\, \mathrm{tr}_E\big(f(x)\, a_k(P)(x)\big)$ and the $a_k(P) = c_i\, \alpha^i_k(P)$ are linear combination with constants c_i of all possible independent invariants $\alpha^i_k(P)$ of dimension k constructed from E, Ω, R and their derivatives (Ω is the curvature of the connection ω, and R is the Riemann curvature tensor). As an example, for $k = 2$, E and s are the only independent invariants.

Point 3) follows since there is no odd-dimension invariant.

If s is the scalar curvature and ';' denote multiple covariant derivative with respect to Levi-Civita connection on M, one finds, using variational methods

$$
\begin{aligned}
a_0(f,P) &= (4\pi)^{-d/2} \int_M dvol_g \, \mathrm{tr}_V(f)\,, \\
a_2(f,P) &= \tfrac{(4\pi)^{-d/2}}{6} \int_M dvol_g \, \mathrm{tr}_V\big[(f(6E+s)\big], \qquad (19)\\
a_4(f,P) &= \tfrac{(4\pi)^{-d/2}}{360} \int_M dvol_g \, \mathrm{tr}_V\big[f(60E_{;kk} + 60Es + 180E^2 + 12R_{;kk} \\
&\qquad + 5s^2 - 2R_{ij}R_{ij} + 2R_{ijkl}R_{ijkl} + 30\Omega_{ij}\Omega_{ij})\big].
\end{aligned}
$$

The coefficient a_6 was computed by Gilkey, a_8 by Amsterdamski, Berkin and O'Connor and a_{10} in 1998 by van de Ven.[108] Some higher coefficients are known in flat spaces.

5.2. *Wodzicki residue and heat expansion*

Wodzicki has proved that, in (17), $a_k(P)(x) = \frac{1}{m}\, c_{P^{(k-d)/m}}(x)$ is true not only for $k=0$ as seen in Theorem 3.2 (where $P \leftrightarrow P^{-1}$), but for all $k \in \mathbb{N}$. In this section, we will prove this result when P is is the inverse of a Dirac operator and this will be generalized in the next section.

Let M be a compact Riemannian manifold of dimension d even, E a Clifford module over M and D be the Dirac operator (definition 4.2) given by a Clifford connection on E. By Theorem 4.1, D is a selfadjoint (unbounded) operator on $\mathcal{H} = L^2(M,S)$.

We are going to use the heat operator e^{-tD^2} since D^2 is related to the Laplacian via the Schrödinger–Lichnerowicz formula (4.2) and since the asymptotics of the heat kernel of this Laplacian is known.

For $t>0$, $e^{-tD^2} \in \mathcal{L}^1$: follows from $e^{-tD^2} = (1+D^2)^{(d+1)/2}\, e^{-tD^2}(1+D^2)^{-(d+1)/2}$, since $(1+D^2)^{-(d+1)/2} \in \mathcal{L}^1$ and $\lambda \to (1+\lambda^2)^{(d+1)/2} e^{-t\lambda^2}$ is bounded. So $\mathrm{Tr}\big(e^{-tD^2}\big) = \sum_n e^{-t\lambda_n^2} < \infty$.

Moreover, the operator e^{-tD^2} has a smooth kernel since it is regularizing ([70]) and the asymptotics of its kernel is, see (16): $k_t(x,y) \underset{t\downarrow 0^+}{\sim} \frac{1}{(4\pi t)^{d/2}} \sqrt{det\, g_x} \sum_{j\geq 0} k_j(x,y)\, t^j\, e^{-d_g(x,y)^2/4t}$ where k_j is a smooth section on $E^* \otimes E$. Thus

$$
\mathrm{Tr}\big(e^{-tD^2}\big) \underset{t\downarrow 0^+}{\sim} \sum_{j\geq 0} t^{(j-d)/2}\, a_j(D^2) \qquad (20)
$$

with $a_{2j}(D^2) = \frac{1}{(4\pi)^{d/2}} \int_M \mathrm{tr}\big(k_j(x,x)\big) \sqrt{det\, g_x}\, |dx|$, $\quad a_{2j+1}(D^2) = 0$, for $j \in \mathbb{N}$.

Theorem 5.1. *For any integer p, $0 \le p \le d$, $D^{-p} \in \Psi DO^{-p}(M,E)$ and*

$$\begin{aligned} WRes(D^{-p}) &= \tfrac{2}{\Gamma(p/2)} a_{d-p}(D^2) \\ &= \tfrac{2}{(4\pi)^{d/2}\Gamma(p/2)} \int_M tr\big(k_{(d-p)/2}(x,x)\big)\, dvol_g(x). \end{aligned}$$

A few remarks are in order:

1) If $p = d$, $WRes(D^{-d}) = \frac{2}{\Gamma(p/2)}\, a_0(D^2) = \frac{2}{\Gamma(p/2)} \frac{Rank(E)}{(4\pi)^{d/2}} \mathrm{Vol(M)}$. Since $\mathrm{Tr}(e^{-tD^2}) \underset{t\downarrow 0^+}{\sim} a_0(D^2)\, t^{-d/2}$, the Tauberian theorem used in Example 3.1 implies that $D^{-d} = (D^{-2})^{d/2}$ is measurable and we obtain Connes' trace Theorem 3.2

$$\mathrm{Tr}_{Dix}(D^{-d}) = \mathrm{Tr}_\omega(D^{-d}) = \tfrac{a_0(D^2)}{\Gamma(d/2+1)} = \tfrac{1}{d}\, WRes(D^{-d}).$$

2) When $D = \not{D}$ and E is the spinor bundle, the Seeley-deWit coefficient $a_2(\mathcal{D}^2)$ (see (19) with $f = 1$) can be easily computed (see[43,49]): if s is the scalar curvature,

$$a_2(\not{D}^2) = -\tfrac{1}{12(4\pi)^{d/2}} \int_M s(x)\, dvol_g(x). \qquad (21)$$

So $WRes(\not{D}^{-d+2}) = \frac{2}{\Gamma(d/2-1)}\, a_2(\not{D}^2) = c \int_M s(x)\, dvol_g(x)$. This is a quite important result since this last integral is nothing else but the Einstein–Hilbert action. In dimension 4, this is an example of invariant by diffeomorphisms, see (9).

6. Noncommutative integration

The Wodzicki residue is a trace so can be viewed as an integral. But of course, it is quite natural to relate this integral to zeta functions used in (7): with notations of Section 2.4, let $P \in \Psi DO^{\mathbb{Z}}(M,E)$ and $D \in \Psi DO^1(M,E)$ which is elliptic. The definition of zeta function $\zeta_D^P(s) = \mathrm{Tr}\big(P\,|D|^{-s}\big)$ has been useful to prove that $WRes\, P = \underset{s=0}{\mathrm{Res}}\, \zeta_D^P(s) = -\int_M c_P(x)\, |dx|$.

The aim now is to extend this notion to noncommutative spaces encoded in the notion of spectral triple. References:[24,30,33,36,49]

6.1. *Notion of spectral triple*

The main properties of a compact spin Riemannian manifold M can be recaptured using the following triple $(\mathcal{A} = C^\infty(M), \mathcal{H} = L^2(M,S), \not{D})$. The coordinates $x = (x^1, \cdots, x^d)$ are exchanged with the algebra $C^\infty(M)$, the Dirac operator $\mathcal{D}$ gives the dimension d as seen in Theorem 4.2, but also the metric of M via Connes formula and more generally generates a quantized calculus. The idea is to forget about the commutativity of the algebra and to impose axioms on a triplet $(\mathcal{A}, \mathcal{H}, \mathcal{D})$ to generalize the above one in order to be able to obtain appropriate definitions of other notions: pseudodifferential operators, measure and integration theory, KO-theory, orientability, Poincaré duality, Hochschild (co)homology etc.

Definition 6.1. A spectral triple $(\mathcal{A}, \mathcal{H}, \mathcal{D})$ is the data of an involutive (unital) algebra $\mathcal{A}$ with a faithful representation π on a Hilbert space $\mathcal{H}$ and a selfadjoint operator $\mathcal{D}$ with compact resolvent (thus with discrete spectrum) such that $[\mathcal{D}, \pi(a)]$ is bounded for any $a \in \mathcal{A}$.

We could impose the existence of a C^*-algebra A where $\mathcal{A} := \{\, a \in A \,|\, [\mathcal{D}, \pi(a)] \text{ is bounded} \,\}$ is norm dense in A so $\mathcal{A}$ is a pre-C^*-algebra stable by holomorphic calculus.

When there is no confusion, we will write a instead of $\pi(a)$.
We now give useful definitions:

Definition 6.2. Let $(\mathcal{A}, \mathcal{H}, \mathcal{D})$ be a spectral triple.

It is even if there is a grading operator χ s.t. $\chi = \chi^*$, $[\chi, \pi(a)] = 0$, $\forall a \in \mathcal{A}, \mathcal{D}\chi = -\chi\mathcal{D}$.

It is real of KO-dimension $d \in \mathbb{Z}/8$ if there is an antilinear isometry $J : \mathcal{H} \to \mathcal{H}$ such that $J\mathcal{D} = \epsilon\, \mathcal{D}J$, $J^2 = \epsilon'$, $J\chi = \epsilon''\, \chi J$ with the following table for the signs $\epsilon, \epsilon', \epsilon''$

d	0	1	2	3	4	5	6	7
ϵ	1	-1	1	1	1	-1	1	1
ϵ'	1	1	-1	-1	-1	-1	1	1
ϵ''	1		-1		1		-1	

(22)

and the following commutation rules

$$[\pi(a), \pi(b)^\circ] = 0, \qquad \big[[\mathcal{D}, \pi(a)], \pi(b)^\circ\big] = 0, \ \forall a, b \in \mathcal{A} \tag{23}$$

where $\pi(a)^\circ = J\pi(a^*)J^{-1}$ is a representation of the opposite algebra $\mathcal{A}^\circ$.

It is d-summable (or has metric dimension d) if the singular values of $\mathcal{D}$ behave like $\mu_n(\mathcal{D}^{-1}) = \mathcal{O}(n^{-1/d})$.

It is regular if $\mathcal{A}$ and $[\mathcal{D}, \mathcal{A}]$ are in the domain of δ^n for all $n \in \mathbb{N}$ where $\delta(T) = [|\mathcal{D}|, T]$.

It satisfies the finiteness condition if the space of smooth vectors $\mathcal{H}^\infty = \bigcap_k \mathcal{D}^k$ is a finitely projective left $\mathcal{A}$-module.

It satisfies the orientation condition if there is a Hochschild cycle $c \in Z_d(\mathcal{A}, \mathcal{A} \otimes \mathcal{A}^\circ)$ such that $\pi_\mathcal{D}(c) = \chi$, where $\pi_\mathcal{D}\big((a \otimes b^\circ) \otimes a_1 \otimes \cdots \otimes a_d\big) = \pi(a)\pi(b)^\circ[\mathcal{D}, \pi(a_1)] \cdots [\mathcal{D}, \pi(a_d)]$ and d is its metric dimension.

An interesting example of noncommutative space of non-zero KO-dimension is given by the finite part of the noncommutative standard model.[20,27,30]

Moreover, the reality (or charge conjugation in the commutative case) operator J is related to Tomita theory.[101]

A reconstruction of the manifold is possible, starting only with a spectral triple where the algebra is commutative (see[28] for a more precise formulation, and also[92]):

Theorem 6.1. [28] *Given a commutative spectral triple $(\mathcal{A}, \mathcal{H}, \mathcal{D})$ satisfying the above axioms, then there exists a compact spin^c manifold M such that $\mathcal{A} \simeq C^\infty(M)$ and $\mathcal{D}$ is a Dirac operator.*

The manifold is known as a set, $M = \text{Sp}(\mathcal{A}) = \text{Sp}(A)$. Notice that $\mathcal{D}$ is known only via its principal symbol, so is not unique. J encodes the nuance between spin and spin^c structures. The spectral action selects the Levi-Civita connection so *the* Dirac operator $\not{D}$.

The way, the operator $\mathcal{D}$ recaptures the original Riemannian metric g of M is via the Connes' distance: the map

$$d(\phi_1, \phi_2) = \sup \{\, |\phi_1(a) - \phi_2(a)| \mid \|[\mathcal{D}, \pi(a)]\| \leq 1, a \in \mathcal{A} \,\} \qquad (24)$$

defines a distance (eventually infinite) between two states ϕ_1, ϕ_2 on the C^*-algebra A.

The role of $\mathcal{D}$ is non only to provide a metric by (24), but its homotopy class represents the K-homology fundamental class of the noncommutative space $\mathcal{A}$.

It is known that one cannot hear the shape of a drum since the knowledge of the spectrum of a Laplacian does not determine the metric of the manifold, even if its conformal class is given.[7] But Theorem 6.1 shows that one can hear the shape of a spinorial drum (or better say, of a spectral triple) since the knowledge of the spectrum of the Dirac operator and the volume form, via its cohomological content, is sufficient to recapture the metric and spin structure. See however the more precise refinement made in.[29]

6.2. *Notion of pseudodifferential operators*

Definition 6.3. Let $(\mathcal{A}, \mathcal{H}, \mathcal{D})$ be a spectral triple.
For $t \in \mathbb{R}$ define the map $F_t : T \in \mathcal{B}(\mathcal{H}) \to e^{it|\mathcal{D}|}Te^{-it|\mathcal{D}|}$ and for $\alpha \in \mathbb{R}$
$OP^0 = \{\, T \mid t \to F_t(T) \in C^\infty\big(\mathbb{R}, \mathcal{B}(\mathcal{H})\big) \,\}$ is the set of operators or order ≤ 0,
$OP^\alpha = \{\, T \mid T|\mathcal{D}|^{-\alpha} \in OP^0 \,\}$ is the set of operators of order $\leq \alpha$.

Moreover, we set $\delta(T) = [|\mathcal{D}|, T], \quad \nabla(T) = [\mathcal{D}^2, T]$.

For instance, $C^\infty(M) = OP^0 \bigcap L^\infty(M)$ and $L^\infty(M)$ is the von Neumann algebra generated by $\mathcal{A} = C^\infty(M)$. The spaces OP^α have the expected properties:

Prop 6.1. Assume that $(\mathcal{A}, \mathcal{H}, \mathcal{D})$ is regular so $\mathcal{A} \subset OP^0 = \bigcap_{k\geq 0} \delta^k \subset \mathcal{B}(\mathcal{H})$. Then, for any $\alpha, \beta \in \mathbb{R}$,

$$OP^\alpha\, OP^\beta \subset OP^{\alpha+\beta}, \quad OP^\alpha \subset OP^\beta \text{ if } \alpha \leq \beta, \quad \delta(OP^\alpha) \subset OP^\alpha$$

and

$$\nabla(OP^\alpha) \subset OP^{\alpha+1}.$$

As an example, let us compute the order of $X = a\,|\mathcal{D}|\,[\mathcal{D}, b]\,\mathcal{D}^{-3}$: since the order of a is 0, of $|\mathcal{D}|$ is 1, of $[\mathcal{D}, b]$ is 0 and of $\mathcal{D}^{-3}$ is -3, we get $X \in OP^{-2}$.

Definition 6.4. Let $(\mathcal{A}, \mathcal{H}, \mathcal{D})$ be a spectral triple and $\mathcal{D}(\mathcal{A})$ be the polynomial algebra generated by $\mathcal{A}$, $\mathcal{A}^\circ$, $\mathcal{D}$ and $|\mathcal{D}|$. Define the set of pseudodifferential operators as

$$\Psi(\mathcal{A}) = \{\, T \mid \forall N \in \mathbb{N} \exists P \in \mathcal{D}(\mathcal{A}),\, R \in OP^{-N},\, p \in \mathbb{N} \text{ s.t. } T = P|\mathcal{D}|^{-p} + R \,\}\,.$$

The idea behind this definition is that we want to work modulo the set $OP^{-\infty}$ *of smoothing operators.* This explains the presence of the arbitrary N and R. In the commutative case where $\mathcal{D} \in \mathrm{Diff}^1(M, E)$, we get the natural inclusion $\Psi\big(C^\infty(M)\big) \subset \Psi DO(M, E)$.
The reader should be aware that Definition 6.4 is not exactly the same as in[30,33,49] since it pays attention to the reality operator J when it is present.

6.3. *Zeta-functions and dimension spectrum*

Definition 6.5. For $P \in \Psi^*(\mathcal{A})$, we define the zeta-function associated to P (and $\mathcal{D}$) by

$$\zeta_{\mathcal{D}}^P : s \in \mathbb{C} \to \mathrm{Tr}\big(P|\mathcal{D}|^{-s}\big) \tag{25}$$

which makes sense since for $\Re(s) \gg 1$, $P|\mathcal{D}|^{-s} \in \mathcal{L}^1(\mathcal{H})$.

The dimension spectrum Sd of $(\mathcal{A}, \mathcal{H}, \mathcal{D})$ is the set $\{$poles of $\zeta_{\mathcal{D}}^P(s) \mid P \in \Psi(\mathcal{A}) \cap OP^0\}$. It is said simple if it contains poles of order at most one.

The noncommutative integral of P is defined by $⨍ P = \operatorname*{Res}_{s=0} \zeta_{\mathcal{D}}^P(s)$.

In (25), we assume $\mathcal{D}$ invertible since otherwise, one can replace $\mathcal{D}$ by the invertible operator $\mathcal{D} + P$, P being the projection on $\mathcal{D}$. This change does not modify the computation of the integrals $⨍$ which follow since $⨍ X = 0$ when X is a trace-class operator.

6.4. *One-forms and fluctuations of $\mathcal{D}$One-forms and fluctuations of D*

The unitary group $\mathcal{U}(\mathcal{A})$ of $\mathcal{A}$ gives rise to the automorphism α_u : $a \in \mathcal{A} \to uau^* \in \mathcal{A}$. This defines the inner automorphisms group $Inn(\mathcal{A})$ which is a normal subgroup of the automorphisms $Aut(\mathcal{A}) = \{\, \alpha \in Aut(A) \,|\, \alpha(\mathcal{A}) \subset \mathcal{A} \,\}$. For instance, in case of a gauge theory, the algebra $\mathcal{A} = C^\infty\big(M, M_n(\mathbb{C})\big) \simeq C^\infty(M) \otimes M_n(\mathbb{C})$ is typically used. Then, $Inn(\mathcal{A})$ is locally isomorphic to $\mathcal{G} = C^\infty\big(M, PSU(n)\big)$. Since $Aut\big(C^\infty(M)\big) \simeq \text{Diff}(M)$, we get a complete parallel analogy between following two exact sequences:

$$\begin{array}{ccccccccc} 1 & \to & Inn(\mathcal{A}) & \longrightarrow & Aut(\mathcal{A}) & \longrightarrow & Aut(\mathcal{A})/Inn(\mathcal{A}) & \to & 1, \\ 1 & \to & \mathcal{G} & \longrightarrow & \mathcal{G} \rtimes \text{Diff}(M) & \longrightarrow & \text{Diff}(M) & \to & 1. \end{array}$$

Thus the internal symmetries of physics have to be replaced by the inner automorphisms.

The appropriate framework for inner fluctuations of a spectral triple $(\mathcal{A}, \mathcal{H}, \mathcal{D})$ is Morita equivalence, see.[24,49]

Definition 6.6. Let $(\mathcal{A}, \mathcal{H}, \mathcal{D})$ be a spectral triple.
One-forms are defined as $\Omega^1_{\mathcal{D}}(\mathcal{A}) = \text{span}\,\{\, adb \mid a, b \in \mathcal{A} \,\}$, $db = [\mathcal{D}, b]$ which is a $\mathcal{A}$-bimodule.

The Morita equivalence which does not change neither the algebra $\mathcal{A}$ nor the Hilbert space $\mathcal{H}$, gives a natural hermitean fluctuation of $\mathcal{D}$: $\mathcal{D} \to \mathcal{D}_A = \mathcal{D} + A$ with $A = A^* \in \Omega^1_{\mathcal{D}}(\mathcal{A})$. For instance, in commutative geometries, $\Omega^1_{\not{D}}\big(C^\infty(M)\big) = \{\, c(da) \,|\, a \in C^\infty(M) \,\}$.

When a reality operator J exists, we also want $\mathcal{D}_A J = \epsilon\, J \mathcal{D}_A$, so we choose

$$\mathcal{D}_{\widetilde{A}} = \mathcal{D} + \widetilde{A}, \quad \widetilde{A} = A + \epsilon JAJ^{-1}, \quad A = A^*. \tag{26}$$

The next two results show that, with the same algebra $\mathcal{A}$ and Hilbert space $\mathcal{H}$, a fluctuation of $\mathcal{D}$ still give rise to a spectral triple $(\mathcal{A}, \mathcal{H}, \mathcal{D}_A)$ or $(\mathcal{A}, \mathcal{H}, \mathcal{D}_{\widetilde{A}})$.

Lemma 6.1. *Let $(\mathcal{A}, \mathcal{H}, \mathcal{D})$ be a spectral triple with a reality operator J and chirality χ. If $A \in \Omega^1_{\mathcal{D}}$, the fluctuated Dirac operator $\mathcal{D}_A$ or $\mathcal{D}_{\widetilde{A}}$ is an operator with compact resolvent, and in particular its kernel is a finite dimensional space. This space is invariant by J and χ.*

Note that $\mathcal{U}(\mathcal{A})$ acts on $\mathcal{D}$ by $\mathcal{D} \to \mathcal{D}_u = u\mathcal{D}u^*$ leaving invariant the spectrum of $\mathcal{D}$. Since $\mathcal{D}_u = \mathcal{D} + u[\mathcal{D}, u^*]$ and in a C^*-algebra, any element

a is a linear combination of at most four unitaries, Definition 6.6 is quite natural.

The inner automorphisms of a spectral triple correspond to inner fluctuation of the metric defined by (24).

One checks directly that a fluctuation of a fluctuation is a fluctuation and that the unitary group $\mathcal{U}(\mathcal{A})$ is gauge compatible for the adjoint representation but to be an inner fluctuation is not a symmetric relation. It can append that $\mathcal{D}_A = 0$ with $\mathcal{D} \neq 0$.

Lemma 6.2. *Let $(\mathcal{A}, \mathcal{D}, \mathcal{H})$ be a spectral triple and $X \in \Psi(\mathcal{A})$. Then $⨍ X^* = \overline{⨍ X}$. If the spectral triple is real, then, for $X \in \Psi(\mathcal{A})$, $JXJ^{-1} \in \Psi(\mathcal{A})$ and $⨍ JXJ^{-1} = ⨍ X^* = \overline{⨍ X}$.*

If $A = A^*$ then, for $k, l \in \mathbb{N}$, the integrals $⨍ A^l \mathcal{D}^{-k}$, $⨍ \left(A\mathcal{D}^{-1}\right)^k$, $⨍ A^l\, |\mathcal{D}|^{-k}$, $⨍ \chi A^l\, |\mathcal{D}|^{-k}$ and $⨍ A^l\, \mathcal{D}\, |\mathcal{D}|^{-k}$ are real valued.

We remark that the fluctuations leave invariant the first term of the spectral action (33). This is a generalization of the fact that in the commutative case, the noncommutative integral depends only on the principal symbol of the Dirac operator $\mathcal{D}$ and this symbol is stable by adding a gauge potential like in $\mathcal{D} + A$. Note however that *the symmetrized gauge potential $A + \epsilon JAJ^{-1}$ is always zero in this commutative case for any selfadjoint one-form A.*

Theorem 6.2. *Let $(\mathcal{A}, \mathcal{H}, \mathcal{D})$ be a regular spectral triple which is simple and of dimension d. Let $A \in \Omega^1_{\mathcal{D}}(\mathcal{A})$ be a selfadjoint gauge potential. Then, $\zeta_{D_{\tilde{A}}}(0) = \zeta_D(0) + \sum_{q=1}^{d} \frac{(-1)^q}{q} ⨍ (\tilde{A}\mathcal{D}^{-1})^q$.*

The proof of this result, necessary for spectral action computation, needs few preliminaries.

Definition 6.7. For an operator T, define the one-parameter group and notation

$$\sigma_z(T) = |D|^z T |D|^{-z},\ z \in \mathbb{C}, \qquad \epsilon(T) = \nabla(T) D^{-2},$$

(recall that $\nabla(T) = [\mathcal{D}^2, T]$).

The expansion of the one-parameter group σ_z gives for $T \in OP^q$

$$\sigma_z(T) \sim \sum_{r=0}^{N} g(z,r)\, \varepsilon^r(T) \mod OP^{-N-1+q} \tag{27}$$

where $g(z,r) = \frac{1}{r!}(\frac{z}{2}) \cdots (\frac{z}{2} - (r-1)) = \binom{z/2}{r}$ with the convention $g(z,0) = 1$.

We fix a regular spectral triple $(\mathcal{A},\mathcal{H},\mathcal{D})$ of dimension d and a self-adjoint 1-form A.
Despite previous remark before Lemma 6.2, we pay attention here to the kernel of $\mathcal{D}_A$ since this operator can be non-invertible even if $\mathcal{D}$ is, so we define

$$\mathcal{D}_A = \mathcal{D} + \widetilde{A} \text{ where } \widetilde{A} = A + \varepsilon JAJ^{-1}, \qquad D_A = \mathcal{D}_A + P_A \tag{28}$$

where P_A is the projection on $\mathcal{D}_A$. Remark that $\widetilde{A} \in \mathcal{D}(\mathcal{A}) \cap OP^0$ and $\mathcal{D}_A \in \mathcal{D}(\mathcal{A}) \cap OP^1$.
We denote $V_A = P_A - P_0$ and, as the following lemma shows, V_A is a smoothing operator:

Lemma 6.3. *(i)* $\bigcap_{k\geq 1}(\mathcal{D}_A)^k \subseteq \bigcap_{k\geq 1} |D|^k$.
(ii) $\mathcal{D}_A \subseteq \bigcap_{k\geq 1} |D|^k$.
(iii) For any $\alpha,\beta \in \mathbb{R}$, $|D|^\beta P_A |D|^\alpha$ *is bounded.*
(iv) $P_A \in OP^{-\infty}$.

Let us define $X = \mathcal{D}_A^2 - \mathcal{D}^2 = \widetilde{A}\mathcal{D} + \mathcal{D}\widetilde{A} + \widetilde{A}^2$, $\quad X_V = X + V_A$, thus $X \in \mathcal{D}_1(\mathcal{A}) \cap OP^1$ and by Lemma 6.3, $X_V \sim X \mod OP^{-\infty}$. Now let $Y = \log(D_A^2) - \log(D^2)$ which makes sense since $D_A^2 = \mathcal{D}_A^2 + P_A$ is invertible for any A.
By definition of X_V, we get $Y = \log(D^2 + X_V) - \log(D^2)$.

Lemma 6.4. *Y is a pseudodifferential operator in OP^{-1} with the expansion, for any $N \in \mathbb{N}$,*

$$Y \sim \sum_{p=1}^{N} \sum_{k_1,\cdots,k_p=0}^{N-p} \frac{(-1)^{|k|_1+p+1}}{|k|_1+p} \nabla^{k_p}(X\nabla^{k_{p-1}}(\cdots X\nabla^{k_1}(X)\cdots))D^{-2(|k|_1+p)},$$

mod OP^{-N-1}. For any $N \in \mathbb{N}$ and $s \in \mathbb{C}$,

$$|D_A|^{-s} \sim |D|^{-s} + \sum_{p=1}^{N} K_p(Y,s)|D|^{-s} \mod OP^{-N-1-\Re(s)} \tag{29}$$

with $K_p(Y,s) \in OP^{-p}$. For any $p \in \mathbb{N}$ and $r_1,\cdots,r_p \in \mathbb{N}_0$, $\varepsilon^{r_1}(Y)\cdots\varepsilon^{r_p}(Y) \in \Psi(\mathcal{A})$.

Proof of Theorem 6.2. See.[12] Since the spectral triple is simple, equation (29) entails that $\zeta_{D_A}(0) - \zeta_D(0) = \mathrm{Tr}(K_1(Y,s)|D|^{-s})_{|s=0}$. Thus, with (27), we get $\zeta_{D_A}(0) - \zeta_D(0) = -\frac{1}{2}⨍ Y$.
Now the conclusion follows from $⨍ \log\big((1+S)(1+T)\big) = ⨍ \log(1+S) + ⨍ \log(1+T)$ for $S,T \in \Psi(\mathcal{A}) \cap OP^{-1}$ (since $\log(1+S) = \sum_{n=1}^{\infty} \frac{(-1)^{n+1}}{n} S^n$)

with $S = D^{-1}A$ and $T = AD^{-1}$; so $\fint \log(1 + XD^{-2}) = 2\fint \log(1 + AD^{-1})$ and $-\frac{1}{2}\fint Y = \sum_{q=1}^{d} \frac{(-1)^q}{q} \fint (\widetilde{A}D^{-1})^q$. □

Lemma 6.5. *For any $k \in \mathbb{N}_0$,*

$$\operatorname*{Res}_{s=d-k} \zeta_{D_A}(s) = \operatorname*{Res}_{s=d-k} \zeta_D(s) + \sum_{p=1}^{k} \sum_{r_1,\cdots,r_p=0}^{k-p} \operatorname*{Res}_{s=d-k} h(s,r,p)\, \mathrm{Tr}\big(\varepsilon^{r_1}(Y)\cdots\varepsilon^{r_p}(Y)|D|^{-s}\big),$$

where $h(s,r,p) = (-s/2)^p \int_{0\le t_1\le\cdots\le t_p\le 1} g(-st_1,r_1)\cdots g(-st_p,r_p)\, dt$.

Our operators $|D_A|^k$ are pseudodifferential operators: for any $k \in \mathbb{Z}$, $|D_A|^k \in \Psi^k(\mathcal{A})$.

The following result is quite important since it shows that one can use $\fint$ for D or D_A:

Prop 6.2. If the spectral triple is simple, $\operatorname*{Res}_{s=0} \mathrm{Tr}\big(P|D_A|^{-s}\big) = \fint P$ for any pseudodifferential operator P. In particular, for any $k \in \mathbb{N}_0$ $\fint |D_A|^{-(d-k)} = \operatorname*{Res}_{s=d-k} \zeta_{D_A}(s)$. Moreover,

$$\begin{aligned}
&\fint |D_A|^{-d} = \fint |D|^{-d}, &(30)\\
&\fint |D_A|^{-(d-1)} = \fint |D|^{-(d-1)} - (\tfrac{d-1}{2})\fint X|D|^{-d-1},\\
&\fint |D_A|^{-(d-2)} = \fint |D|^{-(d-2)} + \tfrac{d-2}{2}\big(-\fint X|D|^{-d} + \tfrac{d}{4}\fint X^2|D|^{-2-d}\big).
\end{aligned}$$

6.5. *Tadpole*

In,[30] the following definition is introduced inspired by the quantum field theory.

Definition 6.8. In $(\mathcal{A},\, \mathcal{H},\, \mathcal{D})$, the tadpole $\mathrm{Tad}_{\mathcal{D}+A}(k)$ of order k, for $k \in \{\, d-l \,:\, l \in \mathbb{N}\,\}$ is the term linear in $A = A^* \in \Omega^1_{\mathcal{D}}$, in the Λ^k term of (33) where $\mathcal{D} \to \mathcal{D} + A$.

If moreover, the triple $(\mathcal{A},\, \mathcal{H},\, \mathcal{D},\, J)$ is real, the tadpole $\mathrm{Tad}_{\mathcal{D}+\tilde{A}}(k)$ is the term linear in A, in the Λ^k term of (33) where $\mathcal{D} \to \mathcal{D} + \widetilde{A}$.

Prop 6.3. Let $(\mathcal{A}, \mathcal{H}, \mathcal{D})$ be a spectral triple of dimension d with simple dimension spectrum. Then

$$\mathrm{Tad}_{\mathcal{D}+A}(d-k) = -(d-k) \int\!\!\!\!\!\!- A\mathcal{D}|\mathcal{D}|^{-(d-k)-2}, \quad \forall k \neq d,$$

$$\mathrm{Tad}_{\mathcal{D}+A}(0) = -\int\!\!\!\!\!\!- A\mathcal{D}^{-1}.$$

Moreover, if the triple is real, $\mathrm{Tad}_{\mathcal{D}+\widetilde{A}} = 2\mathrm{Tad}_{\mathcal{D}+A}$.

Corollary 6.1. *In a real spectral triple* $(\mathcal{A}, \mathcal{H}, \mathcal{D})$, *if* $A = A^* \in \Omega^1_{\mathcal{D}}(\mathcal{A})$ *is such that* $\widetilde{A} = 0$, *then* $\mathrm{Tad}_{D+A}(k) = 0$ *for any* $k \in \mathbb{Z}$, $k \leq d$.

The vanishing tadpole of order 0 has the following equivalence (see[12])
$\int\!\!\!\!- A\mathcal{D}^{-1} = 0, \forall A \in \Omega^1_{\mathcal{D}}(\mathcal{A}) \iff \int\!\!\!\!- ab = \int\!\!\!\!- a\alpha(b), \forall a, b \in \mathcal{A}$,
where $\alpha(b) := \mathcal{D}b\mathcal{D}^{-1}$.
The existence of tadpoles is important since, for instance, $A = 0$ is not necessarily a stable solution of the classical field equation deduced from spectral action expansion.[50]

6.6. *Commutative geometry*

Definition 6.9. Consider a commutative spectral triple given by a compact Riemannian spin manifold M of dimension d without boundary and its Dirac operator $\not{D}$ associated to the Levi–Civita connection. This means $\big(\mathcal{A} := C^\infty(M),\, \mathcal{H} := L^2(M, S),\, \not{D}\big)$ where S is the spinor bundle over M. This triple is real since, due to the existence of a spin structure, the charge conjugation operator generates an anti-linear isometry J on $\mathcal{H}$ such that $JaJ^{-1} = a^*, \quad \forall a \in \mathcal{A}$, and when d is even, the grading is given by the chirality matrix $\chi := (-i)^{d/2}\gamma^1\gamma^2\cdots\gamma^d$. Such triple is said to be a commutative geometry.

In the polynomial algebra $\mathcal{D}(\mathcal{A})$ of Definition 6.4, we added $\mathcal{A}^\circ$. In the commutative case, $\mathcal{A}^\circ \simeq J\mathcal{A}J^{-1} \simeq \mathcal{A}$ which also gives $JAJ^{-1} = -\epsilon\, A^*, \quad \forall A \in \Omega^1_{\mathcal{D}}(\mathcal{A})$ or $\widetilde{A} = 0$ when $A = A^*$.

As noticed by Wodzicki, $\int\!\!\!\!- P$ is equal to -2 times the coefficient in $\log t$ of the asymptotics of $\mathrm{Tr}(P\, e^{-t\not{D}^2})$ as $t \to 0$. It is remarkable that this coefficient is independent of $\not{D}$ as seen in Theorem 2.4 and this gives a close relation between the ζ function and heat kernel expansion with *WRes*. Actually, by [47, Theorem 2.7]

$\mathrm{Tr}(P\, e^{-t\not{D}^2}) \underset{t\downarrow 0+}{\sim} \sum_{k=0}^\infty a_k\, t^{(k-ord(P)-d)/2} + \sum_{k=0}^\infty(-a'_k \log t + b_k)\, t^k,$

so $\fint P = 2a'_0$. Since $\fint$, $WRes$ are traces on $\Psi\big(C^\infty(M)\big)$, Corollary 2.2 gives $\fint P = c\, WRes\, P$. Because $\mathrm{Tr}(P\, D\!\!\!/^{-2s}) = \frac{1}{\Gamma(s)}\int_0^\infty t^{s-1}\,\mathrm{Tr}(P\, e^{-t\, D\!\!\!/^2})\,dt$, the non-zero coefficient a'_k, $k \neq 0$ creates a pole of $\mathrm{Tr}(P\, D\!\!\!/^{-2s})$ of order $k+2$ as we get $\int_0^1 t^{s-1}\log(t)^k\,dt = \frac{(-1)^k k!}{s^{k+1}}$ and $\Gamma(s) = \frac{1}{s} + \gamma + s\,g(s)$ where γ is the Euler constant and the function g is also holomorphic around zero.

Prop 6.4. Let $Sp(M)$ be the dimension spectrum of a commutative geometry of dimension d. Then $Sp(M)$ is simple and $Sp(M) = \{\, d-k \mid k \in \mathbb{N}\,\}$.

Remark 6.1. When the dimension spectrum is not simple, the analog of $WRes$ is no longer a trace.

The equation (30) can be obtained via (8) as $\sigma_d^{|\mathcal{D}_A|^{-d}} = \sigma_d^{|\mathcal{D}|^{-d}}$.

In dimension $d = 4$, the computation in (19) of coefficient $a_4(1, \mathcal{D}_A^2)$ gives

$$\zeta_{\mathcal{D}_A}(0) = c_1 \int_M (5R^2 - 8R\mu\nu r^{\mu\nu} - 7R_{\mu\nu\rho\sigma}R^{\mu\nu\rho\sigma}\,dvol + c_2\int_M \mathrm{tr}(F_{\mu\nu}F^{\mu\nu})\,dvol,$$

see Corollary 7.1 to see precise correspondence between $a_k(1, \mathcal{D}_A^2)$ and $\zeta_{\mathcal{D}_A}(0)$. One recognizes the Yang–Mills action which will be generalized in Section 7.1.3 to arbitrary spectral triples.

According to Corollary 6.1, a commutative geometry has no tadpoles.[58]

6.7. *Scalar curvature*

What could be the scalar curvature of a spectral triple $(\mathcal{A}, \mathcal{H}, \mathcal{D})$? Of course, we consider first the case of a commutative geometry of dimension $d = 4$: if s is the scalar curvature and $f \in C^\infty(M)$, we know that $\fint f(x) D\!\!\!/^{-d+2} = \int_M f(x)\,s(x)\,dvol(x)$. This suggests the

Definition 6.10. Let $(\mathcal{A}, \mathcal{H}, \mathcal{D})$ be a spectral triple of dimension d. The scalar curvature is the map $\mathcal{R} : a \in \mathcal{A} \to \mathbb{C}$ defined by $\mathcal{R}(a) = \fint a\,\mathcal{D}^{-d+2}$.

In the commutative case, $\mathcal{R}$ is a trace on the algebra. More generally

Prop 6.5. [30, Proposition 1.153] If $\mathcal{R}$ is a trace on $\mathcal{A}$ and the tadpoles $\fint A\,\mathcal{D}^{-d+1}$ are zero for all $A \in \Omega^1_{\mathcal{D}}$, $\mathcal{R}$ is invariant by inner fluctuations $\mathcal{D} \to \mathcal{D} + A$.

6.8. *Tensor product of spectral triples*

There is a natural notion of tensor for spectral triples which corresponds to direct product of manifolds in the commutative case. Let $(\mathcal{A}_i, \mathcal{D}_i, \mathcal{H}_i)$,

$i = 1, 2$, be two spectral triples of dimension d_i with simple dimension spectrum. Assume the first to be of even dimension, with grading χ_1. The spectral triple $(\mathcal{A}, \mathcal{D}, \mathcal{H})$ associated to the tensor product is defined by

$$\mathcal{A} := \mathcal{A}_1 \otimes \mathcal{A}_2, \quad \mathcal{D} := \mathcal{D}_1 \otimes 1 + \chi_1 \otimes \mathcal{D}_2, \quad \mathcal{H} := \mathcal{H}_1 \otimes \mathcal{H}_2.$$

The interest of χ_1 is to guarantee additivity: $\mathcal{D}^2 = \mathcal{D}_1^2 \otimes 1 + 1 \otimes \mathcal{D}_2^2$.

Lemma 6.6. *Assume that* $\mathrm{Tr}(e^{-t\mathcal{D}_1^2}) \sim_{t\to 0} a_1\, t^{-d_1/2}$ *and* $\mathrm{Tr}(e^{-t\mathcal{D}_2^2}) \sim_{t\to 0} a_2\, t^{-d_2/2}$.

The triple $(\mathcal{A}, \mathcal{D}, \mathcal{H})$ *has dimension* $d = d_1 + d_2$.

Moreover, the function $\zeta_\mathcal{D}(s) = \mathrm{Tr}(|D|^{-s})$ *has a simple pole at* $s = d_1 + d_2$ *with*

$$\mathrm{Res}_{s=d_1+d_2}\big(\zeta_\mathcal{D}(s)\big) = \tfrac{1}{2}\,\tfrac{\Gamma(d_1/2)\Gamma(d_2/2)}{\Gamma(d/2)}\mathrm{Res}_{s=d_1}\big(\zeta_{\mathcal{D}_1}(s)\big)\,\mathrm{Res}_{s=d_2}\big(\zeta_{\mathcal{D}_2}(s)\big).$$

Proof. We get $\zeta_\mathcal{D}(2s) = \sum_{n=0}^{\infty} \mu_n(D_1^2 \otimes 1 + 1 \otimes \mathcal{D}_2^2)^{-s} = \sum_{n,m=0}^{\infty}\big(\mu_n(\mathcal{D}_1^2) + \mu_m(\mathcal{D}_2^2)\big)^{-s}$. Since $\big(\mu_n(\mathcal{D}_1)^2 + \mu_m(\mathcal{D}_2)^2\big)^{-(c_1+c_2)/2} \leq \mu_n(\mathcal{D}_1)^{-c_1}\mu_m(\mathcal{D}_2)^{-c_2}$, $\zeta_\mathcal{D}(c_1+c_2) \leq \zeta_{\mathcal{D}_1}(c_1)\zeta_{\mathcal{D}_2}(c_2)$ if $c_i > d_i$, so $d := \inf\{\, c \in \mathbb{R}^+ \,:\, \zeta_\mathcal{D}(c) < \infty \,\} \leq d_1 + d_2$. We claim that $d = d_1 + d_2$: recall first that $a_i := \underset{s=d_i/2}{\mathrm{Res}}\big(\Gamma(s)\zeta_{\mathcal{D}_i}(2s)\big) = \Gamma(d_i/2)\underset{s=d_i/2}{\mathrm{Res}}\big(\zeta_{\mathcal{D}_i}(2s)\big) = \tfrac{1}{2}\Gamma(d_i/2)\underset{s=d_i}{\mathrm{Res}}\big(\zeta_{\mathcal{D}_i}(s)\big)$. If $f(s) := \Gamma(s)\,\zeta_\mathcal{D}(2s)$, $f(s) = \int_0^1 \mathrm{Tr}\big(e^{-t\mathcal{D}^2}\big)\, t^{s-1}\, dt + g(s) = \int_0^1 \mathrm{Tr}\big(e^{-t\mathcal{D}_1^2}\big)\mathrm{Tr}\big(e^{-t\mathcal{D}_2^2}\big)\, t^{s-1}\, dt + g(s)$ where g is holomorphic since the map $x \in \mathbb{R} \to \int_1^\infty e^{-tx^2} t^{x-1}\, dt$ is in Schwartz space.

Since $\mathrm{Tr}\big(e^{-t\mathcal{D}_1^2}\big)\mathrm{Tr}\big(e^{-t\mathcal{D}_2^2}\big) \sim_{t\to 0} a_1 a_2\, t^{-(d_1+d_2)/2}$, the function $f(s)$ has a simple pole at $s = (d_1 + d_2)/2$. We conclude that $\zeta_\mathcal{D}(s)$ has a simple pole at $s = d_1 + d_2$. Moreover, using a_i, $\tfrac{1}{2}\Gamma((d_1 + d_2)/2)\underset{s=d}{\mathrm{Res}}\big(\zeta_\mathcal{D}(s)\big) = \tfrac{1}{2}\Gamma(d_1/2)\underset{s=d_1}{\mathrm{Res}}\big(\zeta_{\mathcal{D}_1}(s)\big)\,\tfrac{1}{2}\Gamma(d_2/2)\underset{s=d_2}{\mathrm{Res}}\big(\zeta_{\mathcal{D}_2}(s)\big)$. □

7. Spectral action

7.1. *On the search for a good action functional*

We would like to obtain a good action for any spectral triple and for this it is useful to look at some examples in physics. In any physical theory based on geometry, the interest of an action functional is, by a minimization process, to exhibit a particular geometry, for instance, trying to distinguish between different metrics. This is the case in general relativity with the Einstein–Hilbert action (with its Riemannian signature).

7.1.1. *Einstein–Hilbert action*

This action is $S_{EH}(g) = -\int_M s_g(x)\, dvol_g(x)$ where s is the scalar curvature (chosen positive for the sphere). Up to a constant, this is $⨍ \not{D}^{-2}$ in dimension 4 as quoted after (21).

This action is interesting for the following reason: Let $\mathcal{M}_1$ be the set of Riemannian metrics g on M such that $\int_M dvol_g = 1$. By a theorem of Hilbert,[5] $g \in \mathcal{M}_1$ is a critical point of $S_{EH}(g)$ restricted to $\mathcal{M}_1$ if and only if (M, g) is an Einstein manifold (the Ricci curvature R of g is proportional by a constant to g: $R = c\, g$). Taking the trace, this means that $s_g = c \dim(M)$ and such manifold have a constant scalar curvature.

But in the search for invariants under diffeomorphisms, they are more quantities than the Einstein–Hilbert action, a trivial example being $\int_M f\big(s_g(x)\big)\, dvol_g(x)$ and they are others.[42] In this desire to implement gravity in noncommutative geometry, the eigenvalues of the Dirac operator look as natural variables.[69] However we are looking for observables which add up under disjoint unions of different geometries.

7.1.2. *Quantum approach and spectral action*

In a way, a spectral triple fits quantum field theory since $\mathcal{D}^{-1}$ can be seen as the propagator (or line element ds) for (Euclidean) fermions and we can compute Feynman graphs with fermionic internal lines. As glimpsed in section 6.4, the gauge bosons are only derived objects obtained from internal fluctuations via Morita equivalence given by a choice of a connection which is associated to a one-form in $\Omega^1_{\mathcal{D}}(\mathcal{A})$. Thus, the guiding principle followed by Connes and Chamseddine is to use a theory which is pure gravity with a functional action based on the spectral triple, namely which depends on the spectrum of $\mathcal{D}$.[10] They proposed the

Definition 7.1. The spectral action of $(\mathcal{A}, \mathcal{H}, \mathcal{D})$ is defined by $\mathcal{S}(\mathcal{D}, f, \Lambda) := \operatorname{Tr}\big(f(\mathcal{D}^2/\Lambda^2)\big)$ where $\Lambda \in \mathbb{R}^+$ plays the role of a cut-off and f is any positive function (such that $f(\mathcal{D}^2/\Lambda^2)$ is a trace-class operator).

Remark 7.1. We can also define $\mathcal{S}(\mathcal{D}, f, \Lambda) = \operatorname{Tr}\big(f(\mathcal{D}/\Lambda)\big)$ when f is positive and even. With this second definition, $S(\mathcal{D}, g, \Lambda) = \operatorname{Tr}\big(f(\mathcal{D}^2/\Lambda^2)\big)$ with $g(x) = f(x^2)$.

For f, one can think of the characteristic function of $[-1, 1]$, thus $f(\mathcal{D}/\Lambda)$ is nothing else but the number of eigenvalues of $\mathcal{D}$ within $[-\Lambda, \Lambda]$.

When this action has an asymptotic series in $\Lambda \to \infty$, we deal with an effective theory. Naturally, $\mathcal{D}$ has to be replaced by $\mathcal{D}_A$ which is just a decoration. To this bosonic part of the action, one adds a fermionic term $\frac{1}{2}\langle J\psi, \mathcal{D}\psi\rangle$ for $\psi \in \mathcal{H}$ to get a full action. In the standard model of particle physics, this latter corresponds to the integration of the Lagrangian part for the coupling between gauge bosons and Higgs bosons with fermions. Actually, the finite dimension part of the noncommutative standard model is of KO-dimension 6, thus $\langle \psi, \mathcal{D}\psi\rangle$ has to be replaced by $\frac{1}{2}\langle J\psi, \mathcal{D}\psi\rangle$ for $\psi = \chi\psi \in \mathcal{H}$, see.[30]

7.1.3. *Yang–Mills action*

This action plays an important role in physics so recall first the classical situation: let G be a compact Lie group with its Lie algebra $\mathfrak{g}$ and let $A \in \Omega^1(M, \mathfrak{g})$ be a connection. If $F = da + \frac{1}{2}[A, A] \in \Omega^2(M, \mathfrak{g})$ is the curvature (or field strength) of A, then the Yang-Mills action is $S_{YM}(A) = \int_M \mathrm{tr}(F \wedge \star F)\, dvol_g$. In the abelian case $G = U(1)$, it is the Maxwell action and its quantum version is the quantum electrodynamics (QED) since the ungauged $U(1)$ of electric charge conservation can be gauged and its gauging produces electromagnetism.[97] It is conformally invariant when $\dim(M) = 4$.

The study of its minima and its critical values can also been made for a spectral triple $(\mathcal{A}, \mathcal{H}, \mathcal{D})$ of dimension d:[23,24] let $A \in \Omega^1_{\mathcal{D}}(\mathcal{A})$ and curvature $\theta = dA + A^2$; then it is natural to consider $I(A) = \mathrm{Tr}_{Dix}(\theta^2|\mathcal{D}|^{-d})$ since it coincides (up to a constant) with the previous Yang-Mills action in the commutative case: if $P = \theta^2|\mathcal{D}|^{-d}$, then Theorems 2.4 and 3.2 give the claim since for the principal symbol, $\mathrm{tr}\big(\sigma^P(x,\xi)\big) = c\,\mathrm{tr}(F \wedge \star F)(x)$.

There is nevertheless a problem with the definition of dA: if $A = \sum_j \pi(a_j)[\mathcal{D}, \pi(b_j)]$, then $dA = \sum_j [\mathcal{D}, \pi(a_j)][\mathcal{D}, \pi(b_j)]$ can be non-zero while $A = 0$. This ambiguity means that, to get a graded differential algebra $\Omega^*_{\mathcal{D}}(\mathcal{A})$, one must divide by a junk, for instance $\Omega^2_{\mathcal{D}} \simeq \pi(\Omega^2/\pi\big(\delta(\mathrm{Ker}(\pi) \cap \Omega^1)\big)$ where $\Omega^k(\mathcal{A})$ is the set of universal k-forms over $\mathcal{A}$ given by the set of $a_0\delta a_1 \cdots \delta a_k$ (before representation on $\mathcal{H}$: $\pi(a_0\delta a_1 \cdots \delta a_k) = a_0[\mathcal{D}, a_1] \cdots [\mathcal{D}, a_k]$).
Let $\mathcal{H}_k$ be the Hilbert space completion of $\pi(\Omega^k(\mathcal{A}))$ with the scalar product defined by $\langle A_1, A_2\rangle_k = \mathrm{Tr}_{Dix}(A_2^* A_1 |\mathcal{D}|^{-d})$ for $A_j \in \pi(\Omega^k(\mathcal{A}))$.
The Yang–Mills action on $\Omega^1(\mathcal{A})$ is $S_{YM}(V) = \langle \delta V + V^2, \delta V + V^2\rangle$. It is positive, quartic and gauge invariant under $V \to \pi(u)V\pi(u^*) + \pi(u)[\mathcal{D}, \pi(u^*)]$ when $u \in \mathcal{U}(\mathcal{A})$. Moreover, $S_{YM}(V) = \inf\{\, I(\omega) \mid \omega \in \Omega^1(\mathcal{A}),\ \pi(\omega) = V\,\}$ since the above ambiguity disappears when taking the infimum.

This Yang–Mills action can be extended to the equivalent of Hermitean vector bundles on M, namely finitely projective modules over $\mathcal{A}$.

The spectral action is more conceptual than the Yang–Mills action since it gives no fundamental role to the distinction between gravity and matter in the artificial decomposition $\mathcal{D}_A = \mathcal{D} + A$. For instance, for the minimally coupled standard model, the Yang–Mills action for the vector potential is part of the spectral action, as far as the Einstein–Hilbert action for the Riemannian metric.[11]

As quoted in,[16] the spectral action has conceptual advantages:

- Simplicity: when f is a cutoff function, the spectral action is just the counting function.

- Positivity: when f is positive (which is the case for a cutoff function), the action $\mathrm{Tr}\big(f(\mathcal{D}/\Lambda)\big) \geq 0$ has the correct sign for a Euclidean action: the positivity of the function f will insure that the actions for gravity, Yang-Mills, Higgs couplings are all positive and the Higgs mass term is negative.

- Invariance: the spectral action has a much stronger invariance group than the usual diffeomorphism group as for the gravitational action; this is the unitary group of the Hilbert space $\mathcal{H}$.

However, this action is not local but becomes local when replaced by its asymptotic expansion:

7.2. *Asymptotic expansion for $\Lambda \to \infty$Asymptotic expansion for Lambda going to infinity*

The heat kernel method already used in previous sections will give a control of spectral action $S(\mathcal{D}, f, \Lambda)$ when Λ goes to infinity.

Theorem 7.1. *Let $(\mathcal{A}, \mathcal{H}, \mathcal{D})$ be a spectral triple with a simple dimension spectrum Sd.*
We assume that

$$\mathrm{Tr}\big(e^{-t\mathcal{D}^2}\big) \underset{t\downarrow 0}{\sim} \sum_{\alpha \in Sd} a_\alpha\, t^\alpha \quad \text{with } a_\alpha \neq 0. \tag{31}$$

Then, for the zeta function $\zeta_\mathcal{D}$ defined in (25)

$$a_\alpha = \tfrac{1}{2}\mathrm{Res}_{s=-2\alpha}\big(\Gamma(s/2)\zeta_\mathcal{D}(s)\big). \tag{32}$$

(i) If $\alpha < 0$, $\zeta_\mathcal{D}$ has a pole at -2α with $a_\alpha = \frac{1}{2}\Gamma(-\alpha) \underset{s=-2\alpha}{\mathrm{Res}} \zeta_\mathcal{D}(s)$.

(ii) For $\alpha = 0$, we get $a_0 = \zeta_\mathcal{D}(0) + \dim \mathcal{D}$.

(iii) If $\alpha > 0$, $a_\alpha = \zeta(-2\alpha) \underset{s=-\alpha}{\mathrm{Res}}\, \Gamma(s)$.

(iv) The spectral action has the asymptotic expansion over the positive part Sd^+ of Sd:

$$\mathrm{Tr}\big(f(\mathcal{D}/\Lambda)\big) \underset{\Lambda\to+\infty}{\sim} \sum_{\beta\in Sd^+} f_\beta\, \Lambda^\beta \fint |\mathcal{D}|^\beta + f(0)\, \zeta_\mathcal{D}(0) + \cdots \tag{33}$$

where the dependence of the even function f is $f_\beta := \int_0^\infty f(x)\, x^{\beta-1}\, dx$ and $\cdots$ involves the full Taylor expansion of f at 0.

Proof. (i): Since $\Gamma(s/2)\,|\mathcal{D}|^{-s} = \int_0^\infty e^{-t\mathcal{D}^2}\, t^{s/2-1}\, dt = \int_0^1 e^{-t\mathcal{D}^2}\, t^{s/2-1}\, dt + f(s)$, where the function f is holomorphic (since the map $x \to \int_1^\infty e^{-tx^2} x^{s/2-1}\, dt$ is in the Schwartz space), the swap of $\mathrm{Tr}\big(e^{-t\mathcal{D}^2}\big)$ with a sum of $a_\alpha\, t^\alpha$ and $a_\alpha \int_0^1 t^{\alpha+s/2-1}\, dt = \frac{2a_a}{s+2\alpha}$ yields (32).

(ii): The regularity of $\Gamma(s/2)^{-1} \underset{s\to 0}{\sim} s/2$ around zero implies that only the pole part at $s = 0$ of $\int_0^\infty \mathrm{Tr}\big(e^{-t\mathcal{D}^2}\big)\, t^{s/2-1}\, dt$ contributes to $\zeta_{\mathcal{D}}(0)$. This contribution is $a_0 \int_0^1 t^{s/2-1}\, dt = \frac{2a_0}{s}$.

(iii) follows from (32).

(iv): Assume $f(x) = g(x^2)$ where g is a Laplace transform: $g(x) := \int_0^\infty e^{-sx}\, \phi(s)\, ds$. We will see in Section 7.3 how to relax this hypothesis. Since
$g(t\mathcal{D}^2) = \int_0^\infty e^{-st\mathcal{D}^2}\, \phi(s)\, ds$, $\mathrm{Tr}\big(g(t\mathcal{D}^2)\big) \underset{t\downarrow 0}{\sim} \sum_{\alpha\in \mathrm{Sp}^+} a_\alpha\, t^\alpha \int_0^\infty s^\alpha\, g(s)\, ds$.
When $\alpha < 0$, $s^\alpha = \Gamma(-\alpha)^{-1} \int_0^\infty e^{-sy}\, y^{-\alpha-1}\, dy$ and $\int_0^\infty s^\alpha\, \phi(s)\, ds = \Gamma(-\alpha)^{-1} \int_0^\infty g(y)\, y^{-\alpha-1}\, dy$. Thus

$$\mathrm{Tr}\big(g(t\mathcal{D}^2)\big) \underset{t\downarrow 0}{\sim} \textstyle\sum_{\alpha\in \mathrm{Sp}^-} \big[\frac{1}{2} \underset{s=-2\alpha}{\mathrm{Res}}\, \zeta_{\mathcal{D}}(s) \int_0^\infty g(y)\, y^{-\alpha-1}\, dy\big]\, t^\alpha.$$

Thus (33) follows from (i), (ii) and $\frac{1}{2}\int_0^\infty g(y)\, y^{\beta/2-1}\, dy = \int_0^\infty f(x)\, x^{\beta-1}\, dx$. □

It can be useful to make a connection with (20) of Section 5.2:

Corollary 7.1. *Assume that the spectral triple $(\mathcal{A}, \mathcal{H}, \mathcal{D})$ has dimension d. If*

$$\mathrm{Tr}\big(e^{-t\,\mathcal{D}^2}\big) \underset{t\downarrow 0}{\sim} \sum_{k\in\{0,\cdots,d\}} t^{(k-d)/2}\, a_k(\mathcal{D}^2) + \cdots\,, \tag{34}$$

then $\mathcal{S}(\mathcal{D}, f, \Lambda) \underset{t\downarrow 0}{\sim} \sum_{k\in\{1,\cdots,d\}} f_k\, \Lambda^k\, a_{d-k}(\mathcal{D}^2) + f(0)\, a_d(\mathcal{D}^2) + \cdots$ *with* $f_k := \frac{1}{\Gamma(k/2)} \int_0^\infty f(s) s^{k/2-1} ds$. *Moreover,*

$$a_k(\mathcal{D}^2) = \tfrac{1}{2}\, \Gamma(\tfrac{d-k}{2}) {\int\!\!\!\!\!\!-} |\mathcal{D}|^{-d+k} \qquad \textit{for } k = 0, \cdots, d-1, \tag{35}$$
$$a_d(\mathcal{D}^2) = \dim \mathcal{D} + \zeta_{\mathcal{D}^2}(0).$$

The asymptotics (33) use the value of $\zeta_{\mathcal{D}}(0)$ in the constant term Λ^0, so it is fundamental to look at its variation under a gauge fluctuation $\mathcal{D} \to \mathcal{D} + A$ as we saw in Theorem 6.2.

7.3. *Remark on the use of Laplace transform*

The spectral action asymptotic behavior $S(\mathcal{D}, f, \Lambda) \underset{\Lambda\to+\infty}{\sim} \sum_{n=0}^{\infty} c_n \, \Lambda^{d-n} \, a_n(\mathcal{D}^2)$ has been proved for a smooth function f which is a Laplace transform for an arbitrary spectral triple (with simple dimension spectrum) satisfying (31). However, this hypothesis is too restrictive since it does not cover the heat kernel case where $f(x) = e^{-x}$.
When the triple is commutative and $\mathcal{D}^2$ is a generalized Laplacian on sections of a vector bundle over a manifold of dimension 4, Estrada–Gracia-Bondía–Várilly proved in[37] that previous asymptotics is

$$\mathrm{Tr}\big(f(\mathcal{D}^2/\Lambda^2)\big) \sim \tfrac{1}{(4\pi)^2}\Big[\mathrm{rk(E)} \int_0^\infty x f(x)\, dx\, \Lambda^4 + b_2(\mathcal{D}^2) \int_0^\infty f(x)\, dx\, \Lambda^2 + \sum_{m=0}^{\infty} (-1)^m\, f^{(m)}(0)\, b_{2m+4}(\mathcal{D}^2)\, \Lambda^{-2m}\Big], \quad \Lambda \to \infty$$

where $(-1)^m b_{2m+4}(\mathcal{D}^2) = \frac{(4\pi)^2}{m!}\, \mu_m(\mathcal{D}^2)$ are suitably normalized, integrated moment terms of the spectral density of $\mathcal{D}^2$.

The main point is that *this asymptotics makes sense in the Cesàro sense* (see[37] for definition) for f in $\mathcal{K}'(\mathbb{R})$, which is the dual of $\mathcal{K}(\mathbb{R})$. This latter is the space of smooth functions ϕ such that for some $a \in \mathbb{R}$, $\phi^{(k)}(x) = \mathcal{O}(|x|^{a-k})$ as $|x| \to \infty$, for each $k \in \mathbb{N}$. In particular, the Schwartz functions are in $\mathcal{K}(\mathbb{R})$ (and even dense).

Of course, the counting function is not smooth but is in $\mathcal{K}'(\mathbb{R})$, so the given asymptotic behavior is wrong beyond the first term, but is correct in the Cesàro sense. Actually there are more derivatives of f at 0 as explained on examples in [37, p. 243].

7.4. *About convergence and divergence, local and global aspects of the asymptotic expansion*

The asymptotic expansion series (34) of the spectral action may or may not converge. It is known that each function $g(\Lambda^{-1})$ defines at most a unique expansion series when $\Lambda \to \infty$ but the converse is not true since several functions have the same asymptotic series. We give here examples of convergent and divergent series of this kind.

When $M = \mathbb{T}^d$ as in Example 3.1 with $\Delta = \delta^{\mu\nu}\partial_\mu\partial_\nu$, $\mathrm{Tr}(e^{t\Delta}) = \frac{(4\pi)^{-d/2}\,\mathrm{Vol}(\mathbb{T}^d)}{t^{d/2}} + \mathcal{O}(t^{-d/2}\, e^{-1/4t})$, so the asymptotic series $\mathrm{Tr}(e^{t\Delta}) \simeq \frac{(4\pi)^{-d/2}\,\mathrm{Vol}(\mathbb{T}^d)}{t^{d/2}}$, $t \to 0$, has only one term.

In the opposite direction, let now M be the unit four-sphere $\mathbb{S}^4$ and

$\not{D}$ be the usual Dirac operator. By Proposition 6.4, equation (31) yields (see[19]):

$$\mathrm{Tr}(e^{-t\not{D}^2}) = \tfrac{1}{t^2}\big(\tfrac{2}{3} + \tfrac{2}{3}\,t + \sum_{k=0}^{n} a_k\, t^{k+2} + \mathcal{O}(t^{n+3})\big),$$

where

$$a_k := \tfrac{(-1)^k\, 4}{3\, k!}\big(\tfrac{B_{2k+2}}{2k+2} - \tfrac{B_{2k+4}}{2k+4}\big)$$

with Bernoulli numbers B_{2k}. Thus $t^2\mathrm{Tr}(e^{-t\not{D}^2}) \simeq \frac{2}{3} + \frac{2}{3}\,t + \sum_{k=0}^{\infty} a_k\, t^{k+2}$ when $t \to 0$ and this series is not convergent but only asymptotic: $a_k > \frac{4}{3\,k!}\frac{|B_{2k+4}|}{2k+4} > 0$ and $|B_{2k+4}| = 2\,\frac{(2k+4)!}{(2\pi)^{2k+4}}\,\zeta(2k+4) \simeq 4\sqrt{\pi(k+2)}\,\big(\frac{k+2}{\pi e}\big)^{2k+4} \to \infty$ if $k \to \infty$.

More generally, in the commutative case considered above and when $\mathcal{D}$ is a differential operator—like a Dirac operator, the coefficients of the asymptotic series of $\mathrm{Tr}(e^{-t\mathcal{D}^2})$ are locally defined by the symbol of $\mathcal{D}^2$ at point $x \in M$ but this is not true in general: in[45] is given a positive elliptic pseudodifferential such that non-locally computable coefficients especially appear in (34) when $2k > d$. Nevertheless, all coefficients are local for $2k \leq d$.

Recall that a locally computable quantity is the integral on the manifold of a local frame-independent smooth function of one variable, depending only on a finite number of derivatives of a finite number of terms in the asymptotic expansion of the total symbol of $\mathcal{D}^2$. For instance, some nonlocal information contained in the ultraviolet asymptotics can be recovered if one looks at the (integral) kernel of $e^{-t\sqrt{-\Delta}}$: in $\mathbb{T}^1$, with $\mathrm{Vol}(\mathbb{T}^1) = 2\pi$, we get[38]

$$\mathrm{Tr}(e^{-t\sqrt{-\Delta}}) = \tfrac{\sinh(t)}{\cosh(t)-1} = \coth(\tfrac{t}{2}) = \tfrac{2}{t}\sum_{k=0}^{\infty} \tfrac{B_{2k}}{(2k)!}\, t^{2k} = \tfrac{2}{t}[1+\tfrac{t^2}{12} - \tfrac{t^4}{720} + \mathcal{O}(t^6)]$$

so the series converges when $t < 2\pi$ as $\frac{B_{2k}}{(2k)!} = (-1)^{k+1}\,\frac{2\,\zeta(2k)}{(2\pi)^{2k}}$, thus $\frac{|B_{2k}|}{(2k)!} \simeq \frac{2}{(2\pi)^{2k}}$ when $k \to \infty$.

Thus we have an example where $t \to \infty$ cannot be used with the asymptotic series.

Thus the spectral action of Corollary 7.1 precisely encodes these local and nonlocal behavior which appear or not in its asymptotics for different f. The coefficient of the action for the positive part (at least) of the dimension spectrum correspond to renormalized traces, namely the noncommutative integrals of (35). In conclusion, the asymptotics of spectral action may or may not have nonlocal coefficients.

For the flat torus $\mathbb{T}^d$, the difference between $\mathrm{Tr}(e^{t\Delta})$ and its asymptotic series is an term which is related to periodic orbits of the geodesic flow on $\mathbb{T}^d$. Similarly, the counting function $N(\lambda)$ (number of eigenvalues including multiplicities of Δ less than λ) obeys Weyl's law: $N(\lambda) = \frac{(4\pi)^{-d/2}\,\mathrm{Vol}(\mathbb{T}^d)}{\Gamma(d/2+1)}\,\lambda^{d/2} + o(\lambda^{d/2})$ — see[1] for a nice historical review on these fundamental points. The relationship between the asymptotic expansion of the heat kernel and the formal expansion of the spectral measure is clear: the small-t asymptotics of heat kernel is determined by the large-λ asymptotics of the density of eigenvalues (and eigenvectors). However, the latter is defined modulo some average: Cesàro sense as reminded in Section 7.3, or Riesz mean of the measure which washes out ultraviolet oscillations, but also gives informations on intermediate values of λ.[38]

In[16,76] are examples of spectral actions on commutative geometries of dimension 4 whose asymptotics have only two terms. In the quantum group $SU_q(2)$, the spectral action itself has only 4 terms, independently of the choice of function f. See[62] for more examples.

7.5. *On the physical meaning of the asymptotics of spectral action*

The spectral action is non-local. Its localization does not cover all situations: consider for instance the commutative geometry of a spin manifold M of dimension 4. One adds a gauge connection $A \in \Gamma^\infty\big(M, End(S)\big)$ to $\not{D}$ such that $\mathcal{D} = i\gamma^\mu(\partial_\mu + A_\mu)$, thus with a field strength $F_{\mu\nu} = \partial_\mu A_\nu - \partial_\nu A_\mu + [A_\mu, A_\nu]$. We apply (18) with $P = \mathcal{D}^2$ and find the coefficients $a_i(1, P)$ of (19) with $i = 0, 2, 4$. The asymptotic expansion corresponds to a weak field expansion.

Moreover a commutative geometry times a finite one where the finite one is algebra is a sum of matrices has been deeply and intensively investigated for the noncommutative approach to standard model of particle physics, see.[20,30] This approach offers a lot of interesting perspectives, for instance, the possibility to compute the Higgs representations and mass (for each noncommutative model) is particularly instructive.[10,15,17,57,63,64,71,79] Of course, since the first term in (33) is a cosmological term, one may be worried by its large value (for instance in the noncommutative standard model where the cutoff is, roughly speaking the Planck scale). At the classical level, one can work with unimodular gravity where the metric (so the Dirac operator) $\mathcal{D}$ varies within the set $\mathcal{M}_1$ of metrics which preserve the volume as in Section 7.1.1. Thus it remains only (!) to control the inflaton:

see.[13]

The spectral action has been computed in[60] for the quantum group $SU_q(2)$ which is not a deformation of $SU(2)$ of the type considered on the Moyal plane. It is quite peculiar since (33) has only a finite number of terms.

Due to the difficulties to deal with non-compact manifolds, the case of spheres $\mathbb{S}^4$ or $\mathbb{S}^3 \times \mathbb{S}^1$ has been investigated in[16,19] for instance in the case of Robertson–Walker metrics.

All the machinery of spectral geometry as been recently applied to cosmology, computing the spectral action in few cosmological models related to inflation, see.[66,76–78,81,95]

Spectral triples associated to manifolds with boundary have been considered in.[14,18,18,58,59,61] The main difficulty is precisely to put nice boundary conditions to the operator $\mathcal{D}$ to still get a selfadjoint operator and then, to define a compatible algebra $\mathcal{A}$. This is probably a must to obtain a result in a noncommutative Hamiltonian theory in dimension 1+3.

The case of manifolds with torsion has also been studied in,[53,84,85] and even with boundary in.[61] These works show that the Holst action appears in spectral actions and that torsion could be detected in a noncommutative world.

8. The noncommutative torus

The aim of this section is to compute the spectral action of the noncommutative torus. Due to a fundamental appearance of small divisors, the number theory is involved via a Diophantine condition. As a consequence, the result which essentially says that the spectral action of the noncommutative torus coincide with the action of the ordinary torus (up few constants) is awfully technical and this shows how life can be hard in noncommutative geometry!

Reference:[36]

8.1. *Definition of the nc-torus*

Let $C^\infty(\mathbb{T}^n_\Theta)$ be the smooth noncommutative n-torus associated to a non-zero skew-symmetric deformation matrix $\Theta \in M_n(\mathbb{R})$. It was introduced by Rieffel[93] and Connes[21] to generalize the n-torus $\mathbb{T}^n$. This means that $C^\infty(\mathbb{T}^n_\Theta)$ is the algebra generated by n unitaries u_i, $i = 1, \dots, n$ subject to the relations $u_l\, u_j = e^{i\,\Theta_{lj}}\, u_j\, u_l$, and with Schwartz coefficients: an element $a \in C^\infty(\mathbb{T}^n_\Theta)$ can be written as $a = \sum_{k\in\mathbb{Z}^n} a_k\, U_k$, where $\{a_k\} \in \mathcal{S}(\mathbb{Z}^n)$ with the Weyl elements defined by $U_k = e^{-\frac{i}{2}k.\chi k}\, u_1^{k_1} \cdots u_n^{k_n}$, $k \in \mathbb{Z}^n$, the constraint relation reads

$$U_k U_q = e^{-\frac{i}{2}k.\Theta q}\, U_{k+q}, \text{ and } U_k U_q = e^{-ik.\Theta q}\, U_q U_k$$

where χ is the matrix restriction of Θ to its upper triangular part. Thus unitary operators U_k satisfy $U_k^* = U_{-k}$ and $[U_k, U_l] = -2i\,\sin(\frac{1}{2}k.\Theta l)\, U_{k+l}$.

Let τ be the trace on $C^\infty(\mathbb{T}^n_\Theta)$ defined by $\tau\big(\sum_{k\in\mathbb{Z}^n} a_k\, U_k\big) = a_0$ and $\mathcal{H}_\tau$ be the GNS Hilbert space obtained by completion of $C^\infty(\mathbb{T}^n_\Theta)$ with respect of the norm induced by the scalar product $\langle a, b\rangle = \tau(a^* b)$.

On $\mathcal{H}_\tau = \{\sum_{k\in\mathbb{Z}^n} a_k\, U_k \mid \{a_k\}_k \in l^2(\mathbb{Z}^n)\}$, we consider the left and right regular representations of $C^\infty(\mathbb{T}^n_\Theta)$ by bounded operators, that we denote respectively by $L(.)$ and $R(.)$.

Let also δ_μ, $\mu \in \{1, \ldots, n\}$, be the n (pairwise commuting) canonical derivations, defined by $\delta_\mu(U_k) = ik_\mu U_k$.

We need to fix notations: let $\mathcal{A}_\Theta = C^\infty(\mathbb{T}^n_\Theta)$ acting on $\mathcal{H} = \mathcal{H}_\tau \otimes \mathbb{C}^{2^m}$ with $n = 2m$ or $n = 2m+1$ (i.e., $m = \lfloor \frac{n}{2} \rfloor$ is the integer part of $\frac{n}{2}$), the square integrable sections of the trivial spin bundle over $\mathbb{T}^n$ and each element of $\mathcal{A}_\Theta$ is represented on $\mathcal{H}$ as $L(a) \otimes 1_{2^m}$. The Tomita conjugation $J_0(a) = a^*$ satisfies $[J_0, \delta_\mu] = 0$ and we define $J = J_0 \otimes C_0$ where C_0 is an operator on $\mathbb{C}^{2^m}$. The Dirac-like operator is given by

$$\mathcal{D} = -i\, \delta_\mu \otimes \gamma^\mu,$$

where we use hermitian Dirac matrices γ. It is defined and symmetric on the dense subset of $\mathcal{H}$ given by $C^\infty(\mathbb{T}^n_\Theta) \otimes \mathbb{C}^{2^m}$ and $\mathcal{D}$ denotes its selfadjoint extension. Thus $C_0\gamma^\alpha = -\varepsilon\gamma^\alpha C_0$, and $\mathcal{D}\; U_k \otimes e_i = k_\mu U_k \otimes \gamma^\mu e_i$, where (e_i) is the canonical basis of $\mathbb{C}^{2^m}$. Moreover, $C_0^2 = \pm 1_{2^m}$ depending on the parity of m. Finally, the chirality in the even case is $\chi = id \otimes (-i)^m \gamma^1 \cdots \gamma^n$. This yields a spectral triple:

Theorem 8.1.[24,49] *The 5-tuple $(\mathcal{A}_\Theta, \mathcal{H}, \mathcal{D}, J, \chi)$ is a real regular spectral triple of dimension n. It satisfies the finiteness and orientability conditions of Definition 6.2. It is n-summable and its KO-dimension is also n.*

For every unitary $u \in \mathcal{A}$, $uu^* = u^*u = U_0$, the perturbed operator $V_u\, \mathcal{D}\, V_u^*$ by the unitary $V_u = \big(L(u) \otimes 1_{2^m}\big) J \big(L(u) \otimes 1_{2^m}\big) J^{-1}$, must satisfy condition $J\mathcal{D} = \epsilon \mathcal{D} J$. The yields the necessity of a symmetrized covariant Dirac operator $\mathcal{D}_A = \mathcal{D} + A + \epsilon J\, A\, J^{-1}$ since $V_u\, \mathcal{D}\, V_u^* = \mathcal{D}_{L(u)\otimes 1_{2^m}[\mathcal{D}, L(u^*)\otimes 1_{2^m}]}$. Moreover, we get the gauge transformation $V_u \mathcal{D}_A V_u^* = \mathcal{D}_{\gamma_u(A)}$ where the gauged transform one-form of A is $\gamma_u(A) = u[\mathcal{D}, u^*] + uAu^*$, with the shorthand $L(u) \otimes 1_{2^m} \longrightarrow u$. So the spectral action is gauge invariant: $\mathcal{S}(\mathcal{D}_A, f, \Lambda) = \mathcal{S}(\mathcal{D}_{\gamma_u(A)}, f, \Lambda)$.

Any selfadjoint one-form $A \in \Omega^1_\mathcal{D}(\mathcal{A})$, is written as $A = L(-iA_\alpha) \otimes \gamma^\alpha$, $A_\alpha = -A_\alpha^* \in \mathcal{A}_\Theta$, thus $\mathcal{D}_A = -i\,\big(\delta_\alpha + L(A_\alpha) - R(A_\alpha)\big) \otimes \gamma^\alpha$. Defining

$\tilde{A}_\alpha = L(A_\alpha) - R(A_\alpha)$, we get
$\mathcal{D}_A^2 = -g^{\alpha_1\alpha_2}(\delta_{\alpha_1} + \tilde{A}_{\alpha_1})(\delta_{\alpha_2} + \tilde{A}_{\alpha_2}) \otimes 1_{2^m} - \frac{1}{2}\Omega_{\alpha_1\alpha_2} \otimes \gamma^{\alpha_1\alpha_2}$ where

$$\begin{aligned}
\gamma^{\alpha_1\alpha_2} &= \tfrac{1}{2}(\gamma^{\alpha_1}\gamma^{\alpha_2} - \gamma^{\alpha_2}\gamma^{\alpha_1}),\\
\Omega_{\alpha_1\alpha_2} &= [\delta_{\alpha_1} + \tilde{A}_{\alpha_1}, \delta_{\alpha_2} + \tilde{A}_{\alpha_2}] = L(F_{\alpha_1\alpha_2}) - R(F_{\alpha_1\alpha_2})\\
F_{\alpha_1\alpha_2} &= \delta_{\alpha_1}(A_{\alpha_2}) - \delta_{\alpha_2}(A_{\alpha_1}) + [A_{\alpha_1}, A_{\alpha_2}].
\end{aligned}$$

8.2. *Kernels and dimension spectrum*

Since the dimension of the kernel appears in the coefficients (35) of the spectral action, we now compute the kernel of the perturbed Dirac operator:

Prop 8.1. $\mathcal{D} = U_0 \otimes \mathbb{C}^{2^m}$, so $\dim \mathcal{D} = 2^m$. For any selfadjoint one-form A, $\mathcal{D} \subseteq \mathcal{D}_A$ and or any unitary $u \in \mathcal{A}$, $\mathcal{D}_{\gamma_u(A)} = V_u\, \mathcal{D}_A$.

One shows that $\mathcal{D}_A = \mathcal{D}$ in the following cases:

(i) $A = A_u = L(u) \otimes 1_{2^m}[\mathcal{D}, L(u^*) \otimes 1_{2^m}]$ when u is a unitary in $\mathcal{A}$.
(ii) $||A|| < \frac{1}{2}$.
(iii) The matrix $\frac{1}{2\pi}\Theta$ has only integral coefficients.

Conjecture: $\mathcal{D} = \mathcal{D}_A$ *at least for generic* Θ*'s.*
We will use freely the notation (28) about the difference between $\mathcal{D}$ and D. Since we will have to control small divisors, we give first some Diophantine condition:

Definition 8.1. Let $\delta > 0$. A vector $a \in \mathbb{R}^n$ is said to be δ-badly approximable if there exists $c > 0$ such that $|q.a - m| \geq c\,|q|^{-\delta}$, $\forall q \in \mathbb{Z}^n \setminus \{\,0\,\}$ and $\forall m \in \mathbb{Z}$.
We note $\mathcal{BV}(\delta)$ the set of δ-badly approximable vectors and $\mathcal{BV} := \cup_{\delta>0}\mathcal{BV}(\delta)$ the set of badly approximable vectors.

A matrix $\Theta \in \mathcal{M}_n(\mathbb{R})$ (real $n \times n$ matrices) will be said to be badly approximable if there exists $u \in \mathbb{Z}^n$ such that ${}^t\Theta(u)$ is a badly approximable vector of $\mathbb{R}^n$.

Remark. A result from Diophantine approximation asserts that for $\delta > n$, the Lebesgue measure of $\mathbb{R}^n \setminus \mathcal{BV}(\delta)$ is zero (i.e almost any element of $\mathbb{R}^n$ is $\delta-$badly approximable.)

Let $\Theta \in \mathcal{M}_n(\mathbb{R})$. If its row of index i is a badly approximable vector of $\mathbb{R}^n$ (i.e. if $L_i \in \mathcal{BV}$) then ${}^t\Theta(e_i) \in \mathcal{BV}$ and thus Θ is a badly approximable matrix. It follows that almost any matrix of $\mathcal{M}_n(\mathbb{R}) \approx \mathbb{R}^{n^2}$ is badly approximable.

Prop 8.2. When $\frac{1}{2\pi}\Theta$ is badly approximable, the spectrum dimension of the spectral triple $\big(C^\infty(\mathbb{T}^n_\Theta), \mathcal{H}, \mathcal{D}\big)$ is equal to the set $\{\, n-k \,:\, k \in \mathbb{N}_0 \,\}$ and all these poles are simple. Moreover $\zeta_D(0) = 0$.

To get this result, we must compute the residues of infinite series of functions on $\mathbb{C}$ and the commutation between residues and series works under the sufficient Diophantine condition.

We can compute $\zeta_D(0)$ easily but the main difficulty is precisely to calculate $\zeta_{D_A}(0)$.

8.3. *The spectral action*

We fix a self-adjoint one-form A on the noncommutative torus of dimension n.

Prop 8.3. If $\frac{1}{2\pi}\Theta$ is badly approximable, then the first elements of the spectral action expansion (33) are given by

$$\begin{aligned} ⨍ |D_A|^{-n} &= ⨍ |D|^{-n} = 2^{m+1}\pi^{n/2}\,\Gamma(\tfrac{n}{2})^{-1}, \\ ⨍ |D_A|^{-n+k} &= 0 \text{ for } k \text{ odd}, \\ ⨍ |D_A|^{-n+2} &= 0. \end{aligned}$$

Here is the main result of this section.

Theorem 8.2. *Consider the noncommutative torus* $\big(C^\infty(\mathbb{T}^n_\Theta), \mathcal{H}, \mathcal{D}\big)$ *of dimension* $n \in \mathbb{N}$ *where* $\frac{1}{2\pi}\Theta$ *is a real* $n \times n$ *real skew-symmetric badly approximable matrix, and a selfadjoint one-form* $A = L(-iA_\alpha) \otimes \gamma^\alpha$*. Then, the full spectral action of* $\mathcal{D}_A = \mathcal{D} + A + \epsilon J A J^{-1}$ *is*
(i) for $n = 2$, $\mathcal{S}(\mathcal{D}_A, f, \Lambda) = 4\pi\, f_2\, \Lambda^2 + \mathcal{O}(\Lambda^{-2})$,
(ii) for $n = 4$, $\mathcal{S}(\mathcal{D}_A, f, \Lambda) = 8\pi^2\, f_4\, \Lambda^4 - \frac{4\pi^2}{3}\, f(0)\, \tau(F_{\mu\nu}F^{\mu\nu}) + \mathcal{O}(\Lambda^{-2})$,
(iii) More generally, in $\mathcal{S}(\mathcal{D}_A, f, \Lambda) = \sum_{k=0}^{n} f_{n-k}\, c_{n-k}(A)\, \Lambda^{n-k} + \mathcal{O}(\Lambda^{-1})$, $c_{n-2}(A) = 0$, $c_{n-k}(A) = 0$ *for* k *odd. In particular,* $c_0(A) = 0$ *when* n *is odd.*

This result (for $n = 4$) has also been obtained in[41] using the heat kernel method. It is however interesting to get the result via direct computations of (33) since it shows how this formula is efficient.

Remark 8.1. Note that all terms must be gauge invariants, namely invariant by $A_\alpha \longrightarrow \gamma_u(A_\alpha) = uA_\alpha u^* + u\delta_\alpha(u^*)$. A particular case is $u = U_k$ where $U_k\delta_\alpha(U_k^*) = -ik_\alpha U_0$.

In the same way, note that there is no contradiction with the commutative case where, for any selfadjoint one-form A, $\mathcal{D}_A = \mathcal{D}$ (so A is equivalent to 0!), since we assume in Theorem 8.2 that Θ is badly approximable, so $\mathcal{A}$ cannot be commutative.

Conjecture 8.1. *The constant term of the spectral action of $\mathcal{D}_A$ on the noncommutative n-torus is proportional to the constant term of the spectral action of $\mathcal{D} + A$ on the commutative n-torus.*

Remark 8.2. The appearance of a Diophantine condition for Θ has been characterized in dimension 2 by Connes [22, Prop. 49] where in this case, $\Theta = \theta\left(\begin{smallmatrix} 0 & 1 \\ -1 & 0 \end{smallmatrix}\right)$ with $\theta \in \mathbb{R}$. In fact, the Hochschild cohomology $H(\mathcal{A}_\Theta, {\mathcal{A}_\Theta}^*)$ satisfies $\dim H^j(\mathcal{A}_\Theta, {\mathcal{A}_\Theta}^*) = 2$ (or 1) for $j = 1$ (or $j = 2$) if and only if the irrational number θ satisfies a Diophantine condition like $|1 - e^{i2\pi n\theta}|^{-1} = \mathcal{O}(n^k)$ for some k.

The result of Theorem 8.2 without this Diophantine condition is unknown.

Recall that when the matrix Θ is quite irrational (the lattice generated by its columns is dense after translation by $\mathbb{Z}^n$, see [49, Def. 12.8]), then the C*-algebra generated by $\mathcal{A}_\Theta$ is simple.

It is possible to go beyond the Diophantine condition: see.[41]

Acknowledgments

I would like to thank Driss Essouabri, Victor Gayral, José Gracia-Bondía, Cyril Levy, Pierre Martinetti, Thierry Masson, Thomas Schücker, Andrzej Sitarz, Jo Várilly and Dmitri Vassilevich, for our discussions during our collaborations along years. Certain results presented here are directly extracted from these collaborations. I also benefitted from the questions of participants (professors and students) of the school during the lectures, which helped clarify a few points.

I would like to emphasize the huge kindness of the organizers of this school, Alexander Cardona, Hermán Ocampo, Sylvie Paycha and Andrés Reyes which seems to be a natural colombian quality. I strongly encourage all students or professors to participate to this school in a wonderful country which deserves to be visited despite difficulties.

References

1. W. Arendt, R. Nittka, W. Peter and F. Steiner, "Weyl's law: Spectral properties of the Laplacian in mathematics and physics", in *Mathematical Analysis of Evolution, Information, and Complexity*, W. Arendt and W. Schleich ed., Wiley-VCH Verlag GmbH & Co. KGaA, Weinheim, 2009.
2. C. Bär, "Conformal structures in noncommutative geometry", J. Noncomm. Geom. **1** (2007), 385–395.
3. R. Beals and P. Greiner, *Calculus on Heisenberg Manifolds*, Princeton University Press, Princeton, 1988.
4. N. Berline, E. Getzler and M. Vergne, *Heat Kernels and Dirac Operators*, Springer-Verlag, Berlin Heidelberg New York, 1992.
5. A. Besse, *Einstein Manifolds*, Springer-Verlag, Berlin Heidelberg New York, 2002.
6. J.-P. Bourguignon and P. Goduchon, "Spineurs, opérateurs de Dirac et variations", Commun. Math. Phys. **144** (1992), 581–599.
7. R. Brooks and C. Gordon, "Isospectral families of conformally equivalent Riemannian metric", Bull. Amer. Math. Soc. **23** (1990), 433–436.
8. A. Carey, V. Gayral, A. Rennie and F. Sukochev, "Integration on locally compact noncommutative spaces", J. Funct. Anal. **263** (2012), 383–414.
9. A. Carey, J. Phillips and F. Sukochev, "Spectral flow and Dixmier traces", Adv. Math. **173** (2003), 68–113.
10. A. Chamseddine and A. Connes, "The spectral action principle", Commun. Math. Phys. **186** (1997), 731–750.
11. A. Chamseddine and A. Connes, "Universal formula for noncommutative geometry actions: Unification of gravity and the standard model", Phys. Rev. Letters **77** (1996), 4868–4871.
12. A. Chamseddine and A. Connes, "Inner fluctuations of the spectral action", J. Geom. Phys. **57** (2006), 1–21.
13. A. Chamseddine and A. Connes, "Scale invariance in the spectral action", J. Math. Phys. **47** (2006), 063504.
14. A. Chamseddine and A. Connes, "Quantum gravity boundary terms from the spectral action on noncommutative space", Phys. Rev. Lett. **99** (2007), 071302.
15. A. Chamseddine and A. Connes, "Why the Standard Model", J. Geom. Phys. **58** (2008), 38–47.
16. A. Chamseddine and A. Connes, "The uncanny precision of the spectral action", Commun. Math. Phys. **293** (2010), 867–897.
17. A. Chamseddine and A. Connes, "Noncommutative geometry as a framework for unification of all fundamental interactions including gravity. Part I", Fortschritte der Physik **58** (2010), 553–600.
18. A. Chamseddine and A. Connes, "Noncommutative geometric spaces with boundary: spectral action", J. Geom. Phys. **61** (2011), 317–332.
19. A. Chamseddine and A. Connes "Spectral Action for Robertson–Walker metrics", arXiv:1105.4637v1 [hep-th].
20. A. Chamseddine, A. Connes and M. Marcolli, "Gravity and the standard model with neutrino mixing", Adv. Theor. Math. Phys. **11** (2007), 991–

1090.
21. A. Connes, "C^*-algèbres et géométrie différentielle", C. R. Acad. Sci. Paris **290** (1980), 599–604.
22. A. Connes, "Noncommutative differential geometry", Pub. Math. IHÉS **39** (1985), 257–360.
23. A. Connes, "The action functional in non-commutative geometry", Commun. Math. Phys. **117** (1988), 673–683.
24. A. Connes, *Noncommutative Geometry*, Academic Press, London and San Diego, 1994.
25. A. Connes, "Noncommutative geometry and reality", J. Math. Phys. **36** (1995), 6194–6231.
26. A. Connes, "Gravity coupled with matter and the foundation of non commutative geometry", Commun. Math. Phys. **182** (1996), 155–177.
27. A. Connes, "Noncommutative geometry and the standard model with neutrino mixing", JHEP **0611**:081 (2006).
28. A. Connes, "On the spectral characterization of manifolds", arXiv:0810.2088v1.
29. A. Connes, "A unitary invariant in Riemannian geometry", Inter. Journ. Geom. Meth. Mod. Phys. **5** (2008), 1215–1242.
30. A. Connes and M. Marcolli, *Noncommutative Geometry, Quantum Fields and Motives*, Colloquium Publications, Vol. 55, American Mathematical Society, 2008.
31. A. Connes and H. Moscovici, "Transgression du caractère de Chern et cohomologie cyclique", C. R. Acad. Sci. Paris, Série **I 303** (1986), 913–918.
32. A. Connes and H. Moscovici, "Transgression and Chern character of finite-dimensional K-cycles", Commun. Math. Phys. **155** (1993), 103–122.
33. A. Connes and H. Moscovici, "The local index formula in noncommutative geometry", Geom. Funct. Anal. **5** (1995), 174–243.
34. J. Dixmier, "Existence de traces non normales", C. R. Acad. Sci. Paris **262A** (1966), 1107–1108.
35. A. Edery, "Multidimensional cut-off technique, odd-dimensional Epstein zeta functions and Casimir energy of massless scalar fields", J. Phys. A: Math. Gen. **39** (2006), 678–712.
36. D. Essouabri, B. Iochum, C. Levy and A. Sitarz, "Spectral action on noncommutative torus", J. Noncommut. Geom. **2** (2008), 53–123.
37. R. Estrada, J. M. Gracia-Bondía and J. C. Várilly, "On summability of distributions and spectral geometry", Commun. Math. Phys. **191** (1998), 219–248.
38. S. Fulling, "Some properties of Riesz means and spectral expansions", Electron. J. Diff. Equ. **6**(1999), 1–39.
39. V. Gayral, J. M. Gracia-Bondía, B. Iochum, T. Schücker and J. C. Várilly, "Moyal planes are spectral triples", Commun. Math. Phys. **246** (2004), 569–623.
40. V. Gayral and B. Iochum, "The spectral action for Moyal plane", J. Math. Phys. **46** (2005), 043503.
41. V. Gayral, B. Iochum and D. Vassilevich, "Heat kernel and number theory

on NC-torus", Commun. Math. Phys. **273** (2007), 415–443.
42. S. Giddings, D. Marlof and J. Hartle, "Observables in effective gravity", Phys. Rev. D **74** (2006) 064018.
43. P. B. Gilkey, *Invariance Theory, the Heat equation, and the Atiyah–Singer Index Theory*, CRC Press, Boca Raton, 1995.
44. P. B. Gilkey, *Asymptotic Formulae in Spectral Geometry*, Chapman & Hall, 2003.
45. P. B. Gilkey and G. Grubb, "Logarithmic terms in asymptotic expansions of heat operator traces", Comm. Partial Diff. Equ. **23** (1998), 777–792.
46. N. Ginoux, *The Dirac spectrum*, Lecture Notes in Mathematics 1976, Springer-Verlag, Berlin Heidelberg 2009.
47. G. Grubb and R. Seeley, "Weakly parametric pseudodifferential operators and the Atiyah–Patodi–Singer boundary problem", Invent. Math. **121** (1995), 481–529.
48. J. M. Gracia-Bondía and J. C. Várilly, "Algebras of distributions suitable for phase-space quantum mechanics I", J. Math. Phys. **29** (1988), 869–879.
49. J. M. Gracia-Bondía, J. C. Várilly and H. Figueroa, *Elements of Noncommutative Geometry*, Birkhäuser Advanced Texts, Birkhäuser, Boston, 2001.
50. H. Grosse and R. Wulkenhaar, "8D-spectral triple on 4D-Moyal space and the vacuum of noncommutative gauge theory", J. Geom. Phys. **62** (2012), 1583–1599.
51. V. Guillemin, "A new proof of Weyl's formula on the asymptotic distribution of eigenvalues", Adv. Math. **55** (1985), 131–160.
52. V.W. Guillemin, S. Sternberg and J. Weitsman, "The Ehrhart function for symbols", arXiv:math.CO/06011714.
53. F. Hanisch, F. Pfäffle and C. Stephan, "The spectral action for Dirac operators with skew-symmetric torsion", Commun. Math. Phys. **300** (2010, 877–888.
54. G. Hardy, *Divergent Series*, Clarendon Press, Oxford, 1949.
55. G. H. Hardy and E. M. Wright, *An Introduction to the Theory of Numbers*, Clarendon, Oxford, 1979.
56. O. Hijazi, "Spectral properties of the Dirac operator and geometrical structures", Lecture notes for the Summer School in Villa de Leyva, Columbia, 1999. http://www.iecn.u-nancy.fr/~hijazi/HTML/l2.pdf
57. B. Iochum, D. Kastler and T. Schücker, "Fuzzy mass relations for the Higgs", J. Math. Phys. **36** (1995), 6232–6254.
58. B. Iochum and C. Levy, "Tadpoles and commutative spectral triples", J. Noncommut. Geom. **5** (2011), 299–329.
59. B. Iochum and C. Levy, "Spectral triples and manifolds with boundary", J. Func. Anal. **260** (2011), 117–134.
60. B. Iochum, C. Levy and A. Sitarz, "Spectral action on $SU_q(2)$", Commun. Math. Phys. **289** (2009), 107–155.
61. B. Iochum, C. Levy and D. Vassilevich, "Spectral action for torsion with and without boundaries", Commun. Math. Phys., **310** (2012), 367–382.
62. B. Iochum, C. Levy and D. Vassilevich, "Global and local aspects of spectral actions", J. Phys. A: Math. Theor. **45** (2012), 374020.

63. J.-H. Jureit and C. Stephan, "On a classification of irreducible almost commutative geometries IV", J. Math. Phys. **49** (2008), 033502.
64. J.-H. Jureit and C. Stephan, "Finding the standard model of particle physics, a combinatorial problem", Computer Phys. Commun. **178** (2008), 230–247.
65. T. Kato, *Perturbation Theory For Linear Operators*, Springer–Verlag, Berlin-Heidel-berg-New York, 1980.
66. D. Kolodrubetz and M. Marcolli, "Boundary conditions of the RGE flow in the noncommutative geometry approach to particle physics and cosmology", Phys. Lett. **B 693** (2010), 166–174.
67. T. Krajewski, "Classification of finite spectral triples", J. Geom. Phys. **18** (1997), 1–30.
68. G. Landi, *An Introduction to Noncommutative Spaces and Their Geometries*, Springer-Verlag, Berlin Heidelberg 1997.
69. G. Landi and C. Rovelli, "General relativity in terms of Dirac eigenvalues", Phys. Rev. Lett. **78** (1997), 3051–3054.
70. H. B. Lawson and M.-L. Michelsohn, *Spin Geometry*, Princeton Univ. Press, Princeton, 1989.
71. F. Lizzi, G. Mangano, G. Miele and G. Sparano, "Constraints on unified gauge theories from noncommutative geometry", Mod. Phys. Lett. **A11** (1996), 2561–2572.
72. S. Lord, D. Potapov and F. Sukochev, "Measures from Dixmier traces and zeta functions", J. Funct. Anal. **259** (2010), 1915–1949.
73. S. Lord, A. Sedaev and F. Sukochev, "Dixmier traces as singular symmetric functionals and applications to measurable operators", J. Funct. Anal. **224** (2005), 72–106.
74. S. Lord and F. Sukochev, "Noncommutative residues and a characterization of the noncommutative integral", Proc. Amer. Math. Soc. **139** (2011), 243–257.
75. S. Lord and F. Sukochev, "Measure Theory in Noncommutative Spaces", in "Symmetry, Integrability and Geometry: Methods and Applications", SIGMA **6** (2010), 072, 36 pages.
76. M. Marcolli, E. Pierpaoli and K. Teh, "The spectral action and cosmic topology", Commun. Math. Phys. **304** (2011), 125–174.
77. M. Marcolli, E. Pierpaoli and K. Teh, "The coupling of topology and inflation in noncommutative cosmology", Commun. Math. Phys. **309** (2012), 341–360.
78. M. Marcolli and E. Pierpaoli, "Early universe models from noncommutative geometry", Adv. Theor. Math. Phys. **14** (2010), 1373–1432.
79. C. P. Martin, J. M. Gracia-Bondía and J. C. Várilly, "The standard model as a noncommutative geometry: the low-regime", Phys. Rep. **294** (1998), 363–406.
80. J. E. Moyal, "Quantum mechanics as a statistical theory", Proc. Camb. Philos. Soc. **45** (1949), 99–124.
81. W. Nelson and M. Sakellariadou, "Inflation mechanism in asymptotic noncommutative geometry ", Phys. Lett. **B 680** (2009), 263–266.

82. F. Nicola and L. Rodino, *Global Pseudodifferential Calculus on Euclidean Spaces* Birkäuser, Springer Basel, 2010.
83. S. Paycha, "Regularized integrals, sums and traces. An analytic point of view (Preliminary draft)", 2010.
84. F. Pfäffle and C. Stephan, "On gravity, torsion and the spectral action principle", J. Funct. Anal. **261** (2012), 1529–1565).
85. F. Pfäffle and C. Stephan, "The Holst action by the spectral action principle", Commun. Math. Phys. **307** (2011), 261–273.
86. R. Ponge, "Schwartz kernels and noncommutative residue (Preliminary draft)", http://www.ms.u-tokyo.ac.jp/~ponge/`R. Ponge's homepage`
87. R. Ponge, "Calcul hypoelliptique sur les variétés de Heisenberg, résidu non commutatif et géométrie pseudo-hermitienne", PhD Thesis, Univ. Paris-Sud (Orsay), Dec. 2000, http://www.ms.u-tokyo.ac.jp/~ponge/theses/these.pdf
88. M. Reed and B. Simon, *Methods of Modern Mathematical Physics, II: Fourier Analysis, Self-Adjointness*, Academic Press, New York, 1975.
89. A. Rennie, "Commutative geometries are spin manifolds", Rev. Math. Phys. **13** (2001), 409–464.
90. A. Rennie, "Poincaré duality and spin^c structures for complete noncommutative manifolds", math-ph/0107013.
91. A. Rennie, "Smoothness and locality for nonunital spectral triples", K-Theory, **28** (2003), 127–165.
92. A. Rennie and J. Várilly, "Reconstruction of manifolds in noncommutative geometry", arXiv:math/0610418, (2006).
93. M. A. Rieffel, "C^*-algebras associated with irrational rotations", Pac. J. Math. **93** (1981), 415–429.
94. M. A. Rieffel, *Deformation Quantization for Actions of* $\mathbb{R}^d$, Memoirs of the Amer. Math. Soc. **506**, Providence, RI, 1993.
95. M. Sakellariadou, "Cosmology within noncommutative spectral geometry", arXiv:1101.2174v1 [hep-th].
96. S. Schopka, "Noncommutative Einstein manifolds", Shaker Verlag GmbH, Germany, 2007.
97. T. Schücker, "Forces from Connes' geometry", in 'Topology and Geometry in Physics', Eds: E. Bick and F. Steffen Lecture Notes in Physics **659** (2005), 285–350.
98. L. Schwartz, *Méthodes mathématiques pour les sciences physiques*, Hermann, Paris, 1979.
99. M. Shubin, *Pseudodifferential Operators and Spectral Theory*, Springer Verlag 1987, 2nd ed. 2001.
100. B. Simon, *Trace Ideals and Their Applications*, Second Edition, AMS Mathematical Surveys and Monographs 120, Providence, 2005.
101. M. Takesaki, *Theory of Operator Algebras II*, Springer-Verlag, Berlin Heidelberg New York 2003.
102. M. Taylor, *Pseudodifferential Operators*, Princeton Univ. Press, Princeton, NJ, 1981.
103. F. J. Vanhecke, "On the product of real spectral triples", Lett. Math. Phys.

50 (1999), 151–162.
104. J. Várilly, *Dirac Operators and Spectral Geometry*, http://www.impan.pl/Manuals/Varilly_dosg.pdf
105. J. Várilly, *An Introduction to Noncommutative Geometry*, European Math. Soc. Publishing House (EMS Series of Lectures in Mathematics), 2006.
106. J. C. Várilly and J. M. Gracia-Bondía, "Algebras of distributions suitable for phase-space quantum mechanics II: Topologies on the Moyal algebra", J. Math. Phys. **29** (1988), 880–887.
107. D. Vassilevich, "Heat kernel expansion: user's manual", Phys. Rep. **388** (2003), 279–360.
108. A. van de Ven, "Index-free heat kernel coefficients", Class. Quant. Grav. **15** (1998), 2311–2344.
109. M. Wodzicki, "Local invariants of spectral asymmetry", Invent. Math. **75** (1984), 143–177.
110. M. Wodzicki, "Noncommutative residue. Chapter I: Fundamentals", 320–399, in *K-theory, Arithmetic and Geometry*, Yu. I. Manin, ed., Lecture Notes in Math. **1289**, Springer, Berlin 1987.

INDEX THEORY FOR NON-COMPACT G-MANIFOLDS

Maxim Braverman*

Department of Mathematics
Northeastern University
Boston, MA 02115
USA
E-mail: ab_author@university.com

Leonardo Cano

Mathematics Department
Universidad de los Andes
Bogotá, Colombia
E-mail: an_author@laboratory.com

The index theorem, discovered by Atiyah and Singer in 1963,[6] is one of most important results in the twentieth century mathematics. It has found numerous applications in analysis, geometry and physics. Since it was discovered numerous generalizations have been made, see for example[3–5,13,17] to mention a few; some of these generalizations gave rise to new very productive areas of mathematics. In these lectures we first review the classical Atiyah-Singer index theorem and its generalization to so called transversally elliptic operators[3] due to Atiyah and Singer. Then we discuss the recent developments aimed at generalization of the index theorem for transversally elliptic operators to non-compact manifolds,[11,25]

Keywords: Index theorem; non-compact manifolds.

1. The Fredholm Index

In this section we define the index of an operator A and discuss its main properties. The index is useful and nontrivial for operators defined on an infinite dimensional vector space. To explain the main idea of the definition, let us start with the finite dimensional case.

*Supported in part by the NSF grant DMS-1005888.

1.1. *Finite dimensional case*

Consider a linear operator $A : \mathcal{H}_1 \to \mathcal{H}_2$ between two *finite dimensional* vector spaces $\mathcal{H}_1$ and $\mathcal{H}_2$. Then the *index* of A is defined to be

$$\mathrm{Ind}(A) \;:=\; \dim \mathrm{Ker}(A) \;-\; \dim \mathrm{Coker}(A) \;\in\; \mathbb{Z}, \tag{1}$$

where $\mathrm{Coker}(A) := \mathcal{H}_2/\,\mathrm{Im}(A)$. Notice that, although $\dim \mathrm{Ker}(A)$ and $\dim \mathrm{Coker}(A)$ depend on A, the index

$$\mathrm{Ind}(A) \;=\; \dim \mathcal{H}_1 \;-\; \dim \mathcal{H}_2 \tag{2}$$

depends only on the spaces $\mathcal{H}_1$ and $\mathcal{H}_2$.

1.2. *The Fredholm index*

Now suppose the spaces $\mathcal{H}_1$ and $\mathcal{H}_2$ are infinite dimensional Hilbert spaces. Then, in general, both $\mathrm{Ker}(A)$ and $\mathrm{Coker}(A)$ may be infinite dimensional and the index of A cannot be defined.

Exercise 1.3. Show that if the dimension of $\mathrm{Coker}(A)$ is finite, then $\mathrm{Im}(A)$ is a closed subspace of $\mathcal{H}_2$.

Definition 1.4. Let $\mathcal{H}_1, \mathcal{H}_2$ be two Hilbert spaces and let $A : \mathcal{H}_1 \to \mathcal{H}_2$ be a bounded linear operator. We say that A is a *Fredholm operator* iff $\dim \mathrm{Ker}(A) < \infty$, $\dim \mathrm{Coker}(A) < \infty$.

If $A : \mathcal{H}_1 \to \mathcal{H}_2$ is a Fredholm operator, then its index $\mathrm{Ind}(A)$ can be defined by (1). The notion of index of a Fredholm operator was introduced by Fritz Noether[20a], who also established the following "stability" property of the index:

Theorem 1.5. a. *If* $A(t)$ *is a Fredholm operator which depends continuously on the parameter* t*, then* $\mathrm{Ind}(A(t))$ *is constant.*
b. *If* K *is compact and* A *Fredholm, then* $\mathrm{Ind}(A+K) = \mathrm{Ind}(A)$.

The following example illustrates the notion of index.

---

[a]Fritz Noether was a fine mathematician with a very interesting and tragic biography. He was a son of Max Noether and a younger brother of Emmy Noether. We refer the reader to[21,22] for a short description of his biography.

1.6. *An example*

Let

$$\mathcal{H} = l_2 := \{(x_1, x_2, \dots \,|\, x_j \in \mathcal{C}, \sum_{j=1}^{\infty} |x_j|^2 < \infty\}, \qquad (3)$$

and consider the linear operator $T : \mathcal{H} \to \mathcal{H}$ defined by

$$T(x_1, x_2, x_3, ...) := (x_3, x_4, ...). \qquad (4)$$

We have

$$\mathrm{Ker}(T) = \{(x_1, x_2, 0, 0, ...) \in \mathcal{H} \,\big|\, x_1, x_2 \in \mathbb{C}\}.$$

Thus $\dim \mathrm{Ker}(T) = 2$. Since the image of A is the whole space $\mathcal{H}$, we conclude that $\mathrm{Coker}(T) = \{0\}$. Thus $\mathrm{Ind}(T) = 2$.

Remark 1.7. If $\mathcal{H}$ is a finite dimensional vector space then it follows from (2) that $\mathrm{Ind}(A) = 0$ for every linear operator $A : \mathcal{H} \to \mathcal{H}$. The example above shows that this is not true if $\dim \mathcal{H} = \infty$.

Exercise 1.8. Suppose $A : \mathcal{H}_1 \to \mathcal{H}_2$ and $B : \mathcal{H}_2 \to \mathcal{H}_3$ are Fredholm operators. Show that the operator $BA : \mathcal{H}_1 \to \mathcal{H}_3$ is Fredholm and that

$$\mathrm{Ind}(BA) = \mathrm{Ind}(A) + \mathrm{Ind}(B).$$

Exercise 1.9. Let $A : \mathcal{H}_1 \to \mathcal{H}_2$ be a Fredholm operator. Show that the adjoint operator $A^* : \mathcal{H}_2 \to \mathcal{H}_1$ is Fredholm and that

$$\mathrm{Ind}(A) = -\,\mathrm{Ind}(A^*) = \dim \mathrm{Ker}(A) - \dim \mathrm{Ker}(A^*).$$

1.10. *Application of the index*

One of the most common applications of the index is based on Theorem 1.5. Suppose for example that $\mathrm{Ind}(A) > 0$. This means that $\mathrm{Ker}(A) \neq \{0\}$, i.e., the equation $Ax = 0$ has a non-trivial solution. Moreover, for any compact operator K we have $\mathrm{Ind}(A + K) = \mathrm{Ind}(K) > 0$. Hence, the equation

$$(A + K)\,x = 0 \qquad (5)$$

also has a non-trivial solution. In applications to compute the index of an operator B we often compute the index of a simpler operator A, for which the kernel and cokernel can be explicitly computed, and then show that $B = A + K$ for K a compact operator. If the index of A is positive, we conclude that equation (5) has a non-trivial solution, even though we cannot find this solution explicitly.

1.11. *Connected components of the set of Fredholm operators*

Let $\mathrm{Fred}(\mathcal{H}_1, \mathcal{H}_2)$ denote the set of Fredholm operators $A : \mathcal{H}_1 \to \mathcal{H}_2$. It is a metric space with the distance defined by $d(A, B) := \|A - B\|$. Theorem 1.5 implies that if operators A and B belong to the same connected component of $\mathrm{Fred}(\mathcal{H}_1, \mathcal{H}_2)$, then $\mathrm{Ind}(A) = \mathrm{Ind}(B)$. In other words, $\mathrm{Ind}(A)$ is an invariant of the connected component of $\mathrm{Fred}(\mathcal{H}_1, \mathcal{H}_2)$. In fact, $\mathrm{Ind}(A)$ determines the connected component of A in $\mathrm{Fred}(\mathcal{H}_1, \mathcal{H}_2)$ (see [19, Proposition 7.1]).

1.12. *The group action*

The index $\mathrm{Ind}(A)$ of a Fredholm operator A is an integer. If a compact group G acts on a Hilbert space $\mathcal{H}$ one can define a richer invariant, as we shall now explain.

Recall that a G-representation V is called *irreducible* if it has no non-trivial G-invariant subspaces. We denote by $\mathrm{Irr}\,(G)$ the set of irreducible representations of G. The next theorem, cf.,[12] shows that they are the building-blocks of all the other representations.

Theorem 1.13. *Any finite dimensional representation U of G has a unique decomposition into a sum of irreducible representations:*

$$U = \bigoplus_{V \in \mathrm{Irr}\,(G)} m_V V.$$

Here $m_V V$ stands for the direct sum $V \oplus V \oplus \cdots \oplus V$ of m_V copies of V.

The numbers $m_V \in \mathbb{N}$ are called the *multiplicities of the irreducible representation V in U.*

Definition 1.14. Let G be a compact group acting on the vector spaces $\mathcal{H}_1$ and $\mathcal{H}_2$. A linear transformation $A : \mathcal{H}_1 \to \mathcal{H}_2$ is *G-equivariant* iff it commutes with the action of the group, i.e.,

$$gAx = Agx, \qquad \text{for all} \quad x \in \mathcal{H}_1.$$

Exercise 1.15. Let G be a compact group acting on Hilbert spaces $\mathcal{H}_1$ and $\mathcal{H}_2$. Let $A : \mathcal{H}_1 \to \mathcal{H}_2$ be a G-equivariant Fredholm operator.

a. Show that $\mathrm{Ker}(A)$ is invariant under the action of G, i.e. for any $x \in \mathrm{Ker}(A)$, $g \in G$ we have $gx \in \mathrm{Ker}(A)$. Thus the restriction of the action of G on $\mathcal{H}_1$ to $\mathrm{Ker}(A)$ defines a G-action on $\mathrm{Ker}(A)$.

b. Define a natural action of G on $\mathrm{Coker}(A)$.

The exercise above and Theorem 1.13 show that there is a unique decomposition

$$\mathrm{Ker}(A) = \bigoplus_{V \in \mathrm{Irr}\,(G)} m_V^+ V; \qquad \mathrm{Coker}(A) = \bigoplus_{V \in \mathrm{Irr}\,(G)} m_V^- V.$$

Remark 1.16. Note the though the set $\mathrm{Irr}\,(G)$ is infinite both sums above are actually finite, since only finitely many numbers m_+, m_- can be non-zero.

1.17. *The ring of characters*

Under the direct sum $\oplus$ the set of finite dimensional representations of the group G form an abelian semigroup with identity. Thus we can associate to it the *Grothendieck group* $R(G)$ by considering formal differences $V - U$ of finite dimensional representations of G. Let us give a formal definition:

Definition 1.18. Consider the set of formal differences, $V_1 - V_2$, of finite dimensional representations V_1 and V_2 of G. The quotient of this set by the equivalence relation[b]

$$V - U \;\sim\; A - B \iff V \oplus B \simeq A \oplus U,$$

is an abelian group denoted by $R(G)$ and called the *group of characters of* G. Formally,

$$R(G) \;:=\; \big\{(V,U)\big|\, V, U \ \textit{are finite dimensional representations of}\ G\,\big\}/\sim.$$

The tensor product of representations defines a ring structure $R(G)$, which is why $R(G)$ is called the *ring of characters of* G. We will not use the ring structure on $R(G)$ in these lectures and will not give a formal definition.

Using Theorem 1.13 one easily obtains the following alternative description of $R(G)$:

Proposition 1.19. *As an abelian group, the ring of characters* $R(G)$ *is isomorphic to the free abelian group generated by the set* $\mathrm{Irr}\,(G)$ *of irreducible*

[b]In general, the definition of the equivalence relation in the Grothendieck group is slightly more complicated, but in our case it is equivalent to the one given here.

representations of G*, i.e., to the group of expressions*

$$\Big\{ \bigoplus_{V \in \mathrm{Irr}\,(G)} m_V V \,\Big|\, m_V \in \mathbb{Z}, \text{ only finitely many numbers } m_V \neq 0 \Big\}. \qquad (6)$$

Exercise 1.20. Introduce a ring structure on the group (6) and show that the isomorphism of Proposition 1.19 is an isomorphism of rings.

Remark 1.21. If one allows infinite formal sums in (6) then one obtains the definition of the *completed ring of characters* of G:

$$\widehat{R}(G) = \Big\{ \bigoplus_{V \in \mathrm{Irr}\,(G)} m_V V \,\Big|\, m_V \in \mathbb{Z} \Big\}. \qquad (7)$$

This ring will play an important role in Sections 4 and 6.

1.22. *The equivariant index*

Suppose a compact group G acts on Hilbert spaces $\mathcal{H}_1$ and $\mathcal{H}_2$ and let $A : \mathcal{H}_1 \to \mathcal{H}_2$ be a G-equivariant Fredholm operator.

Definition 1.23. The *equivariant index* $\mathrm{Ind}_G(A)$ of A is defined to be the formal difference between $\mathrm{Ker}(A)$ and $\mathrm{Coker}(A)$ in $R(G)$:

$$\mathrm{Ind}_G(A) := \mathrm{Ker}(A) - \mathrm{Coker}(A) = \bigoplus_{V \in \mathrm{Irr}\,(G)} (m_V^+ - m_V^-)\, V \in R(G). \qquad (8)$$

By Remark 1.16 the sum in the right hand side of (8) is finite.

Example 1.24. Let V_0 denote the trivial representation. If the inequality $m_{V_0}^+ - m_{V_0}^- > 0$ holds, that means that there exist a nontrivial solution of the equation $Ax = 0$ which is invariant under the action of G.

Exercise 1.25. Let

$$G := \mathbb{Z}_2 = \{1, -1\}.$$

To avoid confusion we denote by q the non-trivial element of G (thus $q = -1$). We define an action of G on l^2 by

$$q(x_1, x_2, \ldots) = (x_2, x_1, x_4, x_3, \ldots).$$

Observe that G has two irreducible representations, namely: $V_0 = \mathbb{C}$ with the trivial action, and $V_1 = \mathbb{C}$ with the action $qz = -z$.

Show that the operator (4) is G-equivariant and that $\mathrm{Ker}(T) = V_0 \oplus V_1$. Use this result to compute $\mathrm{Ind}_G(T)$.

2. Differential operators

In this section we introduce differential operators on manifolds and discuss their main properties. We also define elliptic differential operators. If the manifold is compact, then any elliptic operator on it is Fredholm. In the next section we will discuss the index of elliptic differential operators and the Atiyah-Singer theorem, which computes this index in terms of topological information of the manifold.

2.1. *Differential operators*

We recall the definition of linear differential operators in $\mathbb{R}^n$. Let $D_j : C^\infty(\mathbb{R}^n) \to C^\infty(\mathbb{R}^n)$ denote the operator

$$D_j f \;=\; \frac{1}{i}\frac{\partial f}{\partial x_j}, \qquad j = 1, \ldots, n.$$

A *multi-index* is an n-tuple of non-negative integers $\alpha = (\alpha_1, \ldots, \alpha_n) \in \mathbb{Z}^n_{\geq 0}$. For a multiindex α we set

$$|\alpha| \;:=\; \alpha_1 + \cdots + \alpha_n$$

and

$$D^\alpha \;:=\; D_1^{\alpha_1} D_2^{\alpha_1} \cdots D_n^{\alpha_n} : C^\infty(\mathbb{R}^n) \;\longrightarrow\; C^\infty(\mathbb{R}^n).$$

Definition 2.2. A *linear differential operator of order* k, is a linear operator $\mathcal{D}$ of the form

$$\mathcal{D} \;=\; \sum_{|\alpha|\leq k} a_\alpha(x)\, D^\alpha : C^\infty(\mathbb{R}^n) \;\longrightarrow\; C^\infty(\mathbb{R}^n), \tag{1}$$

where $a_\alpha(x) \in C^\infty(\mathbb{R}^n)$.

Let $C_c^\infty(\mathbb{R}^n)$ denote the set of smooth functions with compact support on $\mathbb{R}^n$ and let $\langle\cdot,\cdot\rangle$ denote the scalar product on $C_c^\infty(\mathbb{R}^n)$ defined by

$$\langle f, g\rangle := \int_{\mathbb{R}^n} f(x)\overline{g}(x)dx.$$

Definition 2.3. For $k \in \mathbb{N}$ define the scalar product on $C_c^\infty(\mathbb{R}^n)$ by the formula

$$\langle f, g\rangle_k \;:=\; \sum_{|\alpha|=k} \langle D^\alpha f, D^\alpha g\rangle. \tag{2}$$

The corresponding norm

$$\|f\|_k \;:=\; \sqrt{\langle f, f\rangle_k} \tag{3}$$

is called the *k-th Sobolev norm.*

The *Sobolev space* $H^k(\mathbb{R}^n) \subset L^2(\mathbb{R}^n)$ is the completion of the space $C_c^\infty(\mathbb{R}^n)$ with respect to the norm (3).

The following theorem [1, Corollary 3.19] gives an alternative description of the Sobolev space $H^k(\mathbb{R}^n)$:

Theorem 2.4. *The Sobolev space $H^k(\mathbb{R}^n)$ is equal to the space of square integrable functions whose distributional derivatives up to order k are in $L^2(\mathbb{R}^n)$. More explicitly*

$$H^k(\mathbb{R}^n) = \{f : \mathbb{R}^n \to \mathcal{C} \,|\, D^\alpha f \in L^2(\mathbb{R}^n) \text{ for all } \alpha \in \mathbb{Z}_+^n \text{ with } |\alpha| \le k\}.$$

Remark 2.5. If $\Omega \subset \mathbb{R}^n$ is an open set, one can define the Sobolev space $H^k(\Omega)$ of functions on Ω. An analogue of Theorem 2.4 still holds if the boundary of Ω is sufficiently nice, see [1, Theorem 3.18].

Exercise 2.6. Let $K \subset M$ be a compact set. Denote by

$$H_K^k(\mathbb{R}^n) = \{ f \in H^k(\mathbb{R}^n) \,|\, \operatorname{supp}(f) \subset K \}.$$

For every $m \geq k$ a differential operator $\mathcal{D} : C_c^\infty(\mathbb{R}^n) \to C_c^\infty(\mathbb{R}^n)$ of order k can be extended continuously to a bounded operator $\mathcal{D} : H_K^m(\mathbb{R}^n) \to L^2(\mathbb{R}^n)$.

With a bit more work one can show the following result, cf. [1, Theorem 6.2].

Theorem 2.7. (Rellich's lemma) *If $K \subset \mathbb{R}^n$ is a compact set, then the embedding $H_K^m(\mathbb{R}^n) \hookrightarrow H^k(\mathbb{R}^n)$ is a compact operator for any $m > k$.*

Exercise 2.8. Use Rellich's lemma to show that the restriction of a differential operator $\mathcal{D}$ of order $k < m$ to $H_K^m(\mathbb{R}^n)$ defines a compact operator $H_K^m(\mathbb{R}^n) \to L^2(\mathbb{R}^n)$.

2.9. *Matrix-valued differential operators*

We extend Definition 2.2 to operators acting on vector valued functions on $\mathbb{R}^n$. Let $C^\infty(\mathbb{R}^n, \mathbb{R}^N)$ denote the set of smooth vector valued functions

$$f = (f_1, \ldots, f_N) : \mathbb{R}^n \longrightarrow \mathbb{R}^N, \qquad f_j \in C^\infty(\mathbb{R}^n).$$

Using the scalar product of $\mathbb{R}^N$ we can generalize the constructions of the previous section to define the spaces $L^2(\mathbb{R}^n, \mathbb{R}^N)$, $H^k(\mathbb{R}^n, \mathbb{R}^N)$ and $H_K^k(\mathbb{R}^n, \mathbb{R}^N)$, etc.

An operator $\mathcal{D} : C^\infty(\mathbb{R}^n, \mathbb{R}^{N_1}) \to C^\infty(\mathbb{R}^n, \mathbb{R}^{N_2})$ is a (matrix-valued) differential operator of order k if it is given by the formula (1) with

$$a_\alpha(x) = \{a_{\alpha,ij}(x)\}_{\substack{1\le i\le N_2\\ 1\le j\le N_1}} \in \mathrm{Mat}_{N_1\times N_2}, \qquad a_{\alpha,ij}(x) \in C^\infty(\mathbb{R}^n).$$

As above, a matrix-valued differential operator $\mathcal{D}$ of order k defines a bounded operator

$$\mathcal{D} : H^k(\mathbb{R}^n, \mathbb{R}^N) \longrightarrow L^2(\mathbb{R}^n, \mathbb{R}^N)$$

and a compact operator

$$\mathcal{D} : H^m_K(\mathbb{R}^n, \mathbb{R}^N) \longrightarrow L^2(\mathbb{R}^n, \mathbb{R}^N), \qquad \text{for } m > k.$$

2.10. *Vector bundles*

Consider a smooth manifold M and let E be an N-dimensional complex vector bundle over M. We refer the reader to[27] for the definition and the basic properties of vector bundles. One should remember that E is itself a smooth manifold which is equipped with a *projection map* $\pi : E \to M$. For each $x \in M$ the preimage $\pi^{-1}(x)$ is called the *fiber of* E *over* x and is denoted by E_x. By definition each fiber E_x is a linear space of dimension N.

Example 2.11. a. The *trivial bundle* $E = M \times \mathcal{C}^N$. Here $\pi : E \to M$ is the projection on the first factor. In this case each fiber E_x is canonically isomorphic with $\mathcal{C}^N$.

b. The *tangent bundle* TM. This is a vector bundle whose fiber at each point $x \in M$ is the tangent space T_xM.

A *smooth section* of a vector bundle E is a smooth map $s : M \to E$ such that

$$\pi\big(s(x)\big) = x, \qquad \text{for all } x \in M.$$

In other words, for each $x \in M$ we have an element $s(x) \in E_x$. Notice that for the case of the trivial bundle $E = M \times \mathcal{C}^N$, defining a section s is the same as defining a vector valued function $M \to \mathcal{C}^N$. Thus smooth sections generalize smooth vector valued functions on M.

We denote the set of smooth sections of E by $C^\infty(M, E)$.

2.12. *Differential operators on manifolds*

Let M be a smooth manifold and let E and F be vector bundles over M. We say that a linear operator

$$\mathcal{D}: C^\infty(M,E) \longrightarrow C^\infty(M,F) \tag{4}$$

is *local* if for every section $f \in C^\infty(M,E)$, we have $\operatorname{supp}(\mathcal{D}f) \subset \operatorname{supp}(f)$. In this case, for any open cover $\{U_i\}_{i=1}^m$ of M the operator $\mathcal{D}$ is completely determined by its restriction to functions with supports in one U_i. If the sets U_i are sufficiently small, we can fix coordinates

$$\phi_i : U_i \overset{\approx}{\longrightarrow} V_i \subset \mathbb{R}^n \qquad (i = 1, \ldots, m)$$

in U_i and also trivializations of E and F over U_i. Then the restriction of $\mathcal{D}$ to sections supported in U_i can be identified with a map

$$\mathcal{D}_i : C^\infty(V_i, \mathbb{R}^{N_1}) \longrightarrow C^\infty(V_i, \mathbb{R}^{N_2}).$$

Definition 2.13. A local linear operator $\mathcal{D}: C^\infty(M,E) \to C^\infty(M,F)$ is called a *differential operator of order* k if one can choose a cover $\{U_i\}_{i=1}^m$ of M and coordinate systems on U_i such that for every $i = 1, \ldots, m$, the operator $\mathcal{D}_i$ has the form (1).

2.14. *Sobolev spaces of sections*

We now introduce the Sobolev spaces of sections of a vector bundle E. Choose a measure on M and a scalar product on the fibers of E. Then we can consider the space $L^2(M,E)$ of square integrable section of E. It is not hard to show that if the manifold M is closed (i.e., compact and without boundary), the space $L^2(M,E)$ is independent of the choices of the measure and of the scalar product on E.

One can define the spaces $H^k(M,E)$ of Sobolev sections of E in a way similar to Definition 2.3. Roughly, one uses a partition of unity to define a scalar product $\langle \cdot, \cdot \rangle_k$ on $C_k^\infty(M,E)$ as a combination of scalar products (2) on each coordinate neighborhood. Then the Sobolev space $H^k(M,E)$ is the completion of $C_c^\infty(M,E)$ with respect to the norm defined by this scalar product. We refer to[27] for details. As in the case $M = \mathbb{R}^n$ we have the following:

Theorem 2.15. *Let $\mathcal{D}: C^\infty(M,E) \to C^\infty(M,F)$ be a differential operator of order k.*

a. *For every $m \geq k$ and every compact $K \subset M$, the operator $\mathcal{D}$ extends continuously to an operator*

$$\mathcal{D} : H^m_K(M, E) \longrightarrow L^2(M, F).$$

b. *If $m > k$ then for every compact $K \subset M$ the operator $\mathcal{D} : H^m_K(M, E) \to L^2(M, F)$ is compact.*

2.16. *The symbol of a differential operator*

We now define the notion of the leading symbol of a differential operator, which is crucial for the discussion of the Atiyah-Singer index theorem.

Exercise 2.17. a. Let $\mathcal{D} : C^\infty(M, E) \to C^\infty(M, F)$ be a differential operator of order k. Fix $x_0 \in M$, a cotangent vector $\xi \in T^*_{x_0}M$ and a vector $e \in E_{x_0}$. Let $f \in C^\infty(M)$ be a smooth function, such that $df_{x_0} = \xi$ and let $s \in C^\infty(M, E)$ be a section, such that $s(x_0) = e$. Set

$$\sigma_L(\mathcal{D})(x_0, \xi)\, e \; := \; \lim_{t \to \infty} t^{-k}\, \mathcal{D}\big(e^{itf(x)} s(x)\big)\big|_{x = x_0}. \tag{5}$$

Show that (5) is independent of the choice of f and s and that $\sigma_L(\mathcal{D})(x_0, \xi)\, e$ is linear in e. Thus we can view $\sigma_L(\mathcal{D})(x_0, \xi)$ as a linear map $E_{x_0} \to F_{x_0}$.

b. Suppose that $\mathcal{D}$ is given in local coordinates by the formula (1). Using local coordinates we can identify $T_{x_0}M$ with $\mathbb{R}^n$. Then

$$\sigma_L(\mathcal{D})(x_0, \xi) \;=\; \sum_{|\alpha| = k} a_\alpha(x_0)\, \xi_1^{\alpha_1} \cdots \xi_n^{\alpha_n}. \tag{6}$$

Notice that the exercise above implies that as a function on T^*M, the right hand side of (6) is independent of the choice of a coordinate system near x_0.

Definition 2.18. Let $\mathcal{D} : C^\infty(M, E) \to C^\infty(M, F)$ be a differential operator of order k. The *leading symbol* $\sigma_L(\mathcal{D})(x_0, \xi)$ ($x_0 \in M$, $\xi \in T^*_{x_0}M$) of $\mathcal{D}$ is the element of $\mathrm{Hom}(E_x, F_x)$ defined by (5).

Remark 2.19. Using the local coordinate representation (1) of $\mathcal{D}$ on can also define the *full symbol*

$$\sigma(\mathcal{D})(x_0, \xi) \;=\; \sum_{|\alpha| \leq k} a_\alpha(x_0)\, \xi_1^{\alpha_1} \cdots \xi_n^{\alpha_n}.$$

As opposed to the leading symbol, the full symbol depends on the choice of the local coordinates.

2.20. *The leading symbol as a section of the pullback bundle*

Let $\pi : T^*M \to M$ denote the projection. For a vector bundle E we denote by π^*E the *pullback* bundle over T^*M. This is a vector bundle over T^*M whose fiber over $(x,\xi) \in T^*M$ is the fiber E_x of E, see [27, §5.1] for a precise definition of the pullback bundle.

Let $\mathcal{D} : C^\infty(M,E) \to C^\infty(M,F)$ be a differential operator. Then for $(x,\xi) \in T^*M$, the leading symbol $\sigma_L(\mathcal{D})(x,\xi)$ is a linear map

$$\sigma_L(\mathcal{D})(x,\xi) : \pi^*E_{(x,\xi)} \to \pi^*F_{(x,\xi)}.$$

Hence, we can view $\sigma_L(\mathcal{D})$ as a section of the vector bundle $\mathrm{Hom}(\pi^*E, \pi^*F)$.

2.21. *Elliptic differential operators*

Definition 2.22. A differential operator $\mathcal{D} : C^\infty(M,E) \to C^\infty(M,F)$ is called *elliptic* if $\sigma_L(\mathcal{D})(x,\xi)$ is invertible for all $\xi \neq 0$.

Example 2.23. Consider the *Laplace operator*

$$\Delta = -\frac{\partial^2}{\partial x_1^2} - \frac{\partial^2}{\partial x_2^2} - \cdots - \frac{\partial^2}{\partial x_n^2}$$

on $\mathbb{R}^n$. Its leading symbol $\sigma_L(\Delta)$, given by the formula

$$\sigma_L(\Delta)(\xi_1, \xi_2, \ldots, \xi_n) = \xi_1^2 + \xi_2^2 + \cdots + \xi_n^2,$$

is invertible for $\xi = (\xi_1, \ldots, \xi_n) \neq 0$. Hence, Δ is elliptic.

Elliptic operators play a very important role in analysis and in index theory. The following results, cf.,[26] describes some of the main properties of elliptic operators.

Theorem 2.24. *Suppose $\mathcal{D} : C^\infty(M,E) \to C^\infty(M,F)$ is an elliptic differential operator.*
a. *(Elliptic regularity) If $\mathcal{D}f = u$ and $u \in C^\infty(M,F)$ then $f \in C^\infty(M,E)$.*
b. *If the manifold M is closed, then $\mathcal{D}$ is a Fredholm.*

Part b. of this theorem implies that the index $\mathrm{Ind}(\mathcal{D})$ of an elliptic operator $\mathcal{D}$ on a closed manifold is defined, cf. Section 1. In the next section we will discuss the Atiyah-Singer index theorem, which computes this index. We finish this section with the following

Exercise 2.25. Use Theorem 2.15 to show that $\mathrm{Ind}(\mathcal{D})$ depends only on the leading symbol of $\mathcal{D}$, i.e. if $\sigma_L(\mathcal{D}_1) = \sigma_L(\mathcal{D}_2)$ then $\mathrm{Ind}(\mathcal{D}_1) = \mathrm{Ind}(\mathcal{D}_2)$.

Exercise 2.26. Let $M = S^1$ be the circle and consider the operator

$$\mathcal{D} := -i\frac{d}{dt} + \sin t : C^\infty(M) \to C^\infty(M).$$

Compute $\mathrm{Ind}(\mathcal{D})$.
Hint: Compute the index of a simpler operator with the same leading symbol.

3. The Atiyah-Singer index theorem

In this section we present a K-theoretical formulation of the Atiyah-Singer index theorem.

3.1. *Index as a topological invariant*

At the end of the last section (cf. Exercise 2.25) we saw that the index of an elliptic differential operator on a closed manifold depends only on its leading symbol. Moreover, Theorem 1.5.a implies that the index does not change when we deform the leading symbol. More precisely, let us consider the space $\mathrm{SEll}(E,F)$ of smooth sections of $\sigma(x,\xi) \in \mathrm{Hom}(\pi^*E, \pi^*F)$, which are invertible for $\xi \neq 0$. We refer to $\mathrm{SEll}(E,F)$ as the *space of elliptic symbols*. We endow it with the topology of uniform convergence on compact sets.

Theorem 3.2. *Let E and F be vector bundles over a compact manifold M. The index of an elliptic differential operator $\mathcal{D} : C^\infty(M,E) \to C^\infty(M,F)$ is determined by the connected component of* $\mathrm{SEll}(E,F)$ *in which the leading symbol $\sigma_L(\mathcal{D})$ lies.*

This result shows that $\mathrm{Ind}(\mathcal{D})$ is a topological invariant. About 50 years ago, Israel Gel'fand[14,15] formulated a problem: how to compute the index of an elliptic operator using only its leading symbol. This problem was solved brilliantly by Atiyah and Singer[6,7] (see also, for example,[9,23]). They did more than just a calculation of the index. They associated to each elliptic symbol an element of so-called K-theory (see Section 3.5 below). Then they associated a number – *the topological index* $\text{t-Ind}(\mathcal{D})$ – to each element of the K-theory. Schematically, their construction can be expressed as

$$\mathcal{D} \rightsquigarrow \sigma_L(\mathcal{D}) \rightsquigarrow \text{an element of } K\text{-theory} \rightsquigarrow \text{t-Ind}(\sigma_L(\mathcal{D}))) \in \mathbb{Z}. \quad (1)$$

The composition of the arrows in the diagram above leads to a map

$$\text{elliptic operators} \to \mathbb{Z}, \qquad \mathcal{D} \mapsto \text{t-Ind}(\sigma_L(\mathcal{D}))$$

which is called the topological index. Note that, as we will explain below, the topological index is constructed using purely topological methods, without any analysis involved.

The following result is the simplest form of the Atiyah-Singer index theorem

Theorem 3.3. (Atiyah-Singer) $\mathrm{Ind}(\mathcal{D}) = \text{t-Ind}(\sigma_L(\mathcal{D}))$.

Remark 3.4. Although by Exercise 2.25 we know that the index of $\mathcal{D}$ could be computed by the leading symbol $\sigma_L(\mathcal{D})$, we emphasize the power of this theorem. The index of $\mathcal{D}$ gives information about the kernel and the cokernel of $\mathcal{D}$, i.e., about the spaces of solutions of the differential equations $\mathcal{D}f = 0$ and $\mathcal{D}^*u = 0$. The index theorem allows one to obtain at least part of this information by purely topological methods, *without solving the differential equations.*

We shall now explain the meaning of the terms in (1). First, we will briefly review the notion of the (topological) K-theory.

3.5. *K-theory*

The direct sum $\oplus$ defines a structure of a semigroup on the set of vector bundles ovex X. If X is a compact manifold, then the K-theory of X is just the Grothendieck group of this semigroup. In general, the definition is a little bit more complicated. This is, essentially, because we are interested in the K-theory *with compact support.* In fact, one of the definitions of the K-theory of a non-compact manifold X is the K-theory of the *one-point compactification of* X. It will be more convenient for us to use an equivalent definition, which is based on considering bundle maps

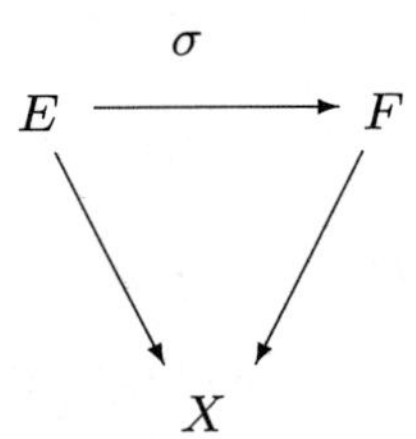

such that the induced map of fibers $\sigma(x) : E_x \to F_x$ is invertible for all x outside of a compact set $K \subset X$. Two such maps $\sigma_1 : E_1 \to F_1$ and $\sigma_2 : E_2 \to F_2$ are said to be equivalent if there exist integers $k_1, k_2 \geq 0$ such that the maps

$$\sigma_1 \oplus \mathrm{Id} : E_1 \oplus \mathcal{C}^{k_1} \to F_1 \oplus \mathcal{C}^{k_1}, \quad \text{and} \quad \sigma_2 \oplus \mathrm{Id} : E_2 \oplus \mathcal{C}^{k_2} \to F_2 \oplus \mathcal{C}^{k_2}, \tag{2}$$

are *homotopic in the class of maps invertible outside of a compact set.* In this case we write $\sigma_1 \sim \sigma_2$.

Definition 3.6. Let X be a topological space. The *K-theory group* $K(X)$ of X is defined by

$$K(X) = \{\sigma : E \to F \mid \sigma(x) \text{ is invertible outside of a compact set}\}/\sim$$

Remark 3.7. If X is compact then any two maps $\sigma_1, \sigma_2 : E \to F$ between the same vector bundles are equivalent with respect to the equivalence relation defined above. Thus $K(X)$ can be described as the set of pairs of vector bundles (E, F) subject to an appropriate equivalence relation. We think about this pair as about *formal difference* of the bundles E and F, and we, usually denote this pair by $E - F$. Thus, as mentioned above, the K-theory group $K(X)$ coincides with the Grothendieck group of the semigroup of vector bundles over X.

The direct sum of vector bundles induces a structure of an abelian group on $K(X)$. The tensor product defines a multiplication. Together these two operations make $K(X)$ a ring.

Example 3.8. If $X = \{pt\}$, then a vector bundle over M is just a vector space and an element of $K(X)$ is a pair of two vector spaces (E, F) up to an equivalence. In this case one easily sees from (2) that the only invariant of the pair (E, F) is the number

$$\dim E - \dim F \in \mathbb{Z}.$$

Thus

$$K(\{pt\}) \cong \mathbb{Z}.$$

Example 3.9. Let $\mathcal{D} : C^\infty(M, E) \to C^\infty(M, F)$ be an elliptic differential operator. Then for each $x \in M$ and each $\xi \in T^*_x M$ we have $\sigma_L(\mathcal{D})(x, \xi) : E_x \to F_x$. In other words, $\sigma(\mathcal{D})$ defines a bundle map

$$\sigma_L(\mathcal{D}) : \pi^\star E \to \pi^\star F. \tag{3}$$

If M is a *compact manifold*, then the ellipticity condition implies that the map (3) is invertible outside of a compact subset of $X = T^*M$. Hence, (3) defines an element of $K(T^*M)$.

This example shows the relevance of the K-theory to computing the index.

3.10. *The pushforward map in K-theory*

One of the most important facts about the K-theory is that it has all the properties of a cohomology theory with compact supports. In particular, given an inclusion $j : Y \hookrightarrow X$ one can define a ring homomorphism

$$j_! : K(Y) \ \to \ K(X),$$

cf.,[7] called *pushforward map.*

The following special case of the *Bott periodicity* theorem[2] plays a crucial role in the definition of the topological index.

Theorem 3.11. (Bott periodicity) *Let $i : \{pt\} \hookrightarrow \mathcal{C}^N$ be an embedding of a point into $\mathcal{C}^N$. Then the pushforward map*

$$i_! : K(pt) \ \longrightarrow \ K(\mathcal{C}^N)$$

is an isomorphism.

3.12. *The topological index*

Consider an embedding $M \hookrightarrow \mathbb{R}^N$ (such an embedding always exists for large enough N by the Whitney embedding theorem). This embedding induces an embedding of the cotangent bundle of M

$$j : T^*M \ \hookrightarrow \mathbb{R}^N \oplus \mathbb{R}^N \ \simeq \ \mathcal{C}^N.$$

We will denote by $i : \{pt\} \hookrightarrow \mathcal{C}^N$ an embedding of a point into $\mathcal{C}^N$. Using the pushforward in K-theory introduced above, we obtain the diagram

$$\begin{array}{ccc} K(T^*M) & \xrightarrow{\ \ j_!\ \ } & K(\mathbb{C}^n) \\ & & \big\uparrow{\scriptstyle i_!} \\ & & K(\{pt\}) \end{array}$$

By Theorem 3.11 we can invert $i_!$ and define the *topological index* as the map

$$\text{t-Ind} \;=\; i_!^{-1} j_! : K(T^*M) \to K(\{pt\}) \;\simeq\; \mathbb{Z}. \tag{4}$$

It is relatively easy to check that this map is independent of the choice of the number N and the embedding $j : M \hookrightarrow \mathbb{R}^N$.

Remark 3.13. Although the definition of the topological index might look abstract, we emphasize that it may be computed explicitly using topological methods as the integrals of certain characteristic classes.[8]

We now introduce a group action into the picture.

3.14. *Equivariant vector bundles*

Suppose a compact group G acts on M. That means that to each $g \in G$ is assigned a diffeomorphism $\phi(g) : M \to M$ such that $\phi(g_1) \circ \phi(g_2) = \phi(g_1 g_2)$.

Definition 3.15. A *G-equivariant vector bundle* over M is a vector bundle $\pi : E \to M$ together with an action

$$g \;\mapsto\; \psi(g) : E \;\to\; E, \qquad g \in G,$$

of G on E such that

$$\pi \circ \psi(g) \;=\; \phi \circ \pi(g).$$

When it does not lead to confusion, we often write $g \cdot x$ and $g \cdot e$ for $\phi(g)(x)$ and $\psi(g)(e)$ respectively.

Definition 3.16. Let E be a G-equivariant vector bundle over M. We define the action of G on the space $C^\infty(M,E)$ of smooth section of E by the formula

$$g \;\mapsto\; l_g : C^\infty(M,E) \;\longrightarrow\; C^\infty(M,E), \qquad l_g(f)(x) \;:=\; g \cdot f(g^{-1} \cdot x). \tag{5}$$

Similarly, we define the action $l_g : L^2(M,E) \to L^2(M,E)$ on the space of square integrable sections of E. In this way $L^2(G)$ becomes a representation of G, called the *left regular representation.*

Suppose E and F are G-equivariant vector bundles over M. A differential operator

$$\mathcal{D} : C^\infty(M,E) \;\longrightarrow\; C^\infty(M,E),$$

is called *G-invariant* if $g \cdot \mathcal{D} = \mathcal{D} \cdot g$.

3.17. *Equivariant K-theory*

If a compact group G acts on M one can define the *equivariant* K*-theory* $K_G(X)$ as the set of the equivalence classes of G-equivariant maps between G-equivariant vector bundles. In particular, G-equivariant K-theory of a point is given by equivalence classes of pairs of finite dimensional representations of G. It should not be a surprise that

$$K_G(\{pt\}) \;=\; R(G), \tag{6}$$

where $R(G)$ is the ring of characters of G, cf. Definition 1.18. The formula (4) generalizes easily to define a G-equivariant topological index

$$\text{t-Ind}_G : K_G(M) \;\rightarrow\; K_G(\{pt\}) \;\simeq\; R(G).$$

All the constructions introduced above readily generalize to the equivariant setting.

3.18. *The Atiyah-Singer index theorem*

Suppose a compact group G acts on a *compact* manifold M. Let E and F be equivariant vector bundles over M and let

$$\mathcal{D} : C^\infty(M, E) \;\rightarrow\; C^\infty(M, F)$$

be a G-invariant elliptic operator. Then the leading symbol defines a G-equivariant map

$$\sigma_L(\mathcal{D}) : \pi^* E \;\longrightarrow\; \pi^* F,$$

which is invertible outside of the compact set $M \subset T^*M$. Thus $\sigma_L \in K_G(T^*M)$ and its topological index $\text{t-Ind}(\sigma_L(\mathcal{D}))$ is defined. The index theorem of Atiyah and Singer is the following result:

Theorem 3.19. (Atiyah-Singer) $\text{Ind}_G(\mathcal{D}) \;=\; \text{t-Ind}_G(\sigma_L(\mathcal{D}))$.

3.20. *The case of an open manifold*

Suppose now that the manifold M is not compact. The topological index map (4) is still defined. However the symbol of an elliptic operator $\mathcal{D}$ does not define an element of $K(T^*M)$, since it is not invertible on the non-compact set $M \subset T^*M$. Thus one can pose the following natural question

Question 1

Assume that a compact group G acts on an open *manifold M. Let E and F be G-equivariant vector bundles over M and let $\sigma : \pi^*E \to \pi^*F$ be an element of $K_G(T^*M)$. Find a G-invariant Fredholm differential operator $\mathcal{D} : C^\infty(M, E) \to C^\infty(M, F)$ such that*

$$\mathrm{Ind}_G(\mathcal{D}) \;=\; \text{t-Ind}_G(\sigma).$$

To the best of our knowledge the answer to this question is unknown. However, in Section 6 we present a partial answer to a certain generalization of this question. For this we will need a generalization of the index theorem to so-called *transversally elliptic* operators, which we will now discuss.

4. Transversal elliptic operators

In this section we discuss a generalization of the index theorem (cf. Theorem 3.19) to transversally elliptic operators due to Atiyah and Singer.[3]

4.1. *A motivating example*

Suppose N is a closed manifold and let $\mathcal{D} : C^\infty(N) \to C^\infty(N)$ be an elliptic operator. Let G be a compact Lie group. Consider the manifold

$$M \;:=\; N \times G$$

and let G act on M by

$$g_1 \cdot (x, g_2) \;:=\; (x, g_1 g_2), \qquad x \in N,\ g_1, g_2 \in G.$$

There is a natural extension of the differential operator $\mathcal{D}$ to an operator

$$\tilde{\mathcal{D}} : C^\infty(M) \;\longrightarrow\; C^\infty(M)$$

defined as follows: Suppose that in a coordinate chart $(x_1, \ldots, x_n)$ on N the operator $\mathcal{D}$ is

$$\mathcal{D} \;=\; \sum_{|\alpha| \le k} a_\alpha(x) \frac{\partial^{\alpha_1}}{\partial x_1^{\alpha_1}} \cdots \frac{\partial^{\alpha_n}}{\partial x_n^{\alpha_n}}.$$

Let $(y_1, \cdots, y_r)$ be a coordinate system on G. Then, $(x_1, \ldots, x_n, y_1, \ldots, y_r)$ is a coordinate system on M. By definition the operator $\tilde{\mathcal{D}}$ takes the form

$$\tilde{\mathcal{D}} \;:=\; \sum_{|\alpha| \le k} a_\alpha(x) \frac{\partial^{\alpha_1}}{\partial x_1^{\alpha_1}} \cdots \frac{\partial^{\alpha_n}}{\partial x_n^{\alpha_n}}.$$

Exercise 4.2. Check that $\tilde{\mathcal{D}}$ is G-equivariant but not elliptic.

Notice that

$$\mathrm{Ker}(\tilde{\mathcal{D}}) \;=\; \mathrm{Ker}(\mathcal{D})\otimes L^2(G), \quad \mathrm{Coker}(\tilde{\mathcal{D}}) \;=\; \mathrm{Coker}(\mathcal{D})\otimes L^2(G). \qquad (1)$$

Thus despite the fact that $\mathrm{Ker}(\tilde{\mathcal{D}})$ and $\mathrm{Coker}(\tilde{\mathcal{D}})$ are infinite dimensional they are "manageable", and the index of $\tilde{\mathcal{D}}$ can be defined as follows:

By the Peter-Weyl theorem,[18] the left regular representation (cf. Definition 3.16) $L^2(G)$ decomposes into direct sum of irreducible representations,

$$L^2(G) \;=\; \bigoplus_{V\in \mathrm{Irr}\,(G)} (\dim V)\, V. \qquad (2)$$

In particular, every irreducible representation occurs in (2) with finite multiplicity equal to $\dim V$.

Equations (1) and (2) suggest that the index of $\tilde{\mathcal{D}}$ can be defined by

$$\mathrm{Ind}_G(\tilde{\mathcal{D}}) \;:=\; \bigoplus_{V\in \mathrm{Irr}\,(G)} (\dim V)\big(\dim \mathrm{Ker}(\mathcal{D}) - \dim \mathrm{Ker}(\mathcal{D})\big)\, V. \qquad (3)$$

Notice that in contrast with (8) the sum in the right hand side of (3) is infinite. Thus $\mathrm{Ind}(\tilde{\mathcal{D}})$ lies in the completed ring of characters $\widehat{R}(G)$, cf. Remark 1.21.

Roughly speaking, the reason we are able to define a version of an index of $\tilde{\mathcal{D}}$ is that although this operator is not elliptic on M, it is elliptic on the quotient $N = M/G$. The purpose of this section is to define the analogue of this situation when M is not a product and the action of G on M is not free.

4.3. *The transversal cotangent bundle*

Suppose a compact group G acts on a smooth manifold M. Recall that an *orbit* of a point $x\in M$ is the set

$$\mathcal{O}(x) \;:=\; \big\{\, g\cdot x \big|\, g\in G \,\big\}.$$

This is a smooth submanifold of M.

We say that a cotangent vector $\xi\in T^*M$ is *perpendicular to the orbits of* G if for $x\in M$ and any tangent vector $v\in T\mathcal{O}(x)$, we have $\xi(v)=0$.

Definition 4.4. The *transversal cotangent bundle* T^*_GM is

$$T^*_GM \;:=\; \big\{\xi\in T^*M \big|\, \xi \textit{ is perpendicular to the orbits of } G\,\big\}. \qquad (4)$$

We set $T^*_{G,x}M := T^*_GM\cap T^*_xM$.

4.5. *The analytical index of transversally elliptic operators*

We now introduce the class of operators which generalizes the example considered in Section 4.1.

Definition 4.6. Suppose E and F are G-equivariant vector bundles over M. A G-invariant differential operator

$$\mathcal{D}: C^\infty(M, E) \longrightarrow C^\infty(M, F)$$

is called *transversally elliptic* if its leading symbol $\sigma_L(\mathcal{D})(x, \xi)$ is invertible for every non-zero $0 \neq \xi \in T^*_G M$.

Notice that every elliptic operator is transversally elliptic. Also, the operator $\tilde{\mathcal{D}}$ from Section 4.1 is transversally elliptic.

Theorem 4.7. (Atiyah-Singer [3, Lemma 2]) *Suppose $\mathcal{D}$ is a transversally elliptic operator on a* compact *manifold M. As representations of G, the spaces* $\mathrm{Ker}(\mathcal{D})$ *and* $\mathrm{Coker}(\mathcal{D})$ *can be decomposed into a direct sum of irreducible representation in which every irreducible representation appears finitely many times:*

$$\mathrm{Ker}(\mathcal{D}) = \bigoplus_{V \in \mathrm{Irr}\,(G)} m_V^+ V, \qquad \mathrm{Coker}(\mathcal{D}) = \bigoplus_{V \in \mathrm{Irr}\,(G)} m_V^- V. \tag{5}$$

Notice that the sums in (5) are, in general, infinite, but the numbers $m_V^\pm \in \mathbb{Z}_{\geq 0}$ are finite.

Definition 4.8. Let $\mathcal{D}$ be a transversally elliptic operator on a *compact* manifold M. The (analytical) index of $\mathcal{D}$ is defined by

$$\mathrm{Ind}_G(\mathcal{D}) := \bigoplus_{V \in \mathrm{Irr}\,(G)} (m_V^+ - m_V^-)\, V \in \widehat{R}(G), \tag{6}$$

where the numbers $m_V^\pm \in \mathbb{Z}_{\geq 0}$ are defined in (5) and $\widehat{R}(G)$ stands for the completed ring of characters, cf. Remark 1.21.

The index (6) possesses many properties of the index of elliptic operators. In particular an analogue of Theorem 3.2 holds.

4.9. *Transversal K-theory and the topological index*

Notice that in general $T^*_G M$ is not a manifold since the dimension of the fibers $T^*_{G,x} M$ might depend on x. But it is a topological space, and one can define the K-theory $K_G(T^*_G M)$ as the set of equivalence classes of pairs of vector bundles over $T^*_G M$ exactly as we did in Section 3.17.

The topological index

$$\text{t-Ind}_G : K_G(T^*_G M) \longrightarrow \widehat{R}(G) \tag{7}$$

is still defined, but the image lies not in the ring of characters and the definition is more involved. In fact, for compact manifold M, for $\sigma \in K_G(T^*_G M)$ one just chooses a G-invariant operator $\mathcal{D}$ with $\sigma_L(D) = \sigma$ and sets

$$\text{t-Ind}_G(\sigma) := \text{Ind}_G(\mathcal{D}). \tag{8}$$

(see[10] for a more topological construction). For non-compact M, the topological index is defined in [24, §3]. As in Section 3.20, if M is not compact the symbol of a transversally elliptic operator does not define an element of $K_G(T^*_G M)$. The index of such an operator is not defined since, in general, the irreducible representations of G appear in the kernel and the cokernel of $\mathcal{D}$ with infinite multiplicities. So as in Section 3.20, a natural question arises:

Question 2

Assume that a compact group G acts on an open *manifold M. Let E and F be G-equivariant vector bundles over M and let $\sigma : \pi^*E \to \pi^*F$ be an element of $K_G(T^*_G M)$. Find a G-invariant differential operator $\mathcal{D} : C^\infty(M,E) \to C^\infty(M,F)$ such that each irreducible representations of G appear in the kernel and the cokernel of $\mathcal{D}$ with finite multiplicities and*

$$\text{Ind}_G(\mathcal{D}) = \text{t-Ind}_G(\sigma) \in \widehat{R}(G).$$

Of course, this question is even harder than the corresponding question for usual K-theory (cf. Question 1 in Section 3.20), and the answer to this question is unknown. However in Section 6 we will present an answer for Question 2 in a special case. To present this answer we need to introduce the notions of Clifford action and Dirac-type operators, which we do in the next section.

5. Dirac-type operators

In this section we define the notions of Clifford bundle and generalized Dirac operator.

5.1. *Clifford action*

Let V be a finite dimensional vector space over $\mathbb{R}$ endowed with a scalar product $\langle\cdot,\cdot\rangle$.

Definition 5.2. A *Clifford action* of V on a complex vector space W is a linear map

$$c : V \ \to \ \mathrm{End}\,(W)$$

such that for any $v \in V$ we have

$$c(v)^2 \ = \ -|v|^2\,\mathrm{Id}\,. \tag{1}$$

Exercise 5.3. Show that a linear map $c : V \to \mathrm{End}\,(W)$ is a Clifford action if and only if

$$c(v)\,c(u) + c(u)\,c(v) \ = \ -2\langle v, u\rangle\ \mathrm{Id}, \qquad \text{for all } u, v \in V.$$

Exercise 5.4. Consider the Pauli matrices:

$$\sigma_1 \ = \ \begin{pmatrix} 0 & 1 \\ 1 & 0 \end{pmatrix}, \qquad \sigma_2 \ = \ \begin{pmatrix} 0 & -i \\ i & 0 \end{pmatrix}, \qquad \sigma_3 \ = \ \begin{pmatrix} 1 & 0 \\ 0 & -1 \end{pmatrix}. \tag{2}$$

Show that the map

$$c : \mathbb{R}^3 \ \longrightarrow \ \mathrm{End}\,(\mathcal{C}^2), \quad c(x_1, x_2, x_3) \ := \ \frac{1}{i}\,\big(\,x_1\sigma_1 + x_2\sigma_2 + x_2\sigma_3\,\big) \tag{3}$$

defines a Clifford action of $\mathbb{R}^3$ on $\mathcal{C}^2$.

Exercise 5.5. Let V be a real vector space endowed with a scalar product and let $V^{\mathcal{C}} = V \otimes_{\mathbb{R}} \mathcal{C}$ be its complexification. Let

$$W \ = \ \Lambda^*(V^{\mathcal{C}})$$

be the exterior algebra of $V^{\mathcal{C}}$. For $v \in V$ we denote $\varepsilon(v) : W \to W$ the exterior multiplication by v

$$\varepsilon(v)\,\alpha \ := \ v \wedge \alpha, \qquad \alpha \in W.$$

Using the scalar product on V, we identify v with an element of the dual space V^* and denote by $\iota_v : W \to W$ the interior multiplication by v. Set

$$c(v) \ := \ \varepsilon_v \ - \ \iota_v : W \ \longrightarrow \ W. \tag{4}$$

Show that the map $c : v \mapsto c(v)$ defines a Clifford action of V on W.

5.6. *A Clifford bundles*

Now let M be a smooth manifold endowed with a Riemannian metric g. Then for every $x \in M$, the tangent space T_xM and the cotangent space T_x^*M are endowed with a scalar product.

Definition 5.7. A *Clifford bundle* over M is a complex vector bundle $E \to M$ over M together with a bundle map

$$c : T^*M \longrightarrow \mathrm{End}\,(E), \tag{5}$$

such that for any $v \in T^*M$ we have $c(v)^2 = -|v|^2\,\mathrm{Id}$.

In other words we assume that for every $x \in M$ there is given a Clifford action

$$c : T_x^*M \to \mathrm{End}\,(E_x)$$

of the cotangent space T_x^*M on the fiber E_x of E and that this action depends smoothly on x.

Example 5.8. Set $M = \mathbb{R}^3$ and $E = \mathbb{R}^3 \times \mathcal{C}^2$. Then the projection on the first factor makes E a vector bundle over M. The action (3) defines a structure of a Clifford bundle on E.

Example 5.9. Let M be a Riemannian manifold and let $E = \Lambda^*(T^*M \otimes \mathcal{C})$. Note that the space of sections of E is just the space $\Omega^*(M)$ of complex valued differential forms on M. The formula (4) defines a Clifford bundle structure on E, such that for $v \in T^*M$ and $\omega \in \Omega^*(M)$ we have

$$c(v)\,\omega \;=\; v \wedge \omega \;-\; \iota_v\,\omega.$$

5.10. *A Clifford connection*

The Riemannian metric on M defines a connection on T^*M called the *Levi-Civita connection* and denoted by ∇^{LC}, cf.[27]

Definition 5.11. A connection $\nabla : C^\infty(M,E) \to \Omega^1(M,E)$ on a Clifford bundle E is called a *Clifford connection* if

$$\nabla_u\big(c(v)s\big) \;=\; c(\nabla_u^{LC} v)\,s \;+\; c(v)\,\nabla_u s, \tag{6}$$

for all $u \in TM,\ v \in T^*M,\ s \in C^\infty(M,E)$.

5.12. *A generalized Dirac operator*

We are now ready to define the notion of a (generalized) Dirac operator.

Definition 5.13. Let E be a Clifford bundle over a Riemannian manifold M and let ∇ be a Clifford connection on E. The *(generalized) Dirac operator* is defined by the formula

$$\mathcal{D} := \sum_{j=1}^{n} c(e_i)\,\nabla_{e_i} : C^\infty(M,E) \longrightarrow C^\infty(M,E), \tag{7}$$

where $e_1,\ldots,e_n$ is an orthonormal basis of TM.[c]

Exercise 5.14. Show that the operator (7) is independent of the choice of the orthonormal basis $e_1,\ldots,e_n$.

Exercise 5.15. a. Show that the leading symbol of a generalized Dirac operator is given by

$$\sigma_L(\mathcal{D})(x,\xi) = i\,c(\xi). \tag{8}$$

b. Prove that the generalized Dirac operator is elliptic.

Example 5.16. In the situation of Example 5.8, let ∇ be the standard connection, i.e.,

$$\nabla_{\partial/\partial x_i} = \frac{\partial}{\partial x_i}.$$

Then (7) is equal to the classical Dirac operator

$$\mathcal{D} = \begin{pmatrix} 0 & 1 \\ 1 & 0 \end{pmatrix} \frac{1}{i}\frac{\partial}{\partial x_1} + \begin{pmatrix} 0 & -i \\ i & 0 \end{pmatrix} \frac{1}{i}\frac{\partial}{\partial x_2} + \begin{pmatrix} 1 & 0 \\ 0 & -1 \end{pmatrix} \frac{1}{i}\frac{\partial}{\partial x_3}. \tag{9}$$

Exercise 5.17. Compute the square of the operator (9).

Example 5.18. Let M be a Riemannian manifold and let $d : \Omega^*(M,E) \to \Omega^{*+1}(M,E)$ denote the de Rham differential. The Riemannian metric induces a scalar product $\langle\cdot,\cdot\rangle$ on the space $\Omega^*(M,E)$ of differential forms. Let

$$d^* : \Omega^*(M,E) \to \Omega^{*-1}(M,E)$$

[c] As above, we use the Riemannian metric to identify TM and T^*M.

be the adjoint on d with respect to this scalar product. Thus we have

$$\langle d\alpha, \beta \rangle \;=\; \langle \alpha, d^*\beta \rangle, \qquad \text{for all} \quad \alpha, \beta \in \Omega^*(M, E).$$

Let $E = \Lambda^*(T^*M \otimes \mathcal{C})$. Then the space of sections $C^\infty(M, E)$ equals $\Omega^*(M, E)$. The Riemannian metric defines a canonical connection on M, called the Levi-Civita connection. One can show, cf. [9, Proposition 3.53], that this is a Clifford connection and that the corresponding Dirac operator (7) is given by

$$\mathcal{D} \;=\; d \,+\, d^*. \tag{10}$$

If M is a compact manifold then, cf.,[29] the kernel of $\mathcal{D}$ is naturally isomorphic to the de Rham cohomology on M:

$$\operatorname{Ker}(\mathcal{D}) \;\simeq\; H^*(M). \tag{11}$$

5.19. *A grading*

In index theory one deals with graded Clifford bundles.

Definition 5.20. A *grading* on a Clifford bundle E is a decomposition

$$E \;=\; E^+ \oplus E^-, \tag{12}$$

such that for every $v \in T^*M$ we have

$$c(v) : E^\pm \;\longrightarrow\; E^\mp.$$

A Clifford bundle with a grading is called a *graded Clifford bundle.*

A Clifford connection on a graded Clifford bundle *preserves the grading* if for each $v \in T^*M$,

$$\nabla_v : C^\infty(M, E^\pm) \;\longrightarrow\; C^\infty(M, E^\pm).$$

When speaking about a Clifford connection ∇ on a graded Clifford bundle, we always assume that it preserves the grading.

Exercise 5.21. Let $E = E^+ \oplus E^-$ be a graded Clifford bundle and let ∇ be a Clifford connection on E which preserves the grading. Show that the corresponding Dirac operator satisfies

$$\mathcal{D} : C^\infty(M, E^\pm) \;\longrightarrow\; C^\infty(M, E^\mp).$$

We denote the restriction of $\mathcal{D}$ to $C^\infty(M, E^+)$ (respectively, $C^\infty(M, E^-)$) by $\mathcal{D}^+$ (respectively, $\mathcal{D}^-$). It follows from the exercise above that with respect to the splitting (12), the Dirac operator $\mathcal{D}$ can be written as

$$\mathcal{D} = \begin{pmatrix} 0 & \mathcal{D}^- \\ \mathcal{D}^+ & 0 \end{pmatrix}. \tag{13}$$

To save space we often write (13) as $\mathcal{D} = \mathcal{D}^+ \oplus \mathcal{D}^-$. When a Dirac operator is presented in this form we refer to it as a *graded Dirac operator*. In index theory one usually considers the index of the operator $\mathcal{D}^+ : C^\infty(M, E^+) \to C^\infty(M, E^-)$.

Example 5.22. Let M be a Riemannian manifold and let $E = \Lambda^*(T^*M \otimes \mathcal{C})$. Set

$$E^+ := \bigoplus_{j \text{ even}} \Lambda^j(T^*M \otimes \mathcal{C}), \qquad E^- := \bigoplus_{j \text{ odd}} \Lambda^j(T^*M \otimes \mathcal{C}).$$

Then $E = E^+ \oplus E^-$, i.e., we obtain a grading on E. The Levi-Civita connection preserves this grading. Thus (10) becomes a graded operator. This graded Dirac operator is called the *de Rham-Dirac operator.*

Exercise 5.23. Suppose that the manifold M is compact. Show that the index $\mathrm{Ind}(\mathcal{D}^+)$ of the de Rham-Dirac operator is equal to the Euler characteristic of M:

$$\mathrm{Ind}(\mathcal{D}^+) = \sum_{j=0}^{n} (-1)^n \dim H^j(M).$$

Example 5.24. Suppose the dimension of M is even, $\dim M = n = 2l$. There is another natural grading on $E = \Lambda^*(T^*M \otimes \mathcal{C})$ defined as follows. Let $* : E \to E$ denote the Hodge star operator [27, §19.1]. Define the *chirality operator* $\Gamma : E \to E$ by

$$\Gamma\omega := i^{p(p-1)+l} * \omega, \qquad \omega \in \Lambda^p(M) \otimes \mathcal{C}.$$

One can show [27, §19.1] that $\Gamma^2 = 1$. It follows that the spectrum of Γ is $\{1, -1\}$. Let E^+ (respectively, E^-) denote the eigenspace of Γ corresponding to the eigenvalue $+1$ (respectively, -1). The sections of E^+ (respectively E^-) are called the *self-dual* (respectively, *anti-selfdual*) differential forms. Then $E = E^+ \oplus E^-$ and the Levi-Civita connection preserves this grading. Hence, (10) becomes a graded operator. This graded Dirac operator is called the *signature operator* and the index of $\mathcal{D}^+$ is called the *signature* of M. The signature of M can also be computed in topological terms. The study of the index of the signature operator by Friedrich Hirzebruch[16] was one of the main motivations for Atiyah and Singer's work on index theory.

Remark 5.25. Exercise 5.23 and Example 5.24 show that such topological invariants of a manifold as the Euler characteristic and the signature can be computed as indices of elliptic operators. By Remark 3.13, they can also be computed as integrals of certain characteristic classes. The Euler characteristic is equal to the integral of so called *Euler class* and the signature is equal to the integral of the *Hirzebruch L-polynomial.* We refer the reader to[8] for details.

5.26. *The group action*

Let $E = E^+ \oplus E^-$ be a graded Clifford bundle over a Riemannian manifold M. Let ∇ be a Clifford connection which preserves the grading, and let $\mathcal{D} = \mathcal{D}^+ \oplus \mathcal{D}^-$ be the corresponding graded Dirac operator. Suppose that a compact group G acts on M and E, preserving the Riemannian metric, the connection ∇, and the grading on E. Then the operators $\mathcal{D}^\pm$ are G-equivariant. In particular, if the manifold M is compact, we can consider the equivariant index

$$\operatorname{Ind}_G(\mathcal{D}^+) \ \in \ R(G).$$

6. Index theory on open G-manifolds

We are now ready to define the class of transversally elliptic symbols for which we are able to answer Question 2 of Section 4.9. The section is based on the results of.[11]

6.1. *The settings*

Recall that a Riemannian manifold M is called *complete* if it is complete as a metric space. Suppose M is a complete Riemannian manifold on which a compact Lie group G acts by isometries. To construct our index theory of certain generalized Dirac operators on M, we need an additional structure on M, namely a G-equivariant map

$$\mathbf{v} : M \ \to \ \mathfrak{g} \ = \ \operatorname{Lie} G. \tag{1}$$

Such a map induces a vector field v on M defined by the formula

$$v(x) \;:=\; \frac{d}{dt}\Big|_{t=0} \exp\big(t\mathbf{v}(x)\big)\cdot x. \tag{2}$$

Example 6.2. Let

$$G \;=\; S^1 \;=\; \big\{\, z\in\mathcal{C}:\, |z|=1\,\big\}$$

be the circle group. Then the Lie algebra $\mathfrak{g} = \operatorname{Lie} S^1$ is naturally isomorphic to $\mathbb{R}$. Suppose G acts on a complete Riemannian manifold M. Let $\mathbf{v}: M\to\mathfrak{g}$ be the constant map $\mathbf{v}(x)\equiv 1$. The corresponding vector field v is called the *generating vector field* for the action of $G = S^1$, since it completely determines the action of G.

Exercise 6.3. Let $G = S^1$ acts on $M = \mathcal{C}$ by multiplications: $(e^{it}, z)\mapsto e^{it}z$. Compute the generating vector field $v(x)$.

6.4. *Tamed G-manifolds*

Throughout this section we will make the following

Assumption

There exists a compact subset $K\subset M$, such that

$$v(x) \;\neq\; 0, \qquad \text{for all} \quad x\notin K. \tag{3}$$

Definition 6.5. A map (1) satisfying (3) is called a *taming map*. The pair $(M,\mathbf{v})$, where M is a complete Riemannian manifold and $\mathbf{v}$ is a taming map, is called a *tamed G-manifold*.

6.6. *An element of $K_G(T^*_GM)$*

Suppose now that $(M,\mathbf{v})$ is a tamed G-manifold and let $E = E^+\oplus E^-$ be a G-equivariant graded Clifford bundle over M. Let T^*_GM denote the transversal cotangent bundle to M, cf. Definition 4.4, and let $\pi: T^*_GM\to M$ denote the projection. Consider the pullback bundle $\pi^*E = \pi^*E^+\oplus\pi^*E^-$ over T^*_GM.

Using the Riemannian metric on M we can identify the tangent and cotangent vectors to M. Thus we can consider the vector $v(x)$ defined in (3) as an element of T^*M. Then for $x\in M$, $\xi\in T^*M$, we can consider the map

$$c\big(\xi+v(x)\big):\, \pi^*E^+_{x,\xi} \;\to\; \pi^*E^-_{x,\xi}. \tag{4}$$

The collection of these maps for all $(x,\xi) \in T^*_G M$ defines a bundle map

$$c(\xi + v) : \pi^* E^+ \rightarrow \pi^* E^-.$$

Exercise 6.7. Show that the map $c\big(\xi + v(x)\big)$ is invertible for all $(x,\xi) \in T^*_G M$ such that $v(x) \neq 0$, $\xi \neq 0$. Conclude that the bundle map (4) defines an element of the K-theory $K_G(T^*_G M)$.

Hence, we can consider the topological index

$$\text{t-Ind}_G \big(c(\xi + v)\big) \in \widehat{R}(G).$$

Definition 6.8. Suppose $E = E^+ \oplus E^-$ is a G-equivariant Clifford bundle over a tamed G-manifold $(M, \mathbf{v})$. We refer to the pair $(E, \mathbf{v})$ as a *tamed Clifford bundle over M*. The topological index $\text{t-Ind}_G(E, \mathbf{v})$ of a tamed Clifford bundle is defined by

$$\text{t-Ind}_G(E, \mathbf{v}) := \text{t-Ind}_G \big(c(\xi + v)\big). \tag{5}$$

This index was extensively studied by M. Vergne[28] and P.-E. Paradan.[24,25] Our purpose is to construct a Fredholm operator, whose analytical index is equal to $\text{t-Ind}_G(E, \mathbf{v})$.

6.9. *The Dirac operator*

Suppose ∇ is a G-equivariant Clifford connection on E which preserves the grading, and let $\mathcal{D} = \mathcal{D}^+ \oplus \mathcal{D}^-$ be the corresponding Dirac operator. By Exercise 5.15 the operator $\mathcal{D}$ is elliptic and its leading symbol is equal to $c(\xi)$. However, if M is not compact, the operator $\mathcal{D}$ does not have to be Fredholm and its index is not defined.

6.10. *A rescaling of v*

Our definition of the index uses a certain rescaling of the vector field v, by which we mean the product $f(x)v(x)$, where $f : M \rightarrow [0, \infty)$ is a smooth positive G-invariant function. Roughly speaking, we demand that $f(x)v(x)$ tends to infinity "fast enough" as x tends to infinity. The precise conditions we impose on f are quite technical, cf. Definition 2.6 of,[11] and depend on the geometry of M, E and ∇. If a function f satisfies these conditions we call it *admissible for the quadruple* $(M, E, \nabla, \mathbf{v})$. The index which we are about to define turns out to be independent of the concrete choice of f. It is important, however, to know that at least one admissible function exists. This is proven in Lemma 2.7 of.[11]

6.11. *The analytic index on non-compact manifolds*

Let f be an admissible function for $(M, E, \nabla, \mathbf{v})$. Consider the *deformed Dirac operator*

$$\mathcal{D}_{fv} = \mathcal{D} + ic(fv), \tag{6}$$

and let $\mathcal{D}_{fv}^{+}$ denote the restriction of $\mathcal{D}_{fv}$ to $C^\infty(M, E^+)$. This operator is elliptic, but since the manifold M is not compact it is not Fredholm. In fact, both the kernel and the cokernel of $\mathcal{D}_{fv}^{+}$ are infinite dimensional. However they have an important property, which we shall now describe. First, recall from Exercise 1.15 that, since the operator $\mathcal{D}_{fv}^{+}$ is G-equivariant, the group G acts on $\mathrm{Ker}(\mathcal{D}_{fv}^{+})$ and $\mathrm{Coker}(\mathcal{D}_{fv}^{+})$.

Theorem 6.12. *Suppose f is an admissible function. Then the kernel and the cokernel of the deformed Dirac operator D_{fv}^{+} decompose into an infinite direct sum*

$$\mathrm{Ker}\,\mathcal{D}_{fv}^{+} = \sum_{V\in \mathrm{Irr}\,(G)} m_V^{+}\cdot V, \quad \mathrm{Coker}\,\mathcal{D}_{fv}^{+} = \sum_{V\in \mathrm{Irr}\,(G)} m_V^{-}\cdot V, \tag{7}$$

where $m_V^{\pm}$ are non-negative integers. In particular, each irreducible representation of G appears in $\mathrm{Ker}\,\mathcal{D}_{fv}^{\pm}$ with finite multiplicity.

This theorem allows us to define the index of the operator $\mathcal{D}_{fv}^{+}$:

$$\mathrm{Ind}_G(\mathcal{D}_{fv}^{+}) = \sum_{V\in \mathrm{Irr}\,(G)} (m_V^{+} - m_V^{-})\cdot V \in \widehat{R}(G). \tag{8}$$

The next theorem states that this index is independent of all choices.

Theorem 6.13. *For an irreducible representation $V \in \mathrm{Irr}\,(G)$, let $m_V^{\pm}$ be defined by (7). Then the differences $m_V^{+} - m_V^{-}$ ($V \in \mathrm{Irr}\,(G)$) are independent of the choices of the admissible function f and the G-invariant Clifford connection on E used in the definition of $\mathcal{D}$.*

Definition 6.14. We refer to the pair $(\mathcal{D}, \mathbf{v})$ as a *tamed Dirac operator*. The analytical index of a tamed Dirac operator is defined by the formula

$$\mathrm{Ind}_G(\mathcal{D}, \mathbf{v}) := \mathrm{Ind}_G(\mathcal{D}_{fv}^{+}), \tag{9}$$

where f is an admissible function for $(M, \mathbf{v})$.

Exercise 6.15. Let $(M, \mathbf{v})$ be the tamed $G = S^1$-manifold defined in Example 6.2. Let

$$E^{\pm} = M \times \mathbb{R}$$

be the trivial line bundles over M. Define the Clifford action of T^*M on $E = E^+ \oplus E^-$ by the formula

$$c(\xi) := \frac{1}{i}\,(\xi_1 \sigma_1 + \xi_2 \sigma_2), \qquad \xi = (\xi_1, \xi_2) \in T^*M \simeq \mathbb{R}^2,$$

where σ_1 and σ_2 are the first two Pauli matrices, cf. (2).

The G action on M lifts to $E^{\pm}$ so that

$$e^{it} \cdot (z, \nu) := (e^{it} z, \nu), \qquad z \in M,\ t, \nu \in \mathbb{R}.$$

We endow the bundle E with the trivial connection.

a. Show that the construction above defines a structure of a G-equivariant graded Clifford module on E.

b. One can show that $f \equiv 1$ is an admissible function for E. Compute the operators $\mathcal{D}_v$ and $\mathcal{D}_v^2$.

c. Find the kernel and cokernel of $\mathcal{D}_v$.

d. Compute $\mathrm{Ind}_G(\mathcal{D}, \mathbf{v})$.

6.16. *The index theorem*

We are now ready to formulate the following analogue of the Atiyah-Singer index theorem for tamed Dirac-type operators on non-compact manifolds:

Theorem 6.17. *Suppose $E = E^+ \oplus E^-$ is a graded G-equivariant Clifford bundle over a tamed G-manifold $(M, \mathbf{v})$. Let ∇ be a G-invariant Clifford connection on E which preserves the grading and let $\mathcal{D}$ be the corresponding Dirac operator. Then for any*

$$\mathrm{Ind}(\mathcal{D}, \mathbf{v}) = \text{t-Ind}(E, \mathbf{v}). \tag{10}$$

Exercise 6.18. Show that if the manifold M is compact, then

$$\mathrm{Ind}_G(\mathcal{D}, \mathbf{v}) = \mathrm{Ind}_G(\mathcal{D}). \tag{11}$$

6.19. *Properties of the index on a non-compact manifold*

The index (9) has many properties similar to the properties of the index of an elliptic operator on a compact manifold. It satisfies an analogue of the Atiyah-Segal-Singer fixed point theorem, Guillemin-Sternberg "quantization commutes with reduction" property, etc. It is also invariant under a certain type of cobordism. The description of all these properties lies beyond the scope of these lectures and we refer the reader to[11] for details. We will finish with mentioning just one property – the gluing formula – which illustrates how the non-compact index can be used in the study of the usual index on compact manifolds.

6.20. *The gluing formula*

Let $(M, \mathbf{v})$ be a tamed G-manifold. Suppose $\Sigma \subset M$ is a smooth G-invariant hypersurface in M such that $M \backslash \Sigma$ is a disjoint union of two open manifolds M_1 and M_2:

$$M \backslash \Sigma \;=\; M_1 \sqcup M_2.$$

For simplicity, we assume that Σ is compact. Assume also that the vector field v induced by $\mathbf{v}$ does not vanish anywhere on Σ. Choose a G-invariant complete Riemannian metric on M_j $(j = 1, 2)$. Let $\mathbf{v}_j$ denote the restriction of $\mathbf{v}$ to M_j. Then $(M_j, \mathbf{v}_j)$ $(j = 1, 2)$ are tamed G-manifolds.

Suppose that $E = E^+ \oplus E^-$ is a G-equivariant graded Clifford module over M. Denote by E_j the restriction of E to M_j $(j = 1, 2)$. Let $\mathcal{D}_j$ $(j = 1, 2)$ denote the restriction of $\mathcal{D}$ to M_j.

Theorem 6.21. *In the situation described above*

$$\mathrm{Ind}_G(\mathcal{D}, \mathbf{v}) \;=\; \mathrm{Ind}_G(\mathcal{D}_1, \mathbf{v}_1) \;+\; \mathrm{Ind}_G(\mathcal{D}_2, \mathbf{v}_2). \tag{12}$$

In view of Exercise 6.18, one can use Theorem 6.21 to study the index of an equivariant Dirac operator on a compact manifold. This is done by cutting a compact manifold M along a G-invariant hypersurface Σ into two non-compact, but topologically simpler manifolds M_1 and M_2. We refer the reader to[11] for examples of different applications of this idea.

Acknowledgments

This paper is based on the lectures given by the first author in the Villa de Leyva summer school on Geometric, Algebraic and Topological Methods for Quantum Field Theory. We are grateful to the students Camilo Orduz and Carlos Pinilla for careful note taking and helping us prepare this manuscript. We thank the referee for very careful reading of the manuscript and suggesting many improvements and corrections.

References

1. R. A. Adams, *Sobolev spaces*, Academic Press, New York-London, 1975, Pure and Applied Mathematics, Vol. 65.
2. M. F. Atiyah, *Bott periodicity and the index of elliptic operators*, Quart. J. Math. Oxford Ser. (2) **19** (1968), 113–140.
3. M. F. Atiyah, *Elliptic operators and compact groups*, Springer-Verlag, Berlin, 1974, Lecture Notes in Mathematics, Vol. 401.
4. M. F. Atiyah, *Elliptic operators, discrete groups and von Neumann algebras*, Asrérisque **32/33** (1976), 43–72.
5. M. F. Atiyah, V. K. Patodi, and I. M. Singer, *Spectral asymmetry and Riemannian geometry. I,II,III*, Math. Proc. Cambridge Philos. Soc., **77** (1975) 43–69, (78) (1975) 405–432, **79** (1976) 71–99.
6. M. F. Atiyah and I. M. Singer, *The index of elliptic operators on compact manifolds*, Bull. Amer. Math. Soc. **69** (1963), 422–433.
7. M. F. Atiyah and I. M. Singer, *The index of elliptic operators I*, Ann. Math. **87** (1968), 484–530.
8. M. F. Atiyah and I. M. Singer, *The index of elliptic operators III*, Ann. Math. **87** (1968), 546–604.
9. N. Berline, E. Getzler, and M. Vergne, *Heat Kernels and Dirac Operators*, Springer-Verlag, 1992.
10. N. Berline and M. Vergne, *The Chern character of a transversally elliptic symbol and the equivariant index*, Invent. Math. **124** (1996), 11–49.
11. M. Braverman, *Index theorem for equivariant Dirac operators on noncompact manifolds*, K-Theory **27** (2002), no. 1, 61–101.
12. T. Bröcker, and T.tom Dieck, *Representations of Compact Lie Groups*, Graduate Texts in Mathematics, **98** (1995).
13. A. Connes and G. Skandalis, *The longitudinal index theorem for foliations*, Publ. Res. Inst. Math. Sci. **20** (1984), no. 6, 1139–1183.
14. I. M. Gel′fand, *Some aspects of functional analysis and algebra*, Proceedings of the International Congress of Mathematicians, Amsterdam, 1954, Vol. 1, Erven P. Noordhoff N.V., Groningen, 1957, pp. 253–276.
15. I. M. Gel′fand, *On elliptic equations*, Russian Math. Surveys **15** (1960), no. 3, 113–123. MR 0123085 (23 #A416)
16. F. Hirzebruch, *Topological methods in algebraic geometry*, Third enlarged edition. Die Grundlehren der Mathematischen Wissenschaften, Band 131, Springer-Verlag New York, Inc., New York, 1966.
17. G. G. Kasparov, *Equivariant KK-theory and the Novikov conjecture*, Invent. Math. **91** (1988), no. 1, 147–201.
18. A. W. Knapp, *Representation Theory of Semisimple Groups*, Princeton Landmarks in Mathematics, Princeton University Press, Princeton, NJ, 2001, An overview based on examples, Reprint of the 1986 original.
19. H.B. Lawson, Jr., and M.-L. Michelsohn, *Spin geometry*, Princeton Mathe-

matical Series, **38** (1989).
20. F. Noether, *Über eine Klasse singulärer Integralgleichungen*, Math. Ann. **82** (1920), 42–63.
21. G. E. Noether, *Fritz Noether (1884–194?)*, Integral Equations Operator Theory **8** (1985), 573–576.
22. G. E. Noether, *Letter to the editor: "Fritz Noether (1884–194?)" [Integral Equations Operator Theory 8 (1985), no. 5, 573–576; MR0813349 (86m:01063a)]*, Integral Equations Operator Theory **13** (1990), 303–305.
23. R. S. Palais, *Seminar on the Atiyah-Singer Index Theorem*, Princeton University Press, Princeton, N.J., 1965, Annals of Mathematics Studies, No. 57.
24. P.-É. Paradan, *Localization of the Riemann-Roch character*, J. Funct. Anal. **187** (2001).
25. P.-É. Paradan, Spin^c*-quantization and the K-multiplicities of the discrete series*, Ann. Sci. École Norm. Sup. (4) **36** (2003).
26. M. A. Shubin, *Pseudodifferential Operators and Spectral Theory*, Springer Verlag, Berlin, New York, 1987.
27. C. H. Taubes, *Differential Geometry. Bundles, Connections, Metrics and Curvature.*, Oxford Graduate Texts in Mathematics 23. Oxford: Oxford University Press., 2011.
28. M. Vergne, *Multiplicities formula for geometric quantization. I, II*, Duke Math. J. **82** (1996), no. 1, 143–179, 181–194.
29. F. W. Warner, *Foundations of Differentiable Manifolds and Lie Groups*, Graduate Texts in Mathematics, Springer-Verlag, New-York, Berlin, Heidelberg, Tokyo, 1983.

GENERALIZED EULER CHARACTERISTICS, GRAPH HYPERSURFACES, AND FEYNMAN PERIODS

Paolo Aluffi

Mathematics Department
Florida State University
Tallahassee FL 32306, U.S.A.
E-mail: aluffi@math.fsu.edu

We give a very informal presentation of background on the Grothendieck group of varieties and on characteristic classes, both viewed as generalizations of the ordinary topological Euler characteristic. We then review some recent work using these tools to study 'graph hypersurfaces'—a topic motivated by the algebro-geometric interpretation of Feynman amplitudes as periods of complements of these hypersurfaces. These notes follow closely, both in content and style, my lectures at the Summer school in Villa de Leyva, July 5–8, 2011.

Keywords: Euler characteristics, graph hypersurfaces, Feynman periods.

1. Introduction

These notes collect the material covered in my lectures at the Summer school in Villa de Leyva, and they preserve the very informal style of my presentations. In the first part, the main character in the story is the notion of 'generalized Euler characteristic'. Two examples are presented in some detail: the Grothendieck ring of varieties, and the theory of Chern classes for singular varieties (the *Chern-Schwartz-MacPherson* classes). In the second part graph hypersurfaces are introduced, with a quick explanation of their relevance in the study of Feynman amplitudes. The topics come together in the third part, which gives a review of some recent work studying Grothendieck classes and Chern classes of graph hypersurfaces.

A gentle warning to the reader: nothing in these notes is new, and much is rough beyond the pale. I have tried to provide enough references to the literature so that a reader who so desires can fill in the large gaps left open here. As my audience consisted of both math and physics students, I was not assuming a specific background in algebraic geometry in my lectures, and I am likewise not assuming the same here; this accounts for the need to include fly-by definitions of standard concepts, aimed at providing a bare minimum to be able to make sense of the rest of the notes, and opting for impressionistic blotches of color over detailed photorealism. It goes without saying that any such attempt is doomed to failure, and that at some point in the notes I will have to rely on more background than suggested at the beginning. Nevertheless, I hope that the few rough definitions given in §2 may tug the memory of someone who has had previous encounters with algebraic geometry, and facilitate the recall needed for the rest.

The more substantial part of the notes (§4) summarizes joint work with Matilde Marcolli (especially[AM11a,AM11b,AM11c]). We define a notion of *algebro-geometric Feynman rules* (Definition 4), i.e., graph invariants determined by the graph hypersurface and satisfying a formalism mimicking part of the 'Feynman rules' commonly used in computations in quantum field theory. We discuss two explicit examples of algebro-geometric Feynman rules, including a polynomial invariant $C_G(t)$ defined by means of the Chern-Schwartz-MacPherson class of a graph hypersurface. Some of the information contained in this invariant may be interpreted in rather transparent ways: for example, the coefficient of t in $C_G(t)$ equals the Euler characteristic of the complement of the graph hypersurface in projective space. We also discuss these invariants from the point of view of deletion-contraction relations (§4.2), and note that while the algebro-geometric Feynman rules we consider are not 'Tutte-Grothendieck' invariants, they share some of the structure carried by Tutte-Grothendieck invariants, such as multiple-edge formulas (Theorems 4.2 and 4.4). Finally, we study the stable birational equivalence class of graph hypersurfaces (§4.3), drawing some consequence in terms of their role as generators of the Grothendieck group of varieties.

A much more detailed review of most of this material, including a more extensive description of the physics context which originally motivated it, may be found in.[Mar10]

Acknowledgments. I thank Alexander Cardona, Hernán Ocampo, Sylvie Paycha, and Andrés Reyes for the invitation to lecture at the summer school on *Geometric, Algebraic, and Topological Methods for Quantum Field Theory*. The collegial and relaxed atmosphere of the school, the interaction with the students and the other lecturers, and the intensely beautiful surroundings made this a wonderful experience. I also thank Ellie Abernethy, whose notes taken at my lectures simplified greatly the task of putting together this material. The author's research is partially supported by a

Simons collaboration grant.

2. Generalized Euler characteristics

2.1. *The Euler characteristic*

Throughout these notes, we will work with complex quasi-projective algebraic varieties. Reminder on terminology:

—An *affine algebraic set* is the set of zeros of a collection of polynomials in $\mathbb{C}[x_1, \dots, x_n]$, viewed as a subset of affine space $\mathbb{A}^n_{\mathbb{C}} (= \mathbb{C}^n)$.

—A *projective algebraic set* is likewise the set of zeros of a collection of homogeneous polynomials in $\mathbb{C}[z_0, \dots, z_n]$, viewed as a subset of projective space $\mathbb{P}^n = \mathbb{P}^n_{\mathbb{C}}$.

Algebraic subsets of a given algebraic set are the closed sets of a topology, called the *Zariski* topology. An algebraic set is a *variety* if it is *irreducible,* that is, it cannot be written as a union of two proper closed subsets (i.e., two proper algebraic subsets). Adhering to an unfortunate but common abuse of language, we may use the term 'variety' for sets which are not necessarily irreducible.

—A *quasi-projective* variety is a Zariski-open subset of a projective variety, i.e., a set which may be written as $Y \smallsetminus Z$ where both Y and Z are projective.

Quasi-projective varieties admit open coverings by affine varieties. Morphisms of varieties are *regular maps,* i.e., function which restrict to polynomial functions on members of an affine cover. A *birational* isomorphism $X \dashrightarrow Y$ is the datum of an isomorphism between dense open subsets of X and Y; X and Y are then said to be 'birationally isomorphic' or 'birationally equivalent'. A *hypersurface* of a variety is an algebraic subset which can be defined locally by a single equation. A variety is *smooth,* or *nonsingular,* if it is a manifold. This can be checked easily in terms of the equations defining the variety. For example, a hypersurface X of $\mathbb{A}^n$ given by $f(x_1, \dots, x_n) = 0$ is nonsingular at a point p if the gradient $\nabla_p f$ is not zero; the linear form defined by $\nabla_p f$ is the *tangent space* to X at p. If a variety X is nonsingular at all of its points, the collection of the tangent spaces forms the *tangent bundle* TX to the variety. If Z is a nonsingular subvariety of a nonsingular variety X, then Z carries a *normal bundle* $N_Z X$, that is, the quotient $TX|_Z/TZ$. The *blow-up* of a variety X along a proper closed subvariety Z is a morphism which restricts to an isomorphism over the complement $X \smallsetminus Z$ (so that blow-ups are birational), and replaces Z by an *exceptional divisor,* which is a copy of the projectivized normal bundle to Z in X if both X and Z are nonsingular (and the projectivized normal *cone* in general). According to H. Hironaka's famous theorem on *resolution of singularities* ([Hir64]), every variety is birationally isomorphic to a nonsingular variety, and the resolution may be achieved by a sequence of blow-ups at smooth centers.

Definition 1. The (topological) *Euler characteristic* of X is the alternating sum $\chi(X) := \sum_i (-1)^i \dim H_c^i(X, \mathbb{Q})$. ⌟

Here H_c denotes cohomology with compact support. As it happens, every complex algebraic variety may be compactified within its homotopy type by giving it a topological boundary with odd-dimensional strata. This implies in particular that the compactly supported Euler characteristic equals the ordinary Euler characteristic. (See e.g.,,[Ful93] p. 95 and 141-2.)

As every projective variety X may be triangulated ([Hir75]), we can also define the Euler characteristic as

$$\chi(X) = s_0 - s_1 + s_2 - \cdots$$

where s_i is the number of real-dimension i simplices in a triangulation. If X is quasi-projective, say realized as $X = Y \smallsetminus Z$ with Y and Z projective, then $\chi(X) = \chi(Y) - \chi(Z)$ independently of the realization.

The following properties are easy consequences of the definition:

(i) If X and X' are isomorphic, then $\chi(X) = \chi(X')$;
(ii) If $Z \subseteq X$ is a closed subvariety, then $\chi(X) = \chi(Z) + \chi(X \smallsetminus Z)$;
(iii) $\chi(X \times Y) = \chi(X) \cdot \chi(Y)$.

For us, a 'generalized Euler characteristic' is an invariant satisfying the same properties. In this section we will look at two examples: invariants obtained from homomorphisms from the Grothendieck group/ring of varieties, and the Chern-Schwartz-MacPherson class.

Note that a formal consequence of (ii) is an inclusion-exclusion principle for the Euler characteristic: if Y and Z are subvarieties of a variety X, then

$$\chi(Y \cup Z) = \chi(Y) + \chi(Z) - \chi(Y \cap Z) \quad .$$

As a consequence of (iii) (and of inclusion-exclusion), the reader can verify that if $X \to Y$ is locally trivial, with fiber F, then $\chi(X) = \chi(Y) \cdot \chi(F)$.

2.2. K_0(Var)

Quasi-projective varieties are objects of a category Var, with regular maps as morphisms. We are still taking the base field to be $\mathbb{C}$, but it is useful to consider other possibilities, such as $\mathbb{Q}$, $\mathbb{F}_q$ (the finite field with a prime power q of elements). In fact, it is very useful to consider the case of $\mathbb{Z}$ as ground ring, and this will occasionally play a role in what follows.

We are going to define a group $K_0(\mathrm{Var})$, called the 'Grothendieck group of varieties'. As a motivation for the name, recall that the Grothendieck group of the category of (say, finitely generated) modules over a ring R is defined as the quotient of the free abelian group on isomorphism classes of R-modules by a relation $[M] = [M'] + [M'']$ for every exact sequence $0 \to M' \to M \to M'' \to 0$. More generally, the same concept may be defined for every small abelian category. It would be very convenient if we could define a similar object for a category such as Var, but we cannot use the same strategy, as Var is not an abelian category. There is a 'difficult' approach to this obstacle: define a suitable enhancement of Var which makes it an abelian category of 'motives', and then consider the Grothendieck group of this category. Defining a good category of motives is in fact an extremely worthwhile goal, but there are very substantial difficulties involved in the construction, and the whole topic is well beyond the scope of these notes. For example, one does not expect the result to be an abelian, rather a *triangulated* category; relations in the corresponding Grothendieck group arise from triangles. The reader may consult[Lev08] for a general review, and Chapter 2 in[Mar10] for a treatment very close to the context of these notes. The situation is more manageable for 'pure' motives, i.e., motives of smooth projective varieties. Also, motives come in different flavors; 'Chow' motives are based on an equivalence that is especially germane to algebraic geometry.

In any case, in this category there ought to be a distinguished triangle with vertices Z, X, and $X \smallsetminus Z$ for every variety X and every closed subvariety $Z \subseteq X$. This suggests an accelerated path to the definition of a 'Grothendieck group of varieties'.

Definition 2.

$$K_0(\mathrm{Var}_k) := \frac{\text{Free abelian group on isom. classes of quasi-proj. } k\text{-varieties}}{\langle [X] = [Z] + [X \smallsetminus Z] \text{ for every closed embedding } Z \subseteq X \rangle}$$

is the *Grothendieck group of k-varieties.* This group is a *ring* with the operation defined on generators by $[X] \cdot [Y] = [X \times_k Y]$.

The *Grothendieck class* (or *naive motive*) of a variety X is its class in $K_0(\mathrm{Var}_k)$. ⌟

I will usually (but not always) take $k = \mathbb{C}$, and omit the subscript k in the notation. The good news now is that this simple-minded object carries at least as much information as the Grothendieck group obtained from the more involved theory of motives:

Fact: There is a ring homomorphism from $K_0(\mathrm{Var})$ to the Grothendieck ring of pure Chow motives, assigning to the Grothendieck class $[X]$ the class of the Chow motive of X.

In this sense we view $[X]$ as a 'naive' counterpart to the motive of X. The fact was proven by H. Gillet and C. Soulé ([GS96]), by F. Guillén and V. Navarro Aznar ([GNA02]), and recovered by F. Bittner from a very useful alternative description of $K_0(\mathrm{Var})$ ([Bit04]) which will be recalled below.

Theorem 2.1. *Let R be a ring and e an R-valued invariant of quasi-projective varieties such that*

- *If p is a point, then $e(p) = 1$;*
- *If X and X' are isomorphic, then $e(X) = e(X')$;*
- *If $Z \subseteq X$ is a closed subvariety, then $e(X) = e(Z) + e(X \smallsetminus Z)$;*
- $e(X \times Y) = e(X)e(Y)$.

Then e is induced by a unique ring homomorphism $K_0(\mathrm{Var}) \to R$.

Theorem 2.1 is essentially tautological: the construction of the Grothendieck ring of varieties parallels the construction of many other objects satisfying a universal requirement, and Theorem 2.1 does nothing but spell out this universal requirement in the case of $K_0(\mathrm{Var})$.

It should now be clear that, by Theorem 2.1, the Euler characteristic χ defined in §2.1 is induced by a unique ring homomorphism $K_0(\mathrm{Var}) \to \mathbb{Z}$, and that the ring-valued invariants satisfying our loose requirement for being 'generalized Euler characteristics' are precisely the invariants induced by ring homomorphisms from $K_0(\mathrm{Var})$. The assignment sending a variety X to its Grothendieck class $[X] \in K_0(\mathrm{Var})$ is a 'universal Euler characteristic' in this sense.

There are other interesting ring homomorphisms naturally defined on $K_0(\mathrm{Var})$: as mentioned above, there is a homomorphism from $K_0(\mathrm{Var})$ to the Grothendieck ring of pure Chow motives: so the class of a variety in this ring is a 'generalized Euler characteristic'. As another example, the 'Hodge-Deligne polynomial', recording ranks of mixed Hodge structures, takes values in a polynomial ring $\mathbb{Z}[u, v]$. Also: for varieties defined over $\mathbb{Z}$ and every prime power q we can define homomorphisms $K(\mathrm{Var}_{\mathbb{Z}}) \to \mathbb{Z}$ by sending $[X]$ to the number of points of X over the finite field $\mathbb{F}_q$ (i.e., the number of solutions of the corresponding equations in the finite field

$\mathbb{F}_q$). This latter type of invariants will be of interest to us in §3.2; see also Example 2.1 below.

The reader is addressed to[Bit04] for an excellent treatment of the universal Euler characteristic. In this paper, F. Bittner gives a very useful alternative presentation for $K_0(\text{Var})$. I have recalled that if X is nonsingular and $Z \subset X$ is a nonsingular subvariety, then there exists a 'blow-up' morphism $\widetilde{X} \to X$ which is an isomorphism over $X \smallsetminus Z$; the inverse image E of Z is the 'exceptional divisor' of this blow-up. It follows that

$$[\widetilde{X}] - [E] = [X] - [Z] \tag{1}$$

in $K_0(\text{Var})$. Bittner proves that $K_0(\text{Var})$ is generated by the classes of *smooth projective varieties,* subject to the relations (1) (where Z may also be taken to be smooth).

Therefore, in order to define a 'generalized Euler characteristic' with values in a ring, it suffices to define it for smooth projective varieties, and prove that the definition is compatible with the blow-up relations (1) (and with multiplicativity). This is often much easier to handle than the scissor-type relations $[X] = [Z] + [X \smallsetminus Z]$ used in Definition 2.

Definition 3. The *Lefschetz (naive) motive* is the Grothendieck class $\mathbb{L} := [\mathbb{A}^1]$ of the affine line. Elements of the subring $\mathbb{Z}[\mathbb{L}] \subseteq K_0(\text{Var})$ are *mixed-Tate (naive) motives.* ⌟

The (not very standard) terminology introduced here reflects the fact that classes in $\mathbb{Z}[\mathbb{L}]$ correspond to *mixed-Tate motives* in the category of motives. (For the sake of simplicity of exposition I am neglecting a *localization* that is in principle needed here.) The reader is addressed to,[Mar10] §§2.5-6, for more information on mixed-Tate motives. In practice, we view a variety whose Grothendieck class is a naive mixed-Tate motive as 'decomposable' as a disjoint union of affine cells. The reader should verify that, for example,

$$[\mathbb{P}^{n-1}] = 1 + \mathbb{L} + \cdots + \mathbb{L}^n = \frac{\mathbb{L}^n - 1}{\mathbb{L} - 1}$$

is in $\mathbb{Z}[\mathbb{L}]$, and that that if X is nonsingular, $Z \subseteq X$ is a closed nonsingular variety, and $\widetilde{X}$ is the blow-up of X along Z, then $[\widetilde{X}]$ is in $\mathbb{Z}[\mathbb{L}]$ if $[X]$ and $[Z]$ are in $\mathbb{Z}[\mathbb{L}]$. As we will see in §3, the question of whether the class of a variety is in $\mathbb{Z}[\mathbb{L}]$ can be very interesting and challenging; 'graph hypersurfaces' provide lots of examples, but as we will see not all graph hypersurfaces are mixed-Tate.

Example 2.1. To get right away a feeling about how 'special' it is for a variety to be mixed-Tate, note that if $[X] \in \mathbb{Z}[\mathbb{L}] \subseteq K_0(\mathrm{Var}_{\mathbb{Z}})$, say

$$[X] = a_0 + a_1\mathbb{L} + \cdots + a_r\mathbb{L}^r \quad , \tag{2}$$

then the number of points of X over the finite field $\mathbb{F}_q$ must be

$$N_q(X) = a_0 + a_1 q + \cdots + a_r q^r \quad , \tag{3}$$

a polynomial in q. We will say that X is *polynomially countable;* we will come back to this in §3.2 in the context of graph hypersurfaces.

It is amusing to observe that as the Euler characteristic of the affine line is 1, if the class of X is given by (2), then necessarily $\chi(X) = \sum_i a_i$: the value obtained by formally replacing q by 1 in (3). The Euler characteristic of a mixed-Tate motive is $[X]$ is the 'number of points of X over $\mathbb{F}_1$'. ⌟

2.3. c_{SM} *classes*

Notwithstanding the universality of the Grothendieck ring of varieties, homomorphisms of $K_0(\mathrm{Var})$ do not exhaust all 'generalized Euler characteristics'. Indeed, the invariants mentioned in Theorem 2.1 have a fixed target R, while there is a notion satisfying an inclusion-exclusion principle and (in a suitable sense) multiplicativity, but with target which itself depends on the variety. This is the *Chern-Schwartz-MacPherson* class, which I will denote by $c_{\mathrm{SM}}(X)$. It takes values in the homology of X, or more properly in the *Chow group* A_*X of X. This is not the place to discuss rational equivalence and the Chow group in any generality, particularly as we will only be interested in Chern-Schwartz-MacPherson classes of specific subsets of projective space; but the reader will bear with us for a moment as we summarize the more general context underlying these classes. (A full treatment of intersection theory is given in;[Ful84a] the hurried reader will benefit from consulting.[Ful84b]) In the case of projective space, $A_*\mathbb{P}^n$ agrees with homology, and equals the free abelian group on generators $[\mathbb{P}^0], \ldots, [\mathbb{P}^n]$. It is unfortunate that the notation for classes in A_*X matches the notation for Grothendieck classes; hopefully the context will clarify the meaning of the notation.

If X is nonsingular, then it carries a *tangent bundle* TX, as already mentioned in §2.1; the rank of this bundle is $r = \dim X$. Now, every vector bundle E on a variety X determines a choice of elements of A_*X, the *Chern classes* of E ([Ful84a] Chapter 3). It is common to collect all Chern classes in a 'total Chern class'

$$c(E) = 1 + c_1(E) + c_2(E) + \cdots + c_{\mathrm{rk}\,E}(E) \quad .$$

For our purposes, we will rely on the intuition that the i-th Chern class of E, denoted $c_i(E) \cap [X]$, is a class in codimension i recording information on the subset of X where $\mathrm{rk}\,E - i + 1$ general sections of E are linearly dependent. *Warning:* Making this statement precise is no simple matter; for example, E may have no nontrivial sections whatsoever; here we are simply capturing the gist of Examples 14.4.1 and 14.4.2 in.[Ful84a] According to this rough interpretation, the 'top' Chern class $c_{\dim X}(TX) \cap [X]$ of the tangent bundle of X is a class in codimension $\dim X$ (i.e., in dimension 0), recording the locus where a general section of TX (i.e., a general vector field) is linearly dependent (i.e., it vanishes). Now, the *Poincaré-Hopf theorem* tells us that the (suitably interpreted) number of zeros of a vector field on a compact nonsingular variety equals the Euler characteristic of the variety. Therefore:

Lemma 2.1 (Poincaré-Hopf).

$$\int c(TX) \cap [X] = \chi(X) \quad . \tag{4}$$

Here, the integral sign $\int$ denotes the degree of the dimension 0 part of the class that follows. Thus, Lemma 2.1 is simply a statement of the Poincaré-Hopf theorem, once we interpret the dimension 0 part of the Chern class of TX as mentioned above. (And again, it should be clear that we are sweeping substantial subtleties under the rug. For example, what if $\chi(X) < 0$?) A similar statement may be given relating all Chern classes of TX and loci at which there arise obstructions to the construction of frames of vector fields.

A natural question is whether the Poincaré-Hopf theorem (and the companion statements for all Chern classes) can be recovered if X is *singular:* in this case, X does not carry a tangent *bundle* (the rank of the tangent space will jump at singular points), so the left-hand side of (4) loses its meaning.

In the 1960s, Marie-Hélène Schwartz answered this question by introducing and studying special vector fields at the singularities of X ([Sch65b,Sch65a]). In independent work, Grothendieck and Deligne proposed a conjectural framework, later confirmed by R. MacPherson ([Mac74]), of which Theorem 2.1 is a tiny aspect. Later still, the classes defined by Schwartz and

MacPherson were found to agree (,[BS81AB08]); they are commonly called *Chern-Schwartz-MacPherson classes,* c_{SM} classes for short.

In MacPherson's approach, one defines a class $c_{\mathrm{SM}}(\varphi)$ in A_*X (MacPherson worked in homology, but the theory may be defined in the Chow group; cf.,[Ful84a] Example 19.1.7) for every *constructible function* on a variety X. A constructible function is a linear combination of indicator functions of subvarieties: $\varphi = \sum m_i 1\!\!1_{Z_i}$. Constructible functions on X form an abelian group $C(X)$, and in fact C is a *covariant functor:* if $f : X \to Y$ is a regular map, we can define a homomorphism $C(X) \to C(Y)$ by setting

$$f_*(1\!\!1_Z)(p) = \chi(f^{-1}(p) \cap Z)$$

for any $p \in Y$ and any subvariety Z, and extending this prescription by linearity. MacPherson's definition has the following properties:

- For any X, $c_{\mathrm{SM}} : C(X) \to A_*X$ is a homomorphism;
- If X is nonsingular and compact, then $c_{\mathrm{SM}}(1\!\!1_X) = c(TX) \cap [X]$;
- If $f : X \to Y$ is a proper morphism and $\varphi \in C(X)$, then $c_{\mathrm{SM}}(f_*\varphi) = f_* c_{\mathrm{SM}}(\varphi)$.

In the third property, f is required to be proper in order to have a good notion of push-forward of classes in the Chow group. Various extensions of the theory, to more general fields, or allowing for non-proper morphisms, have been considered (,[Ken90Alu06]). Resolution of singularities and the covariance property of c_{SM} classes reduce (in principle) the computation of any c_{SM} class to computations for nonsingular varieties, where the c_{SM} class agrees with the class of the tangent bundle. This also implies that the c_{SM} natural transformation is uniquely determined by the properties listed above.

The class $c_{\mathrm{SM}}(1\!\!1_X)$ is usually denoted $c_{\mathrm{SM}}(X)$, and called the *Chern-Schwartz-MacPherson* class of X. Abusing language, we denote by $c_{\mathrm{SM}}(Y)$ the class $c_{\mathrm{SM}}(1\!\!1_Y)$ in A_*X for any subvariety (closed or otherwise) of X.

What the above properties say is that c_{SM} is a natural transformation, specializing to ordinary Chern classes in the nonsingular case. To get a feeling for the type of information packed into this statement, consider the extremely special case of the constant function $\kappa : X \to \{\mathrm{pt}\}$ (where X is assumed to be compact). The push-forward at the level of homology/Chow group,

$$\kappa_* : A_*X \to A_*\{\mathrm{pt}\} = \mathbb{Z} \quad ,$$

is nothing but the 'integral' $\int$ mentioned above. According to the above properties,

$$\int c_{\mathrm{SM}}(X) = \kappa_* c_{\mathrm{SM}}(1\!\!1_X) = c_{\mathrm{SM}}(\kappa_* 1\!\!1_X) = \chi(X) c_{\mathrm{SM}}(1\!\!1_{\mathrm{pt}}) = \chi(X) \qquad (5)$$

as the c_{SM} class of a point is 1 (times the class of the point). Compare (5) with (4): we have recovered for arbitrary (compact) varieties a 'Poincaré-Hopf theorem'.

Equation (5) already shows that $c_{\mathrm{SM}}(X)$ is a direct generalization of the Euler characteristic. Further, if $Z \subseteq X$, then

$$\begin{aligned} c_{\mathrm{SM}}(X) &= c_{\mathrm{SM}}(1\!\!1_X) = c_{\mathrm{SM}}(1\!\!1_Z + 1\!\!1_{X \smallsetminus Z}) \\ &= c_{\mathrm{SM}}(1\!\!1_Z) + c_{\mathrm{SM}}(1\!\!1_{X \smallsetminus Z}) = c_{\mathrm{SM}}(Z) + c_{\mathrm{SM}}(X \smallsetminus Z) \end{aligned}$$

(adopting the notational convention mentioned above). Thus, c_{SM} classes satisfy requirement (ii) for 'generalized Euler characteristics' in §2.1. They also satisfy a form of multiplicativity (requirement (iii)), which we won't describe here, and for which the reader may consult[Kwi92] or.[Alu06] There is a subtlety involved in the first requirement: if X and X' are isomorphic, then of course $c_{\mathrm{SM}}(X)$ and $c_{\mathrm{SM}}(X')$ agree up to the identification $A_*X \cong A_*X'$. However, if X and X' are viewed as subvarieties of (for example) projective space, then $c_{\mathrm{SM}}(1\!\!1_X)$ and $c_{\mathrm{SM}}(1\!\!1_{X'})$ will depend on the embeddings, and hence may differ. Thus, the notation $c_{\mathrm{SM}}(X)$ is ambiguous. This is harmless, as long as the embedding of X is clearly specified.

Example 2.2. —Viewing $\mathbb{P}^1$ as a subvariety of itself, $c_{\mathrm{SM}}(\mathbb{P}^1) = [\mathbb{P}^1] + 2[\mathbb{P}^0]$: indeed, since $\mathbb{P}^1$ is nonsingular we have

$$c_{\mathrm{SM}}(\mathbb{P}^1) = c(T\mathbb{P}^1) \cap [\mathbb{P}^1] = (1 + c_1(T\mathbb{P}^1)) \cap [\mathbb{P}^1] = [\mathbb{P}^1] + \chi(\mathbb{P}^1)[\mathbb{P}^0] = [\mathbb{P}^1] + 2[\mathbb{P}^0].$$

—More generally,

$$c_{\mathrm{SM}}(\mathbb{P}^n) = [\mathbb{P}^n] + \binom{n+1}{1}[\mathbb{P}^{n-1}] + \binom{n+1}{2}[\mathbb{P}^{n-2}] + \cdots + \binom{n+1}{n}[\mathbb{P}^0]. \quad (6)$$

This again follows from the normalization property for nonsingular varieties, a standard exact sequence for $T\mathbb{P}^n$ (,[Ful84a] B.5.8), and the *Whitney formula* for Chern classes. See Example 3.2.11 in.[Ful84a] Note that $\int c_{\mathrm{SM}}(\mathbb{P}^n) = \binom{n+1}{n} = n + 1 = \chi(\mathbb{P}^n)$, as it should. A shorthand for (6) is

$$c_{\mathrm{SM}}(\mathbb{P}^n) = (1 + H)^{n+1} \cap [\mathbb{P}^n]$$

where H is the 'hyperplane class', i.e., H^i stands for $[\mathbb{P}^{n-i}]$. (And in particular $H^{n+1} = 0$.)

—On the other hand, a nonsingular *conic* C in $\mathbb{P}^2$ is easily seen to be abstractly isomorphic to $\mathbb{P}^1$, but

$$c_{\mathrm{SM}}(C) = c_{\mathrm{SM}}(1\!\!1_C) = c(TC) \cap [C] = (1 + c_1(TC)) \cap [C] = 2[\mathbb{P}^1] + 2[\mathbb{P}^0].$$

This agrees with $c_{\mathrm{SM}}(\mathbb{P}^1)$ via the isomorphism, but the result of the push-forward to $\mathbb{P}^2$ differs from the class $c_{\mathrm{SM}}(\mathbb{P}^1)$ recorded above.

—A union X of two distinct lines L_1, L_2 in $\mathbb{P}^2$ is a *singular* conic. By inclusion-exclusion, $c_{\mathrm{SM}}(X) = c_{\mathrm{SM}}(L_1 \cup L_2)$ is given by

$$c_{\mathrm{SM}}(L_1) + c_{\mathrm{SM}}(L_2) - c_{\mathrm{SM}}(L_1 \cap L_2) = 2c_{\mathrm{SM}}(\mathbb{P}^1) - c_{\mathrm{SM}}(\mathbb{P}^0) = 2[\mathbb{P}^1] + 3[\mathbb{P}^0].$$

Notice that the c_{SM} class 'sees' the difference between singular and non-singular curves of the same degree. In general, c_{SM} classes may be used to express the *Milnor number* of a singularity, and more general *Milnor classes,* see e.g.,[PP01] Indeed, one of the main motivations in the study of c_{SM} classes is as a handle on interesting singularity invariants.

—Exercise for the reader: For subsets of projective space obtained as unions/ intersections/ differences of linear subspaces, the information carried by the c_{SM} class precisely matches the information carried by the Grothendieck class. The precise statement may be found in,[AM09] Proposition 2.2.

—In particular, c_{SM} classes for hyperplane arrangements (and their complement) carry the same information as the corresponding Grothendieck classes. It is not hard to see that this information is the *characteristic polynomial* of the arrangement (cf.[Alu12]). For instance, for graphical arrangements c_{SM} classes compute the chromatic polynomial. ⌟

For $X \subseteq \mathbb{P}^n$, the information carried by $c_{\mathrm{SM}}(1\!\!1_X)$ is precisely equivalent to the information of the Euler characteristics of all general linear sections of X ([Alub]).

Work of J.-P. Brasselet, J. Schürmann, S. Yokura treat the c_{SM} class in terms of a relative Grothendieck group ([BSY10]). For projective sets, there is a group homomorphism from a Grothendieck group associated with $\mathbb{P}^\infty$ to the corresponding Chow group, a polynomial ring $\mathbb{Z}[T]$, which yields the c_{SM} class ([Alub]). The multiplicativity properties of this morphism are interesting, and they will play a role behind the scenes later in these notes (Theorem 4.1).

In any case, the true 'motivic' nature of c_{SM} classes has likely not yet been understood completely. One of the motivations in carrying out the computations that will be reviewed in the rest of this paper is to collect raw data in interesting situations, with the aim of clarifying possible relations between motives and Chern-Schwartz-MacPherson classes.

3. Feynman periods and graph hypersurfaces

3.1. *Feynman periods*

In quantum field theory, and more specifically perturbative massless scalar field theories, one is interested in action integrals

$$S(\phi) = \int L(\phi) d^D x$$

in dimension D, where $L(\phi)$ is a 'Lagrangian density'. These integrals may be studied by means of a perturbative expansion, which takes the form of an effective action

$$S_{\text{eff}}(\phi) = \sum_G \langle G \rangle \tag{1}$$

where the sum ranges over a collection of graphs G determined by the Lagrangian, and the contribution $\langle G \rangle$ of a graph is of the form $\frac{G(\phi)}{\# \operatorname{Aut}(G)}$; we will say more about $G(\phi)$ below. Further manipulations may be used to restrict attention to more special graphs, for example connected, '1-particle irreducible' (1PI) graphs—i.e., graphs without edges whose removal causes G to become disconnected (i.e., without 'bridges'); also see §4.1. The reader is addressed to any text in quantum field theory (such as[Zee10]) for information on what all of this really means, and to[Mar10] for a more thorough presentation in the same context as this note.

The contribution $G(\phi)$ due to G is itself an integral, of the form

$$G(\phi) = \frac{1}{N!} \int_{\sum_i p_i = 0} \hat{\phi}(p_1) \cdots \hat{\phi}(p_N) U(G(p_1, \dots, p_N)) dp_1 \cdots dp_N \quad .$$

Here N is the number of external edges attached to G, and p_i denotes a momentum associated with the i-th external edge. The integration locus is $\sum_i p_i = 0$ to enforce momentum conservation; $\hat{\phi}$ denotes Fourier transform, and the integral is taken over 'momentum space'. This is not the only option: an expression for $G(\phi)$ may be given over 'configuration space', and this turns out to be advantageous for the questions of interest here, as has been understood very recently, see.[CM] In the above integral,

$$U(G(p_1, \dots, p_N)) = \int I_G(k_1, \dots, k_\ell, p_1, \dots, p_N) d^D k_1 \cdots d^D k_\ell \quad ,$$

where ℓ is the number of loops in G, and the core contribution I_G may be written out in an essentially automatic fashion from the combinatorics of G, by means of the full set of 'Feynman rules'. We will come back to this in §4, where the guiding theme will indeed be the relation between the combinatorics of a graph and the objects of interest to us. But we have not yet described these objects.

Clever use of various reductions (keywords: 'Schwinger parameters', 'Feynman trick') yields an expression for $U(G)$ of the following form:

$$U(G) = \frac{\Gamma(n - D\ell/2)}{(4\pi)^{\ell D/2}} \int_{\sigma_n} \frac{P_G(\underline{t}, p)^{-n+D\ell/2}\omega_n}{\Psi_G(\underline{t})^{-n+D(\ell+1)/2}} \quad . \tag{2}$$

Here n is the number of (internal) edges of G; the integral is over the (real) simplex $\sigma_n = \{\sum_i t_i = 1\}$; ω_n is the volume form; and P_G, Ψ_G are certain polynomials determined by the graph G. The reader should take heart in the news that of all the terms introduced so far in this section, this polynomial Ψ_G (see §3.2) is *the only one* that will be really relevant to what follows. The more curious reader will find expression (2) as (3.21) in.[Mar10] The polynomial $P_G(\underline{t}, p)$ is given by

$$P_G(\underline{t}, p) = \sum_C s_C \prod_{e \in C} t_e \quad ,$$

where the sum ranges over the *cut sets* C of G, and the coefficient s_C is a quadratic function of the external momenta. For example,

$$U(G) = \frac{\Gamma(2 - D/2)}{(4\pi)^{D/2}} \int_{\sigma_2} \frac{(p^2 t_1 t_2)^{-2+D/2}\omega_2}{(t_1 + t_2)^{-2+D}}$$

for the '2-banana' graph

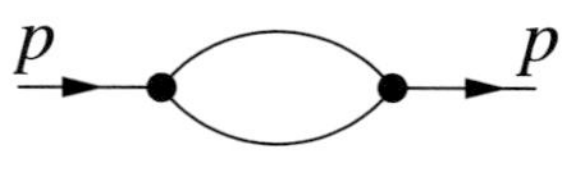

consisting of two vertices joined by two parallel edges, with two external edges carrying a moment p into and out of the graph; this graph has just one cut set, consisting of both edges. Exercise: the above integral agrees with the one given for this graph on p. 23 of.[Mar10]

See Theorem 3.1.9 in[Mar10] for a derivation of (2).

In general, there are issues of convergence with all the integrals mentioned above, with which quantum field theory magically manages to deal. For example, the dimension D is used as a parameter in these integrals (occasionally allowing it to become a complex number); here we will pretend that D is large enough that $D\ell/2 - n \geq 0$, so that P_G is really at numerator and Ψ_G is really at denominator in (2). The Gamma factor $\Gamma(n - D\ell/2)$ contributes to the divergence of $U(G)$, so one can at best aim at dealing with the residue of $U(G)$, which under the above assumptions has the form

$$\int_{\text{cycle } \sigma} \frac{\text{diff. form}}{\Psi_G(t)^M}$$

with $M \geq 0$, an integral of an algebraic differential form over the *complement of the hypersurface* $\hat{X}_G$ *defined by* $\Psi_G = 0$. This hypersurface complement is the main object of interest to us. There are further divergences due to the intersection of $\hat{X}_G$ with the domain of integration σ. Methods have been developed to deal with these, for example sequences of blow-ups which separate the domain from $\hat{X}_G$. This is discussed in work of S. Bloch, H. Esnault, D. Kreimer (,[BEK06BK08]), and I will simply gloss over this important issue, for the sake of simplicity, and on the ground that this should not affect the considerations in this paper.

Integrals of algebraic forms over algebraic domains (all defined over $\overline{\mathbb{Q}}$) are called *periods.* The standard reference for this notion is an extensive review article by M. Kontsevich and D. Zagier,[KZ01] The set of periods forms a ring (often enlarged by including $\frac{1}{2\pi i}$, for technical reasons), and is *countable.* It contains all algebraic numbers, but also numbers such as π; it is surprisingly difficult to prove that a given number is *not* a period, although cardinality ensures that most numbers are not. (For example, e is not believed to be a period, but this is not known.) A period is an expression of the cohomology of a variety, and intuitively varieties with 'simpler' cohomology should have 'simpler' periods. In fact, the periods are determined by the *motive* of the variety on which they are taken, so they may be seen as an avatar of the motive. For example, periods of mixed-Tate motives are known to be (suitable combinations of) *multiple zeta values;* this was a long-standing conjecture, and has recently been proved by F. Brown,[Bro12] More precisely: multiple zeta values are defined by

$$\zeta(n_1, \ldots, n_r) = \sum_{0<k_1<\cdots<k_r} \frac{1}{k_1^{n_1} \cdots k_r^{n_r}}$$

with $n_i \geq 1$, $n_r \geq 2$. Brown was able to confirm that all periods of mixed Tate motives over $\mathbb{Z}$ are $\mathbb{Q}[\frac{1}{2\pi i}]$-linear combinations of multiple zeta values.

Summarizing: the contributions $U(G)$ may be viewed as certain periods for the (motive corresponding to) the complement $\hat{Y}_G$ of the *graph hypersurface* $\hat{X}_G \subseteq \mathbb{A}^n$ defined by $\Psi_G(\underline{t}) = 0$. Again, let me point out that there is an issue of divergence (due to the intersection of $\hat{X}_G$ with the simplex σ over which the integration takes place). In the approach by Bloch et al., this issue is dealt with by means of carefully chosen blow-ups determined by the combinatorics of the graph. While these blow-ups are necessary, their impact on the the motive of the complement should be relatively mild, or so I will assume at any rate. Thus, I will assume that the contributions depend directly on the motive of $\hat{Y}_G$.

Now, there are refined numerical techniques that can closely approximate a contribution $U(G)$ for a given graph G. Thus, a rather extensive catalogue of examples of these 'Feynman periods' has been obtained with high accuracy. If a number is known with sufficiently high precision, there are techniques that can determine very reliably whether the number is, say, algebraic; or, case in point, whether it is a linear combination of multiple zeta values (with coefficients in $\mathbb{Q}[\frac{1}{2\pi i}]$). The reference usually quoted for for these computations is;[BK97] the amazing observation stemming from these very extensive computations is that *Feynman periods appear to be (combinations of) multiple zeta values.* In other words, Feynman periods appear to be periods of mixed Tate motives.

The natural conjecture is then that the complements $\hat{Y}_G$ themselves should be mixed Tate motives. Although, as we will see in a moment, this turns out not to be the case, understanding this situation better has motivated a very substantial amount of work, and my goal in §4 will be to review a tiny portion of this work. I should point out right away that due to work of F. Brown again (,[Bro],[Bro09] etc.), and several others, the available physicists' evidence linking Feynman periods to multiple zeta values has been explained, but the general question of whether Feynman periods are always necessarily combinations of multiple zeta values is still open.

3.2. *Graph hypersurfaces*

The main character of the story has now been identified as the hypersurface given by the vanishing of the 'graph polynomial Ψ_G' mentioned in §3.1. This polynomial is

$$\Psi_G(t_1,\dots,t_n) := \sum_T \prod_{e\notin T} t_e$$

where T ranges over the maximal spanning forests of G (i.e., the unions of spanning trees of the connected components of G), and t_e is a variable associated with the edge e. If G is connected, maximal spanning forests are spanning trees; the reader should note that other references simply set this polynomial to 0 if G is not connected, but we find that it is worth considering Ψ_G as above for arbitrary graphs.

Example 3.1. The spanning trees for the '3-banana graph'

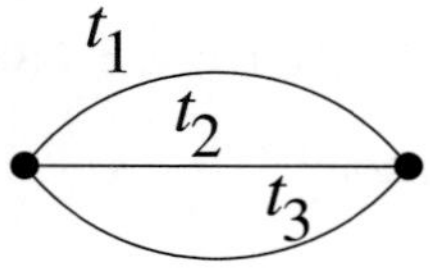

are

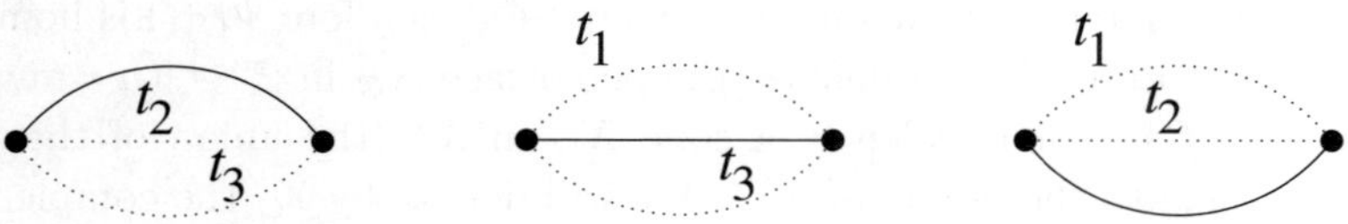

and it follows that

$$\Psi_G(\underline{t}) = t_2 t_3 + t_1 t_3 + t_1 t_2 = t_1 t_2 t_3 \left(\frac{1}{t_1} + \frac{1}{t_2} + \frac{1}{t_3} \right)$$

in this case. Similarly, the graph polynomial for the n-banana graph G is

$$\Psi_G(\underline{t}) = t_1 \cdots t_n \left(\frac{1}{t_1} + \cdots + \frac{1}{t_n} \right)$$

for all $n > 0$. ⌟

Example 3.2. There are eight spanning trees for the graph

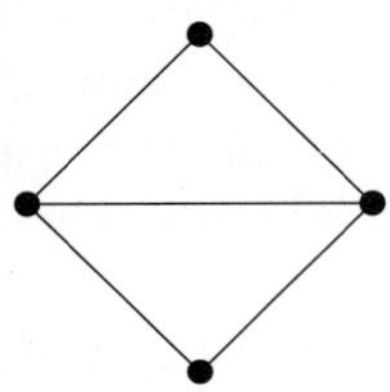

namely

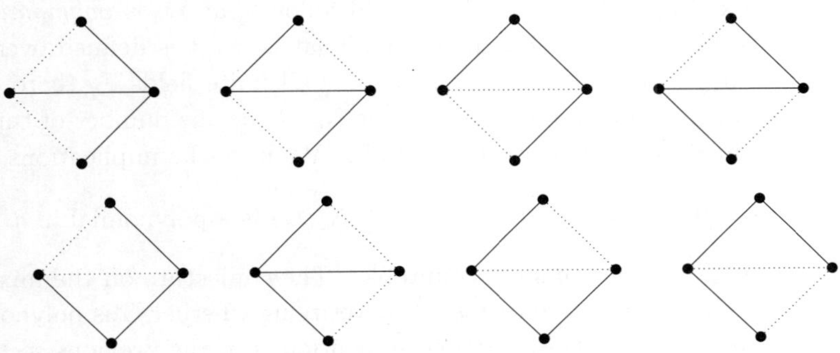

The corresponding graph polynomial is (up to renumbering the edges)

$$t_1t_2 + t_1t_3 + t_1t_5 + t_2t_4 + t_2t_5 + t_3t_4 + t_3t_5 + t_4t_5 \quad ,$$

an irreducible quadratic polynomial. ⌟

In the graph theory literature, Ψ_G is called the *(Kirchoff-Tutte)-Symanzik* polynomial. The number of edges in a maximal spanning forest for G is the same for all such forests (exercise!), therefore $\Psi_G(\underline{t})$ is homogeneous. Thus, its vanishing defines a hypersurface X_G in $\mathbb{P}^{n-1}$ (n =number of edges of G), or equivalently a cone $\hat{X}_G$ in $\mathbb{A}^n$ (the union of the lines corresponding to the points of X_G). We will denote by Y_G the complement of X_G in $\mathbb{P}^{n-1}$, and by $\hat{Y}_G$ the complement of $\hat{X}_G$ in $\mathbb{A}^n$. Of course each of these objects carries essentially the same information. It is clear from the definition that the degree of Ψ_G equals the number of loops $b_1(G)$ of G.

For very small graphs, X_G may be nonsingular: in the example of a 3-banana given above, the corresponding X_G is a nonsingular conic in $\mathbb{P}^2$. But X_G is singular for larger graphs. The reader can verify that the singular locus of X_G is defined by (cones over) graph hypersurfaces determined by smaller graphs $G \smallsetminus e$. The study of singularities of X_G was initiated by E. Patterson ([Pat10]); some of our own work reviewed in §4 is aimed at the computation of global invariants of the singularities of X_G analogous to the Milnor number. We will also see later on that X_G is *rational* when it is irreducible; and it is known (see for example,[SMJŌ77] Proposition 5.2) that Ψ_G is reducible if and only if G can be 'separated' as a union of disjoint subgraphs, or graphs joined at a vertex. Also see,[MA] §5, for a discussion of this question.

The conjecture mentioned at the end of the previous section is that $\hat{Y}_G$ should determine a mixed-Tate motive (over $\mathbb{Z}$). As a consequence, the Grothendieck class $[\hat{Y}_G]$ should belong to the subring $\mathbb{Z}[\mathbb{L}]$ of the Grothendieck ring $K_0(\mathrm{Var}_{\mathbb{Z}})$; and it would follow that $\hat{Y}_G$ is *polynomially countable* (cf. Definition 3, Example 2.1). That is: as it is defined over $\mathbb{Z}$, Ψ_G defines a polynomial in $\mathbb{F}_q[t_1, \ldots, t_n]$ for all finite fields $\mathbb{F}_q$ (here q is of course a prime power); one can then let $N_q(G)$ be the number of tuples $(t_1, \ldots, t_n)$ in $\mathbb{F}_q^n$ such that $\Psi_G(\overline{t}) \neq 0$ in $\mathbb{F}_q$. We have the implications

$$\hat{Y}_G \text{ mixed-Tate} \implies [\hat{Y}_G] \in \mathbb{Z}[\mathbb{L}] \implies N_q(G) \text{ is a polynomial in } q.$$

In this sense G is 'polynomially countable'. The conjecture on the mixed-Tateness of $\hat{Y}_G$ can therefore be tested by verifying whether G is polynomially countable. The physicists' evidence mentioned in the previous section motivated the conjecture (attributed to Kontsevich) that all graphs would be polynomially countable.

The history of this problem is interesting. Initial work by combinatorialists R. Stanley ([Sta98]), J. R. Stembridge ([Ste98]), and others appeared to provide very substantial evidence for the conjecture: all graphs with 12 or fewer edges were found to satisfy it. But the conjecture was soon *disproved* in a remarkable paper by P. Belkale and P. Brosnan ([BB03]), showing that there must be graphs G for which $[\hat{Y}_G]$ is not in the subring $\mathbb{Z}[\mathbb{L}]$ of $K_0(\mathrm{Var}_{\mathbb{Z}})$. Indeed, Belkale and Brosnan show that classes of graph hypersurfaces *generate* the Grothendieck ring $K_0(\mathrm{Var}_{\mathbb{Z}})$, in a sense to which I will come back at the end of §4. In more recent work by D. Doryn and O. Schnetz (,[Dor11Sch11]), it has been shown that all graphs with 13 or fewer edges satisfy the conjecture, and specific graphs G with 14 and 16 edges have been found which are not polynomially countable. The examples with 16 edges fail polynomial countability as the corresponding $N_q(G)$ can be related to the number of points on a K3 in $\mathbb{P}^3$. The appearance of K3s in this context is further studied in the more recent.[BS12]

The polynomial Ψ_G may be written as a determinant of a matrix easily concocted from the graph and recording which edges belong to a fixed basis for $H^1(G)$. Thus, graph hypersurfaces may be viewed as specializations of a determinantal hypersurface, and with suitable care one can write Feynman periods as periods of the complement of a determinant hypersurface. After all, the Feynman periods detect one part of the cohomology of $\hat{Y}_G$; even if $\hat{Y}_G$ is not mixed-Tate, it may be that the part of its cohomology that is responsible for the period is mixed-Tate. Realizing the period as a period over some other potentially simpler locus is a natural approach to proving that it must be a combination of multiple zeta values. This approach is carried out in,[AM10] but hits against a wall: even though determinant hypersurfaces are indeed simpler objects than graph hypersurfaces, the intersection of the hypersurface with the domain of integration becomes much harder to grasp. It may be described in terms of intersections of unions of Schubert cells in flag varieties, and these objects are known to be arbitrarily complex (this is one expression of R. Vakil's "Murphy's law",[Vak06]).

In the following §4 we will focus on general considerations relating the class $[\hat{Y}_G]$ in $K_0(\mathrm{Var})$ with the combinatorics of G. We will also have something to say about the other 'generalized Euler characteristic' considered in §2, that is, the Chern-Schwartz-MacPherson class of a graph hypersurface X_G, and an intriguing polynomial invariant which may be extracted from it. The very fact that the behavior of this invariant is in many ways analogous to the behavior of the Grothendieck class is interesting; and in any case we view $c_{\mathrm{SM}}(X_G)$ as a means to get global information on the singularities of the hypersurface X_G. The issue of polynomial countability and its relation to Kontsevich's conjecture was a strong motivation for the work reviewed in §4, but it will remain as an undercurrent, while the emphasis will simply be on studying these objects for their own sake.

4. Grothendieck classes of graph hypersurfaces

4.1. *Feynman rules*

Recall from §3.1 that the action integrals we are interested in take the form (1): $\sum_{\text{graphs } G} \langle G \rangle$, where $\langle G \rangle$ is a contribution due to G and accounting for a symmetry factor. If G_i are the connected components of G, then

$$\langle G \rangle = \prod_i \langle G_i \rangle / \text{extra symmetries} \quad .$$

The *linked cluster theorem* states that this 'scattering cross section' can then be decomposed as follows:

$$\sum_{\text{graphs } G} \langle G \rangle = \exp\left(\sum_{\textit{connected} \text{ graphs } G} \langle G \rangle \right) \quad .$$

That is, the combinations of the terms due to exponentiating the sum recovers precisely the extra symmetry factors. There is a further reduction: if G is obtained by joining G_1 and G_2 by an edge joining a vertex of G_1 to a vertex of G_2, then

$$\langle G \rangle = \text{'propagator'} \cdot \langle G_1 \rangle \cdot \langle G_2 \rangle \quad ,$$

where the (inverse) 'propagator' term is a multiplicative factor independent of the graphs, allowing for the 'propagation' of momentum from one graph to the other across the bridge joining the two graphs.

These reductions are parts of the lore of 'Feynman rules', a set of recipes that, given a graph, allow a competent physicist to write down the amplitude due to a graph as an integral by extracting information from the combinatorics of the graph. It is natural to ask

> Do invariants such as the Grothendieck or c_{SM} class of a graph hypersurface satisfy a similar structure?

This appears to be a hard and interesting question. In this subsection we are going to observe that the bare-bone essence of the Feynman rules, i.e., the behavior of the invariants with respect to splitting a graph into connected components, or joining two graphs by a bridge, does indeed apply to our invariants in a form that is very similar to the classical one recalled above. This is straightforward for what concerns the Grothendieck class, but requires some work for the $c_{\rm SM}$ invariant. In fact, I was rather surprised when it turned out to work 'on the nose' for our $c_{\rm SM}$ invariant, and I take this instance as good evidence that the invariant is indeed detecting directly some information carried by the combinatorics of the graph.

This material is explored in more detail in.[AM11a] The reader should be warned that the title of this article is more ambitious than the article itself: only the Feynman rules dealing with propagator bridges are really dealt with in this paper. Of course the hope would be that more sophisticated 'algebro-geometric Feynman rules' can be concocted, and eventually allow to write the Grothendieck class or the $c_{\rm SM}$ class for a graph hypersurface by a straightforward combinatorial recipe. This is a good (and difficult) open question, which is not even touched in.[AM11a] (One has to start somewhere...)

The following observation is a direct consequence of the definition of graph polynomials:

Lemma 4.1. *Let G be the disjoint union of two graphs G_1, G_2. Then* $\Psi_G = (\Psi_{G_1})(\Psi_{G_2})$.

Therefore, the polynomial of a graph is the product of the polynomials for its connected components. The same happens for components connected by bridges:

Lemma 4.2. *Let G be obtained by joining G_1 and G_2 by a bridge. Then* $\Psi_G = (\Psi_{G_1})(\Psi_{G_2})$.

The point is that bridges belong to *all* spanning trees, therefore the corresponding variables do not appear in the graph polynomial.

Example 4.1. Consider the graph obtained by joining two polygons by a bridge:

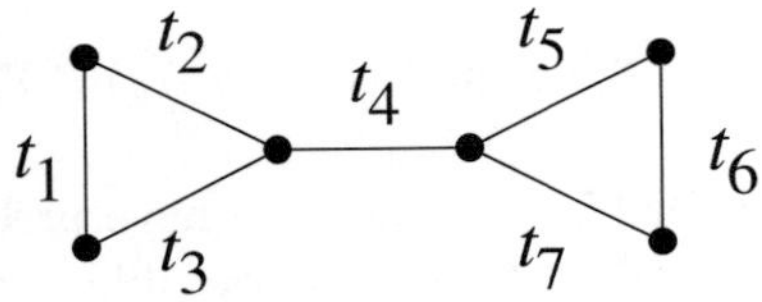

A spanning tree is obtained by joining a spanning tree of the left triangle with a spanning tree of the triangle on the right by the bridge t_4. It follows easily that the graph polynomial equals $(t_1 + t_2 + t_3)(t_5 + t_6 + t_7)$.

The three graphs

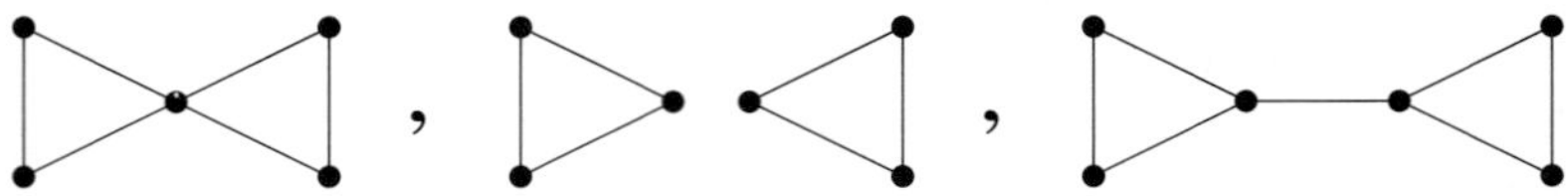

all have the same graph polynomial. But note that the third graph *hypersurface* lives in a space of higher dimension, because of the extra edge. This fact will be responsible for the form of the propagator in our version of the 'Feynman rules'. ⌟

Recall that for a graph G on n edges we are denoting by X_G the hypersurface $\Psi_G(t_1, \dots, t_n) = 0$ in $\mathbb{P}^{n-1}$, by $\hat{X}_G$ the affine cone in $\mathbb{A}^n$, and by $\hat{Y}_G$ the complement $\mathbb{A}^n \smallsetminus \hat{X}_G$.

Prop 4.1. Let G_1, G_2 be graphs.

- If $G = G_1 \amalg G_2$, then $\hat{Y}_G \cong \hat{Y}_{G_1} \times \hat{Y}_{G_2}$.
- If G is obtained by joining G_1 and G_2 by a bridge, then $\hat{Y}_G \cong \hat{Y}_{G_1} \times \hat{Y}_{G_2} \times \mathbb{A}^1$.

Proof. Let $t_1, \dots, t_{n_1}$ be the variables corresponding to the edges of G_1, and let $u_1, \dots, u_{n_2}$ be the variables corresponding to the edges of G_2. If $G = G_1 \amalg G_2$, then $\Psi_G = \Psi_{G_1}\Psi_{G_2}$ as observed above. A point $(t_1, \dots, t_{n_1}, u_1, \dots, u_{n_2})$ of $\mathbb{A}^{n_1+n_2}$ is in $\hat{Y}_G$ if and only if

$$\Psi_G(t_1, \dots, t_{n_1}, u_1, \dots, u_{n_2}) \neq 0 \quad ,$$

if and only if

$$\Psi_{G_1}(t_1, \dots, t_{n_1})\Psi_{G_2}(u_1, \dots, u_{n_2}) \neq 0 \quad ,$$

if and only if

$$\Psi_{G_1}(t_1, \dots, t_{n_1}) \neq 0 \quad \text{and} \quad \Psi_{G_2}(u_1, \dots, u_{n_2}) \neq 0 \quad ,$$

if and only if $(t_1, \dots, t_{n_1}) \in \hat{Y}_{G_1}$ and $(u_1 \dots, u_{n_2}) \in \hat{Y}_2$. The first assertion follows.

If G_1 and G_2 are joined by a bridge, we have precisely the same computation, but in the presence of an extra variable e; this accounts for the extra factor of $\mathbb{A}^1$, giving the second assertion. (In fact, the second assertion follows from the first.) □

Proposition 4.1 says that the very definition of graph hypersurface complement satisfies the 'bare-bone structure' mentioned above. Notice that the extra $\mathbb{A}^1$ appearing in the second formula may be interpreted as the graph hypersurface complement for the added bridge joining the two graphs. For us (and for lack of a better term), 'abstract Feynman rules' consist of any assignment from graphs to rings, $G \mapsto U(G)$, such that

- If $G = G_1 \amalg G_2$ or G is obtained by joining G_1 and G_2 at a single vertex, then $U(G) = U(G_1)U(G_2)$.

In particular, we have that $U(G) = U(G_1)U(G_2)U(e)$ if G is obtained by joining G_1 and G_2 by a bridge e. This identifies $U(e)$ as the (inverse) 'propagator'.

Abstract Feynman rules are of course a dime for a dozen: for example, it suffices to assign any elements of a ring to connected, 1-PI graphs, and define the invariant on more general graphs by adopting the multiplicativity prescribed by the rules. Or we can simply encode into a 'rule' any feature of the graph that is manifestly additive on disjoint unions. For instance, the following example looks rather useless; but we record it anyway.

Example 4.2. Let $\mathcal{SB}(G) \in \mathbb{Z}$ be 0 if G has edges that are not looping edges, and $(-1)^n$ if G has n looping edges and no other edge. ⌟

This is trivially an example of 'abstract Feynman rules' in the sense specified above, and it does not look very interesting (yet).

What is more interesting is to identify invariants that are meaningfully and independently defined for all graphs, and verify that these happen to satisfy the multiplicativity prescribed by the rules. We are interested in those invariants that arise from algebro-geometric properties of graph hypersurfaces.

Definition 4. *Algebro-geometric Feynman rules* are abstract Feynman rules which only depend on the affine hypersurface complement $\hat{Y}_G$. ⌟

It is not difficult to construct 'universal' algebro-geometric Feynman rules, in the form of a Grothendieck ring F of conical immersed affine varieties, so that algebro-geometric Feynman rules with target a ring R correspond precisely to ring homomorphisms $F \to R$. The hypersurface complement $\hat{Y}_G$ determines a class in this Grothendieck ring F.

Two examples are discussed below; there are others, and they await further study. In general, it may be not completely trivial to decide whether given Feynman rules are 'algebro-geometric' in the sense of Definition 4. For instance, is the assignment $\mathcal{SB}(G)$ of Example 4.2 algebro-geometric?

Also, it may be useful to point out that 'meaningful' algebro-geometric invariants need not be Feynman rules. My favorite example is the Euler characteristic $\chi(Y_G)$ of the complement of the graph hypersurface in projective space: if G is the disjoint union of G_1 and G_2, and neither G_1 nor G_2 is a forest, then the class $[Y_G]$ is a multiple of $(\mathbb{L} - 1)$ (exercise!), so $\chi(Y_G) = 0$ no matter what $\chi(Y_{G_1})$ and $\chi(Y_{G_2})$ may be. By itself, this invariant does not seem to have enough 'structure' to qualify as an example of Feynman rules. By the end of this subsection we will see that the information carried by this invariant *can* be incorporated into meaningful Feynman rules, but in a rather nontrivial way.

—*Motivic algebro-geometric Feynman rules:*
Simply send the class of $\hat{Y}_G$ in F to the class $\mathbb{U}(G) := [\hat{Y}_G]$ in the ordinary Grothendieck ring $K_0(\mathrm{Var})$. The content of Proposition 4.1 is precisely that this assignment gives 'abstract Feynman rules' in the sense specified above. The 'motivic propagator' is the class $\mathbb{L} = (\mathbb{T} + 1)$ of $\mathbb{A}^1$. It is a good exercise to establish the effect of some simple graph operations on $\mathbb{U}(G)$; for example, splitting an edge also multiplies $\mathbb{U}(G)$ by the propagator $(\mathbb{T} + 1)$, while adding a looping edge multiplies $\mathbb{U}(G)$ by $\mathbb{T}$.

While this example is very straightforward, the reader should note that variations on this theme—such as $[\hat{X}_G]$, $[X_G]$, or the class $[Y_G]$ of the hypersurface complement in *projective* space, would not work. The class $[\hat{Y}_G]$ of the affine complement is just right.

—*Polynomial algebro-geometic Feynman rules:*
A far less straightforward example may be obtained using c_{SM} classes; the corresponding Feynman rules take values in $A_*\mathbb{P}^n$, so they may be interpreted as (truncated) polynomials in $\mathbb{Z}[T]$.

Consider any affine cone $\hat{X}$ in $\mathbb{A}^n$ as a locally closed subset of the projective space $\mathbb{P}^n$ obtained by adding the hyperplane at infinity. The c_{SM} class of the corresponding indicator function is a combination of classes of projective subspaces, with integer coefficients:

$$c_{\mathrm{SM}}(1\!\!1_{\hat{X}}) = a_0[\mathbb{P}^0] + a_1[\mathbb{P}^1] + \cdots + a_n[\mathbb{P}^n] \quad .$$

Associate with $\hat{X}$ the polynomial $F_{\hat{X}}(T) := a_0 + a_1 T + \cdots + a_n T^n$.

Example 4.3. What is $F_{\mathbb{A}^n}(T)$? Denoting by H the hyperplane class in

$\mathbb{P}^n$, use (6):

$$\begin{aligned}c_{\mathrm{SM}}(1\!\!1_{\mathbb{A}^n}) &= c_{\mathrm{SM}}(1\!\!1_{\mathbb{P}^n} - 1\!\!1_{\mathbb{P}^{n-1}}) \\ &= ((1+H)^{n+1} - (1+H)^n H) \cap [\mathbb{P}^n] = (1+H)^n \cap [\mathbb{P}^n] \\ &= \sum_{i=0}^{n} \binom{n}{i} [\mathbb{P}^i] \rightsquigarrow F_{\mathbb{A}^n}(T) = \sum_{i=0}^{n} \binom{n}{i} T^i \quad .\end{aligned}$$

Therefore, $F_{\mathbb{A}^n}(T) = (1+T)^n$. ⌟

Prop 4.2 (Easy). *F defines a* group *homomorphism* $F \to \mathbb{Z}[T]$.

Proof. Inclusion-exclusion holds for c_{SM} classes. □

Theorem 4.1 (Not so easy). *F defines a* ring *homomorphism* $F \to \mathbb{Z}[T]$.

This is somewhat surprising, if only because the identification of $A_*\mathbb{P}^n$ with truncated polynomials in $\mathbb{Z}[T]$ does not preserve products—the product in $\mathbb{Z}[T]$ does *not* correspond to the intersection product in $A_*\mathbb{P}^n$. *A priori,* there does not seem to be any reasonable sense in which $[\mathbb{P}^i]$ and $[\mathbb{P}^j]$ should 'multiply' to $[\mathbb{P}^{i+j}]$, especially if $i+j$ exceeds the dimension of the ambient space $\mathbb{P}^n$. Proposition 4.1 recovers a way in which this makes sense nevertheless.

Once Theorem 4.1 is established, we can define 'polynomial', or 'c_{SM}' Feynman rules by associating with a graph G the polynomial $C_G(T) := F_{\hat{Y}_G}(T) \in \mathbb{Z}[T]$. The c_{SM} propagator is the polynomial $C_{\mathbb{A}^1}(T) = 1+T$. (Notice the formal similarity with the situation with the motivic Feynman rules, once the variable T is 'interpreted' as the class $\mathbb{T}$ of a torus.)

Example 4.4. If G is an n-sided polygon, then $C_G(T) = T(1+T)^{n-1}$.

To verify this, note that the graph polynomial for an n-sided polygon G is simply $t_1 + \cdots + t_n$. Therefore, $\hat{Y}_G$ consists of the complement of a hyperplane $\mathbb{A}^{n-1}$ in $\mathbb{A}^n$. As verified in Example 4.3, $F_{\mathbb{A}^r}(T) = (1+T)^r$, and it follows that

$$\begin{aligned}C_G(T) &= F_{\mathbb{A}^n \smallsetminus \mathbb{A}^{n-1}}(T) = (1+T)^n - (1+T)^{n-1} \\ &= (1+T-1)(1+T)^{n-1} = T(1+T)^{n-1}\end{aligned}$$

as claimed. ⌟

A number of general properties of $C_G(T)$ may be established easily; the reader may take most of these as good exercises in handling the definitions. A much subtler relation with the combinatorics of G will be mentioned below, in §4.2.

(i) $C_G(T)$ is a monic polynomial, of degree equal to the number n of edges of G;
(ii) The coefficient of T^{n-1} in $C_G(T)$ equals $n - b_1(G)$;
(iii) If G is a forest with n edges, then $C_G(T) = (1+T)^n$;
(iv) If G is not a forest, then $C_G(0) = 0$;
(v) $C'_G(0)$ equals the Euler characteristic of $Y_G = \mathbb{P}^{n-1} \smallsetminus X_G$;
(vi) If G' is obtained from G by attaching an edge or splitting an edge, then $C_{G'}(T) = (1+T)C_G(T)$;
(vii) If G' is obtained from G by attaching a looping edge, then $C_{G'}(T) = TC_G(T)$;
(viii) If G is not 1-particle irreducible, then $C_G(-1) = 0$.

These properties often streamline computations. For example, to obtain (again) $C_G(T)$ for an n-sided polygon, start from a single vertex and no edges; by (3), the invariant is 1 in this case. By (7), the invariant for a single looping edge is T; and then applying (6) $n-1$ times we obtain again that the invariant for an n-sided polygon is $T(1+T)^{n-1}$.

Also: (4) and (5) clarify the Feynman rules role of $\chi(Y_G)$: If G_1 and G_2 are not forests, then $C_{G_1}(T) = \chi(Y_{G_1})T$ + higher order terms and $C_{G_2}(T) = \chi(Y_{G_2})T + \text{h.o.t.}$ By Theorem 4.1,

$$C_{G_1 \amalg G_2}(T) = \chi(Y_{G_1})\chi(Y_{G_2})T^2 + \text{h.o.t} \quad :$$

this confirms that $\chi(Y_{G_1 \amalg G_2}) = 0$, but tells us that the product $\chi(Y_{G_1})\,\chi(Y_{G_2})$ is not lost in the process: it just appears as the coefficient of T^2 in the corresponding c_{SM} Feynman rules.

Incidentally: *In all examples I know, $\chi(Y_G)$ is always* 0 *or* ± 1. (A large number of computations for small graphs are collected in.[Str11]) I have not been able to prove that this holds in general. Based on the above considerations, it is tempting to conjecture the stronger statement that the first nonzero coefficient of $C_G(T)$ should always be ± 1.

If G is not a forest, then one may verify that $C_G(T)$ is the polynomial obtained from $c_{\mathrm{SM}}(1\!\!1_{\mathbb{P}^{n-1} \smallsetminus X_G})$ by replacing $[\mathbb{P}^k]$ with T^{k+1}. Naive variations on this recipe, e.g., using $c_{\mathrm{SM}}(1\!\!1_{X_G})$ instead, simply do not work. The computation behind Theorem 4.1 is somewhat delicate.

Example 4.5. Let G be the 3-banana graph:

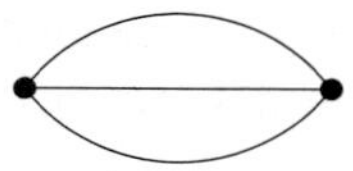

The corresponding graph polynomial is $t_1t_2 + t_1t_3 + t_2t_3$; therefore, the graph hypersurface X_G is a nonsingular conic in $\mathbb{P}^2$, hence $c_{\mathrm{SM}}(1\!\!1_{X_G}) = 2[\mathbb{P}^0]+2[\mathbb{P}^1]$. It follows that $c_{\mathrm{SM}}(1\!\!1_{\mathbb{P}^2 \smallsetminus X_G}) = [\mathbb{P}^0]+[\mathbb{P}^1]+[\mathbb{P}^2]$, and therefore $C_G(T) = T(1+T+T^2)$. ⌟

For the proof of Theorem 4.1, the reader may consult.[AM11a] The surprising role of the product in $\mathbb{Z}[T]$ *vis-a-vis* multiplication of c_{SM} classes is clarified to some extent in.[Alub] Both Jörg Schürmann and Andrzej Weber have remarked that Theorem 4.1 has an interpretation (and perhaps a more natural proof) in an *equivariant* setting; see,[Web12] Proposition 3.

4.2. *Deletion-contraction*

An impressive number of important graph invariants turn out to be specializations of the 'Tutte polynomial' (see e.g.,,[Bol98] Chapter X). This polynomial encodes the basic recursion in graph theory building up a graph by inserting an edge between two vertices or splitting a vertex into two and joining the two resulting vertices by an edge. More precisely: the Tutte polynomial is $T_G(x,y) = x^iy^j$ if G contains i bridges, j looping edges, and no other edge; and in general it is defined by the recurrence relation

$$T_G(x,y) = T_{G/e}(x,y) + T_{G\smallsetminus e}(x,y)$$

where e is an edge of G which is neither a bridge nor a looping edge, and G/e, $G \smallsetminus e$ denote respectively the operations of contracting an edge e and removing it.

In general, a 'Tutte-Grothendieck invariant' is a function τ from the set of graphs to $R[\alpha,\beta,\gamma,x,y]$, where R is a ring (usually taken to be $\mathbb{C}$) such that

- $\tau(G) = \gamma^{\#\text{vertices}}$ if G has no edges;
- $\tau(G) = x\,\tau(G \smallsetminus e)$ if e is a bridge;
- $\tau(G) = y\,\tau(G/e)$ if e is a looping edge;
- $\tau(G) = \alpha\,\tau(G/e) + \beta\,\tau(G \smallsetminus e)$ if e is neither a bridge nor a looping edge.

Examples of Tutte-Grothendieck invariants include the number of spanning trees; the chromatic polynomial; the flow polynomial; the partition function of the Ising model; and many more. Up to a change of variables and minor adjustments, the graph polynomial Ψ_G is itself such an invariant. The Tutte polynomial is the special case $\alpha = \beta = \gamma = 1$.

Fact: Every Tutte-Grothendieck invariant is a specialization of the Tutte polynomial.

This is surprising at first, but rather easy to prove after the fact: just verify that the expression

$$\gamma^{b_0(G)}\alpha^{\#V(G)-b_0(G)}\beta^{b_1(G)}\,T_G\left(\frac{\gamma x}{\alpha},\frac{y}{\beta}\right)$$

satisfies the recursion spelled out above, where $b_0(G)$ is the number of connected components of G, etc. It is clear that the recursion determines the polynomial, so this is all one needs to do.

The very existence of an invariant such as the Tutte polynomial is in itself somewhat magic. The recursive relations allow us to compute it easily for any small graph, for example:

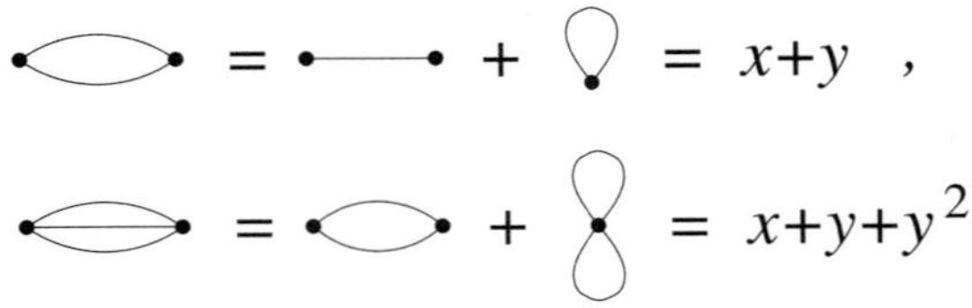

and hence

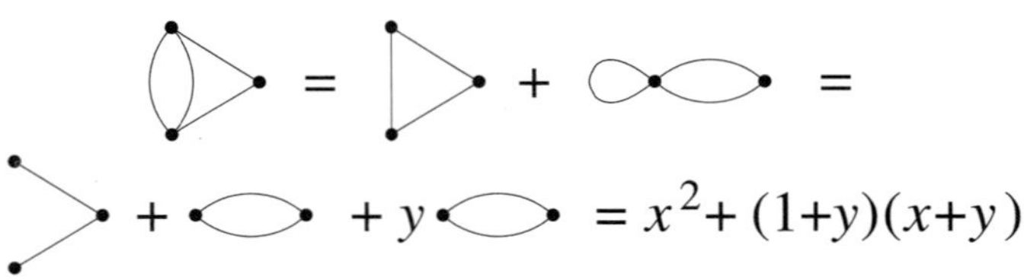

or

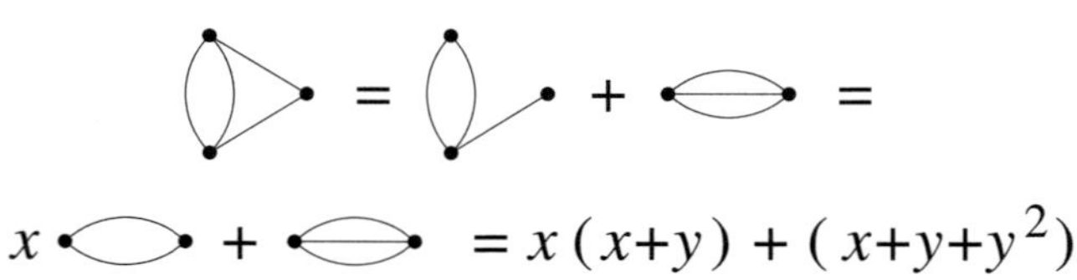

(What is magic is that any choice of sequence of deletions & contractions necessarily gives the same result.) As an illustration of the specialization result mentioned above, the chromatic polynomial of a graph may be recovered from the Tutte polynomial by setting $x = 1 - \lambda$, $y = 0$ (and taking care of a sign and a factor of a power of λ). For the graph shown above, this gives $(1 - \lambda)^2 + (1 - \lambda) = (1 - \lambda)(2 - \lambda)$, which equals the chromatic polynomial up to a factor of λ.

Note that the Tutte polynomial itself is an instance of 'abstract Feynman rules' in the sense of §4.1: it is clear from the recursive definition that

$$T_{G_1 \cup G_2}(x, y) = T_{G_1}(x, y) \cdot T_{G_2}(x, y)$$

if G_1 and G_2 are have at most one vertex in common; the propagator factor is x. Also, the recursive definition of $T_G(x, y)$ implies a formula for doubling a single edge in a graph: if e is an edge of G, denote by G_{2e} the graph obtained by inserting an edge parallel to e in G; then

$$\begin{aligned} T_{G_{2e}}(x, y) &= T_G(x, y) + yT_{G/e}(x, y) \\ &= T_{G \smallsetminus e}(x, y) + (y + 1)T_{G/e}(x, y) \end{aligned}$$

if e is neither a bridge nor a looping edge (this is a good exercise for the reader). For example, this formula produces the Tutte polynomial for the graph (shown above) obtained by doubling an edge of a triangle from the polynomial $x^2 + x + y$ for the triangle and the polynomial $x + y$ for a 2-banana. A little care produces multiple edge formulas, and it is easy to put these together into a generating function,

$$\sum_{m \geq 0} T_{G_{me}}(x, y) \frac{s^m}{m!} = e^s \left(T_{G \smallsetminus e}(x, y) + \frac{e^{(y-1)s} - 1}{y - 1} T_{G/e}(x, y) \right) \quad ,$$

again under the assumption that e is neither a bridge nor a looping edge. The details of these formulas are not important, but the reader is invited to keep in mind their general shape.

At this point it is natural to ask

> Are invariants such as the motivic and $c_{\rm SM}$ Feynman rules discussed in §4.1 Tutte-Grothendieck invariants?

That is: are these Feynman rules specializations of the Tutte polynomial? The quick answer is 'no', as simple examples show. There is a more cogent reason why the answer must be no: if the motivic Feynman rule were a Tutte-Grothendieck invariant, then the Grothendieck class of the complement $\hat{Y}_G$ would recursively be in $\mathbb{Z}[\mathbb{L}]$, while the result of Belkale and Brosnan (;[BB03] see the discussion in §3.2) guarantees that this is not the case. On the other hand, the behavior under the deletion-contraction of these invariants can be understood to some extent, and one can verify that part of the structure implied by the axioms listed above does hold for the algebro-geometric Feynman rules encountered in §4.1: specifically, we can establish doubling edge formulas which have the same flavor as the one shown above for the Tutte polynomial. Details for this material may be found in[AM11b] and.[Alua]

As usual, the result is more straightforward for the motivic Feynman rules. Recall from §4.1 that $\mathbb{U}(G)$ denotes the class $[\hat{Y}_G]$ in $K_0(\mathrm{Var})$.

Theorem 4.2. *Let e be an edge of G, and assume e is neither a bridge nor a looping edge. Then*

- $\mathbb{U}(G) = \mathbb{L} \cdot [\hat{Y}_{G \smallsetminus e} \cup \hat{Y}_{G/e}] - \mathbb{U}(G \smallsetminus e)$,
- $\mathbb{U}(G_{2e}) = (\mathbb{T} - 1) \cdot \mathbb{U}(G) + \mathbb{T} \cdot \mathbb{U}(G \smallsetminus e) + (\mathbb{T} + 1) \cdot \mathbb{U}(G/e)$.

The first formula is a version of a deletion-contraction relation for motivic Feynman rules. In one form or another, this formula has been noticed by anyone working on polynomial countability of graph hypersurfaces, starting at least as early as.[Ste98] The 'problem' with it is that it is not purely combinatorial: the term in the middle involves a locus, $\hat{Y}_{G \smallsetminus e} \cup \hat{Y}_{G/e}$, which does not seem accessible directly in terms of Feynman rules. In a sense it *cannot* be accessible, as mentioned earlier, by the result of.[BB03]

The second formula *is* combinatorial, in the sense that it only involves motivic Feynman rules for graphs obtained from the given one by combinatorial operations. For example, from the knowledge of $\mathbb{U}(G)$ for the graphs in Figure 1, to wit $\mathbb{L}^3 - \mathbb{L}^2 = \mathbb{T}(\mathbb{T}+1)^2$, $(\mathbb{T}+1)^2$, $\mathbb{T}(\mathbb{T}+1)$ (exercise!), we

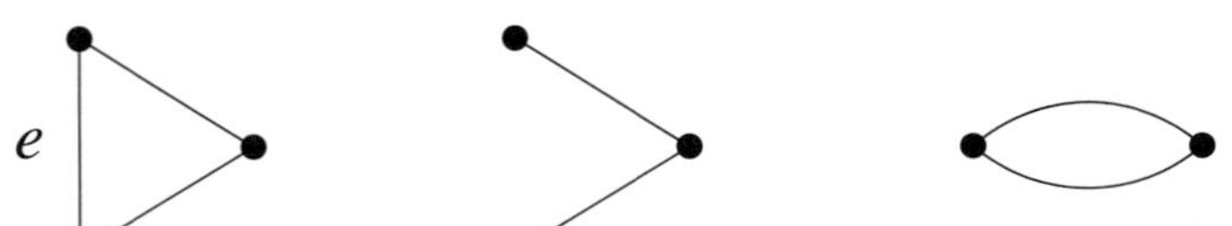

Fig. 1. G, $G \smallsetminus e$, G/e.

get that the motivic class for the triangle with doubled edge is

$$(\mathbb{T} - 1)\mathbb{T}(\mathbb{T}+1)^2 + \mathbb{T}(\mathbb{T}+1)^2 + (\mathbb{T}+1)\mathbb{T}(\mathbb{T}+1) = \mathbb{T}(\mathbb{T}+1)^3 \quad .$$

This is as it should be, as the graph may also be obtained by splitting an edge in a 3-banana, and the class of a 3-banana is $\mathbb{T}(\mathbb{T}+1)^2$. A complete computation of the Grothendieck class for all banana graphs may also be found in.[MA]

In terms of double-edge formulas, the situation is entirely similar to the situation for the Tutte polynomial. As in that case, we get as a formal consequence a generating function for the result of 'multiplying' (bananifying?) the edge e:

$$\sum_{m\geq 0} \mathbb{U}(G_{me})\,\frac{s^m}{m!} = \frac{e^{\mathbb{T}s}-e^{-s}}{\mathbb{T}+1}\mathbb{U}(G) + \frac{e^{\mathbb{T}s}+\mathbb{T}e^{-s}}{\mathbb{T}+1}\mathbb{U}(G\smallsetminus e) + \left(s\,e^{\mathbb{T}s} - \frac{e^{\mathbb{T}s}-e^{-s}}{\mathbb{T}+1}\right)\mathbb{U}(G/e).$$

Our conclusion is that although deletion-contraction does not hold for the motivic Feynman rules, some further consequences do hold 'as if' motivic Feynman rules were Tutte invariants.

Remark 4.1. (i) The second formula in Theorem 4.2 implies that if G, $G\smallsetminus e$, G/e are polynomially countable, then so are all G_{me}.

(ii) The coefficients appearing in the formula for the generating function are functions used in defining Hirzebruch's T_y genus, cf. §11 of Chapter III of.[Hir95]

(iii) The generating function for $\mathbb{U}(G_{me})$ implies that

$$\chi(Y_{G_{me}}) = (-1)^{m-1}(\chi(Y_G) - \chi(Y_{G/e})) \quad ,$$

for $m \geq 2$, still under the assumption that e is neither a bridge nor a looping edge in G, and further that $G\smallsetminus e$ is not a forest. As mentioned in §4.1, $\chi(Y_G)$ is always 0 or ± 1 in all examples I have been able to compute. If this were necessarily the case, then this formula would imply that $\chi(Y_G)$ and $\chi(Y_{G/e})$ cannot be both nonzero and opposite in sign under these assumptions.

(iv) For G =a 2-banana (so that $\mathbb{U}(G) = \mathbb{T}(\mathbb{T}+1)$, $\mathbb{U}(G\smallsetminus e) = \mathbb{T}+1$, and $\mathbb{U}(G/e) = \mathbb{T}$), we obtain a generating function for the class of G_{me} = the $(m+1)$-th banana. This recovers the result of the direct computations in[AM09] and.[MA] ⌟

Proof of Theorem 4.2. The first formula is based on a very useful observation regarding graph polynomials: if e is neither a bridge nor a looping edge, then there is a one-to-one correspondence between the maximal forests of G *not* containing e and the maximal forests of $G \smallsetminus e$, and a one-to-one correspondence between the maximal forests of G *containing* e and the maximal forests of G/e. Denoting by t_e the variable corresponding to e, this shows that

$$\Psi_G = t_e \Psi_{G\smallsetminus e} + \Psi_{G/e} \quad . \tag{1}$$

Now let $\underline{t} \in \mathbb{A}^n$, n =number of edges of G. We have

$$\underline{t} \in Y_G \iff \Psi_G(\underline{t}) \neq 0 \iff \Psi_{G/e}(\underline{t}') \neq -t_e \Psi_{G\smallsetminus e}(\underline{t}') \quad ,$$

where $\underline{t}'$ denotes omission of t_e. Now a simple case-by-case analysis proves the first formula in Theorem 4.2. For example, the case in which $\underline{t}' \notin Y_{G\smallsetminus e}$ and $\underline{t}' \in Y_{G/e}$ accounts for a term

$$\mathbb{L} \cdot [(Y_{G\smallsetminus e} \cup Y_{G/e}) \smallsetminus Y_{G\smallsetminus e}] \quad ,$$

with the $\mathbb{L}$ factor due to the freely varying t_e. Adding the terms obtained from the different possibilities gives the right-hand side of the first formula.

For the second part, assume that e and e' are parallel edges in G_{2e}. Then sorting the maximal forests of G_{2e} according to whether they contain e or e' (they cannot contain both) gives

$$\Psi_{G_{2e}} = t_e t_{e'} \Psi_{G\smallsetminus e} + (t_e + t_{e'}) \Psi_{G/e} \quad .$$

This formula can be used to obtain a relation between the classes $[Y_{G_{2e}/e} \cup Y_{G_{2e}\smallsetminus e}]$ and $[Y_{G/e} \cup Y_{G\smallsetminus e}]$. A serendipitous cancellation gets rid of these 'noncombinatorial' terms, and the second formula follows. The reader is invited to provide details for this argument, which may be found in.[AM11b]

□

There is an analogous story for c_{SM} Feynman rules $C_G(T)$, but it is more complex, and the proofs are substantially harder. There is a deletion-contraction formula:

Theorem 4.3. *Let e be an edge of G, and assume e is neither a bridge nor a looping edge, and satisfies two additional technical conditions. Then*

$$C_G(T) = C_{G\smallsetminus e, G/e}(T) + (T-1) C_{G\smallsetminus e}(T) \quad .$$

Here, $C_{G\smallsetminus e,G/e}(T)$ is a 'non-combinatorial term' obtained from the union $\hat{Y}_{G\smallsetminus e} \cup \hat{Y}_{G/e}$ by the same process (described in §4.1) yielding $C_G(T)$ from $\hat{Y}_G$. The two additional technical conditions are somewhat mysterious, and I address the reader to,[Alua] §2 for a thorough discussion. For example, $\Psi_{G/e}(T)$ is required to belong to the Jacobian ideal of $\Psi_{G\smallsetminus e}$. It is a little surprising that this condition holds for as many graphs as it appears to hold; the smallest example of a graph that does not satisfy it is

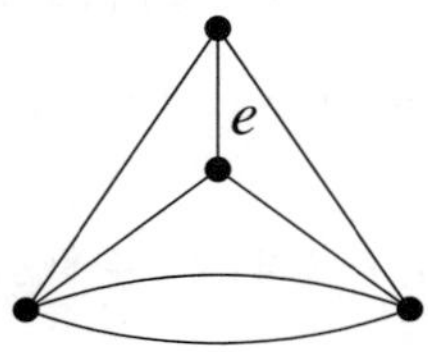

with respect to the vertical edge e. (The second condition is even more technical.) It would be interesting to provide combinatorial interpretations of these algebraic conditions. If the conditions are satisfied, we obtain a 'combinatorial' double-edge formula for c_{SM} Feynman rules:

Theorem 4.4. *Under the same hypotheses of Theorem 4.3,*

$$C_{G_{2e}}(T) = (2T-1)C_G(T) - T(T-1)C_{G\smallsetminus e}(T) + C_{G/e}(T) \quad .$$

The technical conditions are satisfied for sides of a triangle, so knowledge of c_{SM} Feynman rules for the graphs in Figure 1, that is,

$$T(1+T)^2 \quad , \quad (1+T)^2 \quad , \quad T(1+T)$$

(obtained by applying the properties of c_{SM} Feynman rules listed in §4.1) gives

$$C_G(T) = (2T-1)T(1+T)^2 - T(T-1)(1+T)^2 + T(1+T) = T(1+T)(1+T+T^2)$$

for the triangle with a doubled edge. This is as it should be, since this graph may be obtained by splitting an edge of a 3-banana, and $C_{3\text{-banana}}(T) = T(1+T+T^2)$ (Example 4.5).

By a remarkable stroke of luck, the conditions are automatically satisfied *if e is a multiple edge of G*, that is, if its endpoints are adjacent in $G \smallsetminus e$. Thus, the formula given in Theorem 4.4 holds in this case. This is enough to trigger induction, and we get the relation

$$\begin{aligned} C_{G_{(m+3)e}}(T) = (3T-1)\, C_{G_{(m+2)e}}(T) - (3T^2-2T)\, C_{G_{(m+1)e}}(T) \\ +(T^3-T^2)\, C_{G_{me}}(T) \end{aligned}$$

for $m \geq 1$. Note the differentiation relations among the coefficients; I was not able to explain this remarkable feature of the formula to my satisfaction.

The reader may enjoy the task of obtaining a generating function from Theorem 4.4:

$$\sum_{m\geq 0} C_{G_{(m+1)e}}(t)\,\frac{s^m}{m!} = \left(e^{ts} - e^{(t-1)s}\right) C_{G_{2e}}(t)$$
$$- \left((t-1)\,e^{ts} - t\,e^{(t-1)s}\right) C_G(t) + t\left((s-1)\,e^{ts} + e^{(t-1)s}\right) C_{G/e}(t)\,.$$

A proof of Theorem 4.3 is obtained by interpreting the basic relation (1):

$$\Psi_G = t_e \Psi_{G\smallsetminus e} + \Psi_{G/e}$$

(for e neither a bridge nor a looping edge in G) in terms of blow-ups. Assume G has n edges and (for simplicity) that X_G is irreducible. Let $p \in \mathbb{P}^{n-1}$ be the point for which all coordinates except t_e are set to 0. Then one may use (1) to verify that the blow-up $B\ell_p X_G$ of X_G at p is isomorphic to the blow-up of $\mathbb{P}^{n-2}$ along $X_{G\smallsetminus e} \cap X_{G/e}$. (Note that it follows that irreducible graph hypersurfaces are *rational.*) The basic functoriality property of c_{SM} classes may then be used to relate the various classes, and yields Theorem 4.3. This computation requires that the exceptional divisor of the blow-up of $\mathbb{P}^{n-1}$ at p be sufficiently transversal to the proper transform of X_G; this in turn translates into the technical conditions mentioned in the statement of Theorem 4.3.

Again, our conclusion is that while these algebro-geometric Feynman rules are not Tutte-Grothendieck invariants, they share some basic element of the structure of these invariants, which triggers multiple edge formulas. It would be worth formalizing this observation in more technical terms.

4.3. *Stable birational equivalence*

One remarkable application of the deletion-contraction relation in Theorem 4.2 is the complete determination of the *stable birational equivalence* class of graph hypersurfaces.

Definition 5. Two irreducible, nonsingular, compact complex algebraic varieties X, Y are *stably birationally equivalent* if $X \times \mathbb{P}^\ell$ and $Y \times \mathbb{P}^m$ are birationally equivalent for some ℓ, m. ⌟

Recall (§2.1) that X and Y are 'birationally equivalent' if there are dense open sets U in X and V in Y such that U and V are isomorphic. *Stable* birational equivalence is a weakening of this condition. If X_1 is (stably) birationally equivalent to X_2 and Y_1 is (stably) birationally equivalent to Y_2, then $X_1 \times Y_1$ is (stably) birationally equivalent to $X_2 \times Y_2$.

It is not hard to show that two nonsingular compact *curves* are birationally equivalent if and only if they are in fact isomorphic. In higher dimension, birational equivalence is much more interesting: for example, a variety is birationally equivalent to any of its blow-ups; it follows that, in dimension ≥ 2, every variety is birationally equivalent to infinitely many varieties that are mutually non-isomorphic. In fact, blow-ups are the prime example of birational equivalence, in the sense that if X and Y are birationally equivalent, then there exists a sequence of blow-ups and blow-downs at smooth centers which transforms X into Y. This is the content of the 'weak factorization theorem' of,[AKMW02] a fundamental tool in the study of birational equivalence.

A variety is *rational* if it is birationally equivalent to projective space.

We could adopt the definition of stable birational equivalence given above also for *singular*, or *noncompact* varieties; thus, $\mathbb{A}^1$ or a nodal cubic plane curve would be examples of stably rational varieties. In fact, since (by Hironaka) every complex algebraic variety has a nonsingular compact model, every variety would belong to a well-defined stable birational equivalence class, in this naive sense.

However, something much more interesting can be done. Larsen and Lunts define ([LL03]) a *ring* of stable birational equivalence classes as follows:

- Let SB be the monoid of stable birational equivalence classes of smooth irreducible compact varieties, defining the class of $X \times Y$ as the product of the classes of X and Y (this is well-defined, as follows from the facts mentioned above);
- Then let $\mathbb{Z}[SB]$ be the corresponding monoid ring.

Denote by $[X]_{SB}$ the class of X in $\mathbb{Z}[SB]$. Thus, elements of $\mathbb{Z}[SB]$ may be written as formal integer linear combinations of elements $[X]_{SB}$; if X is smooth irreducible compact *and* rational, then $[X]_{SB} = 1$.

Now the main remark is that *every variety* (possibly singular, noncompact) *determines a class in* $\mathbb{Z}[SB]$*;* but for X singular or noncompact, the class of X in $\mathbb{Z}[SB]$ is not necessarily its naive 'stable birational equivalence class' mentioned above. Instead, one can determine the class of X in $\mathbb{Z}[SB]$ by (for example) obtaining X from smooth irreducible compact varieties by ordinary set-theoretic operations, and mapping the result to $\mathbb{Z}[SB]$ by interpreting disjoint union as addition. Of course one must show that this leads to a well-defined class; see Theorem 4.6 below.

Example 4.6. The class of $\mathbb{A}^1$ in $\mathbb{Z}[SB]$ is 0. Indeed, we can view $\mathbb{A}^1$ as $\mathbb{P}^1 \smallsetminus \mathbb{P}^0$, hence map it to $[\mathbb{P}^1]_{SB} - [\mathbb{P}^0]_{SB} = 1 - 1 = 0$.

The class of a *nodal* plane cubic is also 0: the complement of the node may be realized as the complement of 2 points in $\mathbb{P}^1$, so it has class $[\mathbb{P}^1]_{SB} - 2[\mathbb{P}^0]_{SB} = -1$; adding the node back gives 0.

The class of a cuspidal cubic in $\mathbb{Z}[SB]$ is 1: the complement of the cusp is isomorphic to $\mathbb{A}^1$, so has class 0 as seen above; and adding back the cusp gives 1. ⌟

These three examples show that the class of a singular or noncompact *rational* variety may not be 1 in $\mathbb{Z}[SB]$, but sometimes it is—this is a rather sophisticated statement about the type of singularities of the variety. Our graph hypersurfaces X_G in projective space are usually singular and (when irreducible) rational; it is natural to pose the problem:

> Compute the class of X_G in the Larsen-Lunts ring $\mathbb{Z}[SB]$ of stable birational equivalence classes.

Theorem 4.5. *Let G be a graph that is not a forest, and with at least one non-looping edge. Then the projective graph hypersurface X_G has class 1 in $\mathbb{Z}[SB]$.*

This is Corollary 3.3 in;[AM11c] its proof is very easy modulo,[LL03] and is reviewed below. As Example 4.6 hopefully illustrates, the fact that X_G is rational (when irreducible) does *not* suffice to imply that its class in $\mathbb{Z}[SB]$ is 1, because X_G is singular. The fact expressed by Theorem 4.5 is not obvious, even if its proof is very straightforward modulo well-known facts.

Theorem 4.5 follows from the deletion-contraction relation in Theorem 4.2 and from the following key description of the ring $\mathbb{Z}[SB]$:

Theorem 4.6 (Larsen & Lunts:,[LL03] Theorem 2.3, Proposition 2.7). *The assignment $X \mapsto [X]_{SB}$ for X smooth, irreducible, complete descends to an onto ring homomorphism $K_0(\mathrm{Var}) \to \mathbb{Z}[SB]$, whose kernel is the ideal $(\mathbb{L})$.*

This result leads to the description given above for the construction of the class in $\mathbb{Z}[SB]$ determined by a possibly singular or noncompact variety: every such variety has a class in $K_0(\mathrm{Var})$ which may be expressed in terms of irreducible nonsingular compact varieties, by means of basic set-theoretic operations.

One way to prove Theorem 4.6 is to show that stable birational equivalence is compatible with Bittner's relations (cf.,[LL03] Remark 2.4). This is because if B is a nonsingular compact variety and E is the total space of a projective bundle over B, then B and E are stably birationally equivalent: indeed, E is birationally equivalent to $B \times \mathbb{P}^r$ for some r. It follows that if X is nonsingular and compact, and $B \subset X$ is a nonsingular closed subvariety, then

$$[B\ell_B X]_{SB} - [E]_{SB} = [X]_{SB} - [B]_{SB} \quad ,$$

where $B\ell_B X$ is the blow-up of X along B, and E is the exceptional divisor. This shows that there is a homomorphism $K_0(\mathrm{Var}) \to \mathbb{Z}[SB]$ as claimed, and this homomorphism is onto since classes of nonsingular compact irreducible varieties generate $\mathbb{Z}[SB]$. It is clear that $(\mathbb{L})$ is in the kernel of this homomorphism (exercise!); the opposite inclusion may be obtained as a consequence of the weak factorization theorem of:[AKMW02] since birational isomorphisms may be decomposed in terms of blow-ups at nonsingular centers, the key point is that $[B\ell_B X] \equiv [X]$ modulo $(\mathbb{L})$ in $K_0(\mathrm{Var})$.

By Theorem 4.6, we can define 'stable birational Feynman rules' by taking the stable birational equivalence class of the motivic Feynman rules $\mathbb{U}(G)$ of §4.1 modulo $\mathbb{L}$: this defines a homomorphism $F \to \mathbb{Z}[SB]$ from the ring of universal algebro-geometric Feynman rules to the Larsen-Lunts ring. As we will see in a moment, this is essentially the same as my prime example (Example 4.2) of not-so-interesting Feynman rules! (But this shows that that example *is* algebro-geometric.)

Proof of Theorem 4.5. By Theorem 4.6, it suffices to compute the class of X_G modulo $\mathbb{L}$. Equivalently, we can compute the class of $[\hat{Y}_G] = \mathbb{U}(G)$ modulo $\mathbb{L}$. Reading the deletion-contraction formula in Theorem 4.2 modulo $\mathbb{L}$ gives

$$\mathbb{U}(G) \equiv -\mathbb{U}(G \smallsetminus e) \mod (\mathbb{L})$$

if e is neither a bridge nor a looping edge in G. If e is a looping edge, then

$$\mathbb{U}(G) = (\mathbb{L}-1)\mathbb{U}(G \smallsetminus e) \equiv -\mathbb{U}(G \smallsetminus e) \mod (\mathbb{L});$$

and if e is a bridge, then

$$\mathbb{U}(G) = \mathbb{L}\mathbb{U}(G \smallsetminus e) \equiv 0 \mod (\mathbb{L})$$

(as we saw in §4.1). These recursions reduce the computation of $\mathbb{U}(G)$ mod $\mathbb{L}$ to the case in which G has no edges at all; and if G had a non-looping edge, at some stage in the reduction process that edge will become a bridge, killing the class modulo $\mathbb{L}$. It follows that

$$\mathbb{U}(G) = \begin{cases} 0 & \text{mod } (\mathbb{L}) \quad \text{if } G \text{ has edges that are not looping edges} \\ (-1)^n & \text{mod } (\mathbb{L}) \quad \text{if } G \text{ has } n \text{ looping edges and no other edge.} \end{cases}$$

This is of course nothing but the assignment $\mathcal{SB}(G)$ of Example 4.2, followed by the inclusion $\mathbb{Z} \to \mathbb{Z}[SB]$.

The statement for X_G given in Theorem 4.5 is an easy consequence of this, left to the reader. □

The observation that $[\hat{Y}_G]$ is a multiple of $\mathbb{L}$ as soon as G has a non-looping edge may be strengthened: Lemma 15 in[BS12] shows that $[\hat{Y}_G]$ is a multiple of $\mathbb{L}^2$ for all 'physically significant' graphs.

An interesting consequence of Theorem 4.5 is that the classes of graph hypersurfaces in $\mathbb{Z}[SB]$ span $\mathbb{Z}$, and this is a rather small part of the ring of stable birational equivalence classes. By Theorem 4.6, classes of graph hypersurfaces span a small part of $K_0(\text{Var})$: any naive motive M that can be written as a combination of graph hypersurfaces must be of the form $c + \mathbb{L} \cdot M'$ for some $c \in \mathbb{Z}$ and $M' \in K_0(\text{Var})$. This is very special—for example, it implies that the Hodge numbers $h^{p,q}$ of M must be zero if $p = 0, q > 0$ or $p > 0, q = 0$.

The reader may not be very surprised by this observation. As a consequence of Theorem 4.6 (,[LL03] Corollary 2.6) every class of $K_0(\text{Var})$ may be written as a combination of classes of smooth complete varieties that are *uniquely* determined up to stable rational equivalence. On the basis of such results, one might suspect that if the class of a variety is a combination of classes of graph hypersurfaces, then the variety must itself be stably birationally equivalent to a graph hypersurface, and this would give a direct explanation for particular features of its class. However, I do not know if this is true: note again that graph hypersurfaces are *singular* in general, so the quoted result does not apply. Theorem 4.5 bypasses these considerations.

The fact that classes of graph hypersurfaces are 'very special' in $K_0(\text{Var})$ should be contrasted with the result of Belkale and Brosnan disproving Kontsevich's conjecture, mentioned in §3.2, whose essence is that classes of graph hypersurfaces are *as general as possible:* according to,[BB03] p. 149, the complements $\hat{Y}_G$ are *'from the standpoint of their zeta functions, the most general schemes possible'*. The contrast is due to the fact that Belkale and Brosnan work in a *localization* of $K_0(\mathrm{Var}_{\mathbb{Z}})$, in which both $\mathbb{L}^n - 1$ $(n > 0)$ and $\mathbb{L}$ are invertible: in[BB03] it is shown that classes of graph hypersurfaces generate this localization over the localization of $\mathbb{Z}[\mathbb{L}]$. It follows from Theorem 4.5 that the localization at $\mathbb{L}$ is necessary for this result to hold: classes of graph hypersurfaces do *not* generate the localization of $K_0(\mathrm{Var}_{\mathbb{Z}})$ if only $\mathbb{L}^n - 1$ are inverted. (Indeed, these elements are already invertible modulo $\mathbb{L}$.)

Question: Do graph hypersurfaces generate $K(\mathrm{Var}_{\mathbb{Z}})_{\mathbb{L}}$ over $\mathbb{Z}[\mathbb{L}]_{\mathbb{L}}$?

References

AB08. Paolo Aluffi and Jean-Paul Brasselet. Une nouvelle preuve de la concordance des classes définies par M.-H. Schwartz et par R. MacPherson. *Bull. Soc. Math. France*, 136(2):159–166, 2008.

AKMW02. Dan Abramovich, Kalle Karu, Kenji Matsuki, and Jaroslaw Wlodarczyk. Torification and factorization of birational maps. *J. Amer. Math. Soc.*, 15(3):531–572, 2002.

Alua. Paolo Aluffi. Chern classes of graph hypersurfaces and deletion-contraction. arXiv:1106.1447, to appear in Moscow Math. Jour.

Alub. Paolo Aluffi. Euler characteristics of general linear sections and polynomial Chern classes. arXiv:1207.6638.

Alu06. Paolo Aluffi. Classes de Chern des variétés singulières, revisitées. *C. R. Math. Acad. Sci. Paris*, 342(6):405–410, 2006.

Alu12. Paolo Aluffi. Grothendieck classes and Chern classes of hyperplane arrangements. *Int. Math. Res. Not.*, 2012.

AM09. Paolo Aluffi and Matilde Marcolli. Feynman motives of banana graphs. *Commun. Number Theory Phys.*, 3(1):1–57, 2009.

AM10. Paolo Aluffi and Matilde Marcolli. Parametric Feynman integrals and determinant hypersurfaces. *Adv. Theor. Math. Phys.*, 14(3):911–963, 2010.

AM11a. Paolo Aluffi and Matilde Marcolli. Algebro-geometric Feynman rules. *Int. J. Geom. Methods Mod. Phys.*, 8(1):203–237, 2011.

AM11b. Paolo Aluffi and Matilde Marcolli. Feynman motives and deletion-contraction relations. In *Topology of algebraic varieties and singularities*, volume 538 of *Contemp. Math.*, pages 21–64. Amer. Math. Soc., Providence, RI, 2011.

AM11c. Paolo Aluffi and Matilde Marcolli. Graph hypersurfaces and a dichotomy in the Grothendieck ring. *Lett. Math. Phys.*, 95(3):223–232, 2011.

BB03. Prakash Belkale and Patrick Brosnan. Matroids, motives, and a conjecture of Kontsevich. *Duke Math. J.*, 116(1):147–188, 2003.

BEK06. Spencer Bloch, Hélène Esnault, and Dirk Kreimer. On motives associated to graph polynomials. *Comm. Math. Phys.*, 267(1):181–225, 2006.

Bit04. Franziska Bittner. The universal Euler characteristic for varieties of characteristic zero. *Compos. Math.*, 140(4):1011–1032, 2004.

BK97. D. J. Broadhurst and D. Kreimer. Association of multiple zeta values with positive knots via Feynman diagrams up to 9 loops. *Phys. Lett. B*, 393(3-4):403–412, 1997.

BK08. Spencer Bloch and Dirk Kreimer. Mixed Hodge structures and renormalization in physics. *Commun. Number Theory Phys.*, 2(4):637–718, 2008.

Bol98. Béla Bollobás. *Modern graph theory*, volume 184 of *Graduate Texts in Mathematics*. Springer-Verlag, New York, 1998.

Bro. Francis Brown. On the periods of some Feynman integrals. arXiv:0910.0114.

Bro09. Francis Brown. The massless higher-loop two-point function. *Comm. Math. Phys.*, 287(3):925–958, 2009.

Bro12. Francis Brown. Mixed Tate motives over $\mathbb{Z}$. *Ann. of Math. (2)*, 175(2):949–976, 2012.

BS81. Jean-Paul Brasselet and Marie-Hélène Schwartz. Sur les classes de Chern d'un ensemble analytique complexe. In *Caractéristique d'Euler-Poincaré*, volume 83 of *Astérisque*, pages 93–147. Soc. Math. France, Paris, 1981.

BS12. Francis Brown and Oliver Schnetz. A K3 in ϕ^4. *Duke Math. J.*, 161(10):1817–1862, 2012.

BSY10. Jean-Paul Brasselet, Jörg Schürmann, and Shoji Yokura. Hirzebruch classes and motivic Chern classes for singular spaces. *J. Topol. Anal.*, 2(1):1–55, 2010.

CM. Ozgur Ceyhan and Matilde Marcolli. Feynman integrals and periods in configuration spaces. arXiv:1207.3544.

Dor11. Dzmitry Doryn. On one example and one counterexample in counting rational points on graph hypersurfaces. *Lett. Math. Phys.*, 97(3):303–315, 2011.

Ful84a. William Fulton. *Intersection theory*. Springer-Verlag, Berlin, 1984.

Ful84b. William Fulton. *Introduction to intersection theory in algebraic geometry*, volume 54 of *CBMS Regional Conference Series in Mathematics*. Published for the Conference Board of the Mathematical Sciences, Washington, DC, 1984.

Ful93. William Fulton. *Introduction to toric varieties*, volume 131 of *Annals of Mathematics Studies*. Princeton University Press, Princeton, NJ,

1993. The William H. Roever Lectures in Geometry.

GNA02. Francisco Guillén and Vicente Navarro Aznar. Un critère d'extension des foncteurs définis sur les schémas lisses. *Publ. Math. Inst. Hautes Études Sci.*, (95):1–91, 2002.

GS96. H. Gillet and C. Soulé. Descent, motives and K-theory. *J. Reine Angew. Math.*, 478:127–176, 1996.

Hir64. Heisuke Hironaka. Resolution of singularities of an algebraic variety over a field of characteristic zero. I, II. *Ann. of Math. (2) 79 (1964), 109–203; ibid. (2)*, 79:205–326, 1964.

Hir75. Heisuke Hironaka. Triangulations of algebraic sets. In *Algebraic geometry (Proc. Sympos. Pure Math., Vol. 29, Humboldt State Univ., Arcata, Calif., 1974)*, pages 165–185. Amer. Math. Soc., Providence, R.I., 1975.

Hir95. Friedrich Hirzebruch. *Topological methods in algebraic geometry.* Classics in Mathematics. Springer-Verlag, Berlin, 1995. Translated from the German and Appendix One by R. L. E. Schwarzenberger, With a preface to the third English edition by the author and Schwarzenberger, Appendix Two by A. Borel, Reprint of the 1978 edition.

Ken90. Gary Kennedy. MacPherson's Chern classes of singular algebraic varieties. *Comm. Algebra*, 18(9):2821–2839, 1990.

Kwi92. Michal Kwieciński. Formule du produit pour les classes caractéristiques de Chern-Schwartz-MacPherson et homologie d'intersection. *C. R. Acad. Sci. Paris Sér. I Math.*, 314(8):625–628, 1992.

KZ01. Maxim Kontsevich and Don Zagier. Periods. In *Mathematics unlimited—2001 and beyond*, pages 771–808. Springer, Berlin, 2001.

Lev08. Marc Levine. Six lectures on motives. In *Some recent developments in algebraic K-theory*, volume 23 of *ICTP Lect. Notes*, pages 131–227. Abdus Salam Int. Cent. Theoret. Phys., Trieste, 2008.

LL03. Michael Larsen and Valery A. Lunts. Motivic measures and stable birational geometry. *Mosc. Math. J.*, 3(1):85–95, 259, 2003.

MA. Pedro Morales-Almazán. Grothendieck ring class of Banana and Flower graphs. arXiv:1208.4129.

Mac74. Robert D. MacPherson. Chern classes for singular algebraic varieties. *Ann. of Math. (2)*, 100:423–432, 1974.

Mar10. Matilde Marcolli. *Feynman motives.* World Scientific Publishing Co. Pte. Ltd., Hackensack, NJ, 2010.

Pat10. Eric Patterson. On the singular structure of graph hypersurfaces. *Commun. Number Theory Phys.*, 4(4):659–708, 2010.

PP01. Adam Parusiński and Piotr Pragacz. Characteristic classes of hypersurfaces and characteristic cycles. *J. Algebraic Geom.*, 10(1):63–79,

2001.

Sch65a. M.-H. Schwartz. Classes caractéristiques définies par une stratification d'une variété analytique complexe. I. *C. R. Acad. Sci. Paris*, 260:3262–3264, 1965.

Sch65b. M.-H. Schwartz. Classes caractéristiques définies par une stratification d'une variété analytique complexe. II. *C. R. Acad. Sci. Paris*, 260:3535–3537, 1965.

Sch11. Oliver Schnetz. Quantum field theory over $\mathbb{F}_q$. *Electron. J. Combin.*, 18(1):Paper 102, 23, 2011.

SMJŌ77. Mikio Sato, Tetsuji Miwa, Michio Jimbo, and Toshio Ōshima. Holonomy structure of Landau singularities and Feynman integrals. In *Proceedings of the Oji Seminar on Algebraic Analysis and the RIMS Symposium on Algebraic Analysis (Kyoto Univ., Kyoto, 1976)*, volume 12, pages 387–439, 1976/77.

Sta98. Richard P. Stanley. Spanning trees and a conjecture of Kontsevich. *Ann. Comb.*, 2(4):351–363, 1998.

Ste98. John R. Stembridge. Counting points on varieties over finite fields related to a conjecture of Kontsevich. *Ann. Comb.*, 2(4):365–385, 1998.

Str11. Judson Stryker. Chern-Schwartz-MacPherson classes of graph hypersurfaces and Schubert varieties, 2011.

Vak06. Ravi Vakil. Murphy's law in algebraic geometry: badly-behaved deformation spaces. *Invent. Math.*, 164(3):569–590, 2006.

Web12. Andrzej Weber. Equivariant Chern classes and localization theorem. *J. Singul.*, 5:153–176, 2012.

Zee10. A. Zee. *Quantum field theory in a nutshell.* Princeton University Press, Princeton, NJ, second edition, 2010.

GRAVITATION THEORY AND CHERN-SIMONS FORMS

Jorge Zanelli

Centro de Estudios Científicos (CECs)
Casilla 1469, Valdivia, Chile
Universidad Andrés Bello
República 440, Santiago, Chile
E-mail: z@cecs.cl

In the past four decades, the Chern-Simons (**CS**) form has evolved from a curious obstruction in mathematics to a central object in theoretical physics. CS terms play an important role in high temperature superconductivity and in recently discovered topological insulators. In classical physics, the minimal coupling in electromagnetism and to the action for a mechanical system in Hamiltonian form are examples of CS functionals. CS forms are also the natural generalization of the minimal coupling between the electromagnetic field and a point charge when the source is an even-dimensional membrane. A cursory review of the role of CS forms in gravitation theories is presented at an introductory level.*

Keywords: Gravitation, Chern-Simons Forms.

1. Introduction

Chern-Simons forms appeared in mathematics as an accidental discovery: A frustrated attempt to find a formula for the first Pontrjagin number led to the discovery of an unexpected obstruction, a "boundary term", an object that could be locally written as a total derivative, but not globally. As S.-S. Chern and J. Simons described their discovery, ... *This process got stuck by the emergence of a boundary term which did not yield to a simple*

*Lectures presented at the *Villa de Leyva-2011 School.* This is an author-created version of the article published in Classical and Quantum Gravity **29** 133001 (2012). IOP Publishing Ltd is not responsible for any errors or omissions in this version of the manuscript or any version derived from it. The definitive IOP-authenticated version is available online at doi:10.1088/0264- 9381/29/13/133001.

combinatorial analysis. The boundary term seemed interesting in its own right and it and its generalization are the subject of this paper.[1]

In the forty years since their discovery, CS forms have opened new areas of study in mathematics and several excellent books aimed at their many applications in physics have been written,[2,3] for which these notes are no substitute. Our purpose here is to merely collect a few useful observations that could help understand the role of CS forms in physics and the reason for their usefulness in gravitation. This is an elementary introduction, discussing a selection of topics related to gravity, that is expected to be self-contained. There is an abundant literature describing solutions and other applications of CS gravities, like.[4–6] Here, the emphasis is on the construction of the action principles, and in the geometric features that make CS forms particularly suited for the physics of geometric systems.

The next section describes why CS forms are useful in physics. In particular, the gauge invariance of gravity is reviewed, emphasizing the local Lorentz symmetry of spacetime. Section 2 describes the basic ingredients of gravity in the first order formalism, while section 3 takes those fields as building blocks to produce a family of local Lorentz-invariant gravitational actions beyond the modest Einstein-Hilbert form. In section 4 this abundance of theories is reduced by choosing the arbitrary coefficients in the Lagrangian in such a way that, in odd dimensions, the gauge symmetry becomes enlarged. This leads in odd dimensions to three class of theories: those with $SO(D,1)$ (de Sitter), $SO(D-1,2)$ (anti-de Sitter), or $ISO(D-1,1)$ (Poincaré), symmetry groups. Section 5 shows how CS forms can define couplings between non-abelian connections and sources defined by topological structures such as angular defects (naked point like singularities, or zero-branes), or extended localized sources such as $2p$-branes. Section 6 contains a summary and a few comments on future projections.

Note to the readers: I found the wonderful quote of V.I.Arnol'd in,[8] that illustrates the attitude of most physicists, including myself, when confronted with mathematical texts: *It is almost impossible for me to read contemporary mathematicians who, instead of saying, Petya washed his hands, write There is a* $t_1 < 0$ *such that the image of* t_1 *under the natural mapping* $t_1 \to \mathit{Petya}(t_1)$ *belongs to the set of dirty hands, and a* t_2*,* $t_1 < t_2 \leq 0$*, such that the image of* t_2 *under the above-mentioned mappings belongs to the complement of the set defined in the preceding sentence ...* V. I. Arnold

In this spirit, I beg the mathematicians who read these notes compassion with my ineptitude in the mathematical style of high precision, and the resulting lack of rigour. I am sure there is plenty of room for improvement to the notes, but it would take me an exponentially growing amount of energy and time to achieve any finite progress in this direction.

2. Why Chern-Simons forms in physics?

CS forms are useful in physics because they are **not gauge invariant but quasi-invariant**: under a gauge transformation (abelian or not) they transform as an abelian connection. Let us see how this happens. Consider a Yang-Mills connection (vector potential) $\mathbf{A}$ for a nonabelian gauge field theory i.e., the structure group $\mathbb{G}$ of the principal fibre bundle over the space-time manifold M is nonabelian. Under a gauge transformation, it changes as

$$\mathbf{A}_\mu(x) \to \mathbf{A}'_\mu(x) = g^{-1}(x)[\mathbf{A}_\mu(x) + \partial_\mu]g(x), \tag{1}$$

where $\forall x \in M$, $g(x) \in \mathbb{G}$ defines a gauge transformation that can be continuously deformed to the identity throughout M. Then a CS form $\mathcal{C}$, constructed with the connection $\mathbf{A}$ transforms as

$$\mathcal{C}(\mathbf{A}') = \mathcal{C}(\mathbf{A}) + d\Omega, \tag{2}$$

precisely as an abelian connection i.e., a connection on an U(1) principal bundle. Clearly, an abelian connection 1-form is a particular case of a CS form, but in general $\mathcal{C}(\mathbf{A})$ are $(2n+1)$-forms with integer $n \geq 0$.

There are two instances in classical physics where a function that changes by a total derivative –like an abelian gauge field–, gives rise to nontrivial effects:

• The change of a Lagrangian under a symmetry (canonical) transformation,

$$L(q,\dot{q})dt \to L(q',\dot{q}')dt' = L(q,\dot{q})dt + d\Omega(q,t). \tag{3}$$

• The coupling between the electromagnetic potential A_μ and an external current density,

$$\int \mathbf{A}_\mu(x) j^\mu(x) d^n x. \tag{4}$$

These two are not completely independent situations. From Noether's theorem, the existence of a symmetry gives rise to a conserved current, which in turn couples to the dynamical variables as a source for the classical equations. In classical mechanics, the conserved charge is a constant of the Hamiltonian flow in phase space, like the energy-momentum or the angular momentum of an isolated system. In the second case, requiring gauge invariance of the coupling of the electromagnetic potential to an external current leads to the conserved Noether current made up of point electric charges (electric charge conservation).

It is reassuring that not only the coupling, but also the conservation law $\partial_\mu j^\mu = 0$ does not require a metric, since ∂ is the ordinary derivative, and j is a contravariant vector density, which makes the conservation equation valid in any coordinate basis and for any metric. Metric independence ultimately means that the coupling is insensitive to deformations of the worldline of the charge, and of the spacetime metric. Thus, regardless of how the particle twists and turns in it evolution, or the metric properties of spacetime where the interaction takes place, the coupling remains consistently gauge invariant. This fact is crucial for the dynamical consistency of the coupling to membranes or other extended objects.

The lesson one draws from this is that if a $(2n+1)$-form $\mathcal{C}$ changes under a group of transformations ($\mathbb{G}$) by a total derivative $d\Omega$, where Ω is an arbitrary $(2n)$-form, then $\mathcal{C}$ makes a good candidate for the Lagrangian of a gauge invariant theory. This idea has been explored in theoretical physics over the past thirty years, beginning with the pioneering works by Deser, Jackiw and Templeton.[9]

What is less obvious is how the CS forms $\mathcal{C}(\mathbf{A})$ can be used to define a gauge-invariant coupling between the connection $\mathbf{A}$ and a charged brane defined by a conserved current, analogous to j^μ. As we will show by the end of these lectures, the combination

$$I = \int \mathcal{C} \wedge \star j, \tag{5}$$

changes under $\mathbb{G}$ as

$$\delta I = \int_M \delta\mathcal{C} \wedge \star j = \int_M (d\Omega) \wedge \star j = \int_M d(\Omega \wedge \star j) - \int_M \Omega \wedge d \star j, \tag{6}$$

where $\star j$ denotes the Hodge $\star$-dual[a] of j, and M stands for the spacetime manifold. By Stokes' theorem, the first term on the right hand side can be turned into a a surface integral over ∂M, which can be dropped for sufficiently localized Ω and/or j (if they go to zero sufficiently fast at ∂M). The critical term is the last one, which vanishes for any generic Ω if and only if $d\star j = 0$. Hence, the "minimal coupling" (5) is $\mathbb{G}$-invariant provided

[a]For a p-form $\alpha = (p!)^{-1}\alpha_{\mu_1\mu_2\cdots\mu_p} dx^{\mu_1} \wedge dx^{\mu_2} \wedge \cdots \wedge dx^{\mu_p}$, its $\star$-dual is the $(D-p)$-form $\star\alpha = [p!(D-p)!]^{-1}\epsilon_{\nu_1\nu_2\cdots\nu_{D-p}}{}^{\mu_1\mu_2\cdots\mu_p}\alpha_{\mu_1\mu_2\cdots\mu_p} dx^{\nu_1} \wedge dx^{\nu_2} \wedge \cdots \wedge dx^{\nu_{D-p}}$. Unfortunately, here we must use $\star j$ in (6), because in physics we have used j^μ for the current, which turns out to be the dual of the form that couples to the CS form.

$\mathcal{C}$ couples to a conserved gauge-invariant source. If the spacetime M is D-dimensional, a $(2n+1)$-form $\mathcal{C}$ couples to a source j, whose dual $\star j$ is a $(D-2n-1)$ form.

In electromagnetism, the current 1-form that enters in (4) is $j = q\delta(\Gamma)dz$, where $\delta(\Gamma)$ is the Dirac delta with support on the history traced out in spacetime M by a point charge, also known as a *charged zero-brane*. Analogously, the $(2n+1)$-form j in (6) can also be viewed as describing the history of a $2n$-dimensional membrane in the D-dimensional spacetime.

2.1. *Construction of CS forms*

The fundamental object in a gauge theory is the **connection**, a generalization of the abelian vector potential. Typically, the connection is a matrix-valued one-form field,[b]

$$\begin{aligned}\mathbf{A} &= \mathbf{A}_\mu dx^\mu \\ &= A^a_\mu \mathbf{K}_a dx^\mu,\end{aligned} \tag{7}$$

where $\mathbf{K}_a$ $(a = 1, 2, ..., N)$ provide a matrix representation for the generators of the gauge group $\mathbb{G}$. That is, the $\mathbf{K}_a$'s form a basis of the Lie algebra $\mathbb{L}$ associated to $\mathbb{G}$. The connection is defined to transform in the adjoint representation of the gauge group. If $g(x) = \exp[\alpha^a(x)\mathbf{K}_a]$ is an element of the group, it acts on the connection as (1),

$$\mathbf{A}(x) \xrightarrow{g} \mathbf{A}'(x) = g^{-1}(x)\mathbf{A}(x)g(x) + g^{-1}(x)dg(x). \tag{8}$$

This transformation law is prompted by the need to set up a covariant derivative, $D = d - \mathbf{A}$, that generalizes $\partial_\mu - iA_\mu$ in electrodynamics, or $\nabla_\mu = \partial_\mu + \mathbf{\Gamma}_\mu$ in Riemannian geometry.

The connection is a locally defined gauge-dependent and therefore not directly measurable. In contrast with the inhomogeneous transformation (8), the curvature two form (field strength) $\mathbf{F} = d\mathbf{A} + \mathbf{A}\wedge\mathbf{A}$, transforms homogeneously (or covariantly),

$$\mathbf{F} \xrightarrow{g} \mathbf{F}' = g\mathbf{F}g^{-1}. \tag{9}$$

The curvature, like $\vec{E}$ and $\vec{B}$, has directly observable local features. For example, the $2k$-form

$$Tr(\mathbf{F}^k), \tag{10}$$

[b]In what follows, differential forms will be used throughout unless otherwise indicated. We follow the notation and conventions of.[7]

is by construction invariant under (9) and is therefore observable. Invariants of this kind (or more generally, the trace of any polynomial in $\mathbf{F}$), like the Euler or the Pontryagin forms, known as **characteristic classes**, capture the topological nature of the mapping between the spacetime manifold and the Lie algebra $\mathbb{L}$ in which the connection one form $\mathbf{A} : M \mapsto \mathbb{L}$ takes its values.

A **CS form** is defined in association with a characteristic class. Let's denote by $P_{2k}(\mathbf{F}) = \langle \mathbf{F}^k \rangle$ one of those invariants, where $\langle \cdots \rangle$ stands for a symmetric, multilinear operation in the Lie algebra (a generalized trace). To fix ideas, one can take $\langle \cdots \rangle$ to be the ordinary trace in a particular representation, but some Lie algebras could have more than one way to define "Tr" (the algebra of rotations, for example, has two). Then, P_{2k} is a homogeneous polynomial in the curvature $\mathbf{F}$ associated to a gauge connection $\mathbf{A}$ satisfying the following conditions:[2]

i. It is invariant under gauge transformations (8) and (9), which we denote as $\delta_{gauge} P_{2k} = 0$.

ii. It is closed, $dP_{2k} = 0$.

iii. It can be locally expressed as the derivative of a $(2k-1)$-form, $P_{2k} = d\mathcal{C}_{2k-1}$.

iv. Its integral over a $2k$-dimensional compact, orientable manifold without boundary, is a topological invariant, $\int_M P_{2k} = c_{2k}(M) \in \mathbf{Z}$.

Condition (**i**) is satisfied by virtue of the cyclic property of the product of curvature 2-forms. Condition (**ii**) is a consequence of the Bianchi identity which states that the covariant derivative (in the connection $\mathbf{A}$) of the curvature $\mathbf{F}$ vanishes identically, $D\mathbf{F} = d\mathbf{F} + [\mathbf{A}, \mathbf{F}] \equiv 0$. Condition (**iii**) follows from (**ii**). By Poincaré lemma: *If* $d\phi = 0$ *then, locally* ϕ *can be written as* $\phi = d(\textit{something})$.

Finally, (**iv**) means that, although P_{2k} is an exact form in a local chart although globally it may not.

The CS forms $\mathcal{C}_{2n-1}$ identified in[1] are given by the trace of some polynomial in $\mathbf{A}$ and $d\mathbf{A}$ that *cannot* be written as a local function involving only the curvature $\mathbf{F}$. This makes the CS forms rather cumbersome to write, but its exact expression in not needed in order to establish its most important property, as stated in the following

Lemma: *Under a gauge transformation* (8), $\mathcal{C}_{2n-1}$ *changes by a locally exact form (a total derivative in a coordinate patch).*

Proof: The homogeneous polynomial P_{2k} is invariant under gauge transformations (this is easily seen from the transformation (9) if $P_{2k} = Tr[\mathbf{F}^k]$, due to the cyclic property of the trace). Performing a gauge transformation on (**iv**), gives

$$\delta_{gauge} P_{2k} = d(\delta_{gauge} \mathcal{C}_{2k-1}), \tag{11}$$

and since P_{2k} is invariant, one concludes that the right hand side must vanish as well,

$$d(\delta_{gauge} \mathcal{C}_{2k-1} = 0). \tag{12}$$

By Poincaré's lemma, this last equation implies that the gauge variation of $\mathcal{C}_{2k-1}$ can be written locally as an exact form,

$$\delta_{gauge} \mathcal{C}_{2k-1} = d\Omega. \quad \blacksquare \tag{13}$$

This is a nontrivial result: although the nonabelian connection **A** transforms inhomogeneously, as in (8), the CS form transforms in the same way as an abelian connection. This is sufficient to ensure that a CS $(2n-1)$-form defines gauge invariant action in a $(2n-1)$-dimensional manifold,

$$\delta_{gauge} I[A] = \int_{M^{2n-1}} \delta_{gauge} \mathcal{C}_{2n-1} = \int_{M^{2n-1}} d\Omega, \tag{14}$$

which vanishes for an appropriate set of boundary conditions.

CS actions are exceptional in physics because, unlike for most theories, such as Maxwell or Yang-Mills, they do not require a metric structure. The gauge invariance of the action does not depend on the shape of the manifold M^{2n-1}; a metric structure may not even be defined on it. This is a welcome feature in a gravitation theory in which the geometry is dynamical. A particular consequence of this is that in CS gravity theories, the metric is a derived (composite) object and not a fundamental field to be quantized. This in turn implies that concepts such as the energy-momentum tensor and the inertial mass must be regarded as phenomenological constructs of classical or semi-classical nature, an emerging phenomenon.

2.2. *The gauge invariance of gravity*

In order to see how CS forms can enter in gravity, we must identify what is the gauge invariance associated to the gravitational interaction.

2.2.1. *Coordinate transformations*

It is often said that the fundamental symmetry of gravity is the group of general coordinate transformations, usually called the diffeomorphism group (of the underlying spacetime manifold). These transformations do form a group whose action is certainly local. The invariance under diffeomorphisms, however, is not a useful symmetry and much less are they a unique feature of gravity. Indeed, any action, for any physical system whatsoever must be coordinate-invariant, lest one has made a mistake somewhere. Hence, all meaningful statements derived from an action principle must be coordinate-invariant as well.

In fact, *all* well defined physical theories must be invariant under general changes of coordinates, because the coordinates are labels introduced by the physicists in order to describe where and when events occur, together with the units for measuring temperature, pressure, tension, etc., and not part of the physical system. It is a triviality that an objective situation cannot depend on the coordinates we humans employ to describe them. The *representation* may of course change, but the phenomenon itself cannot.

General coordinate invariance is explicitly recognized in Lagrangian mechanics, where the choice of coordinates is left completely arbitrary. In other words, general coordinate transformations are not a distinctive symmetry of gravity, it is the invariance of the laws of Nature under changes in the form humans choose to describe it. We experience this every time we write or Maxwell's or Schrödinger's equations in spherical coordinates, in order to render more transparent the presence of boundaries or sources with spherical symmetry. In such cases, the coordinates are adapted to the symmetry of the physical situation, but that does not imply that coordinates could not be chosen otherwise.

The origin of the confusion seems to be that diffeomorphisms can be conceived as the group of local translations,

$$x^\mu \to x'^\mu = x^\mu + \xi^\mu(x). \tag{15}$$

These are gauge-like transformations in the sense that ξ^μ is an arbitrary function of x, but here the analogy with with gauge transformations stops. Under coordinate transformations, a vector transforms as

$$v'^\mu(x') = L^\mu_\nu(x)v^\nu(x), \text{ where } L^\mu_\nu(x) = \frac{\partial x'^\mu}{\partial x^\nu} = \delta^\mu_\nu + \frac{\partial \xi^\mu}{\partial x^\nu}. \tag{16}$$

In a gauge transformation, the transformed field has the same argument as the original field, whereas the argument of v'^μ in (16) is x' and not x, which is a different point on the manifold. One could try to write (16) in a form similar to a gauge transformation,

$$\begin{aligned} v'^\mu(x) &= L^\mu_\nu(x)v^\nu(x) - \xi^\lambda(x)\partial_\lambda v^\mu(x) \\ &= v^\mu(x) + \frac{\partial \xi^\mu}{\partial x^\nu}v^\nu(x) - \xi^\lambda(x)\partial_\lambda v^\mu(x) \end{aligned} \tag{17}$$

The first two terms on the right of this last expression correspond to the way a vector representation transforms under the action of a gauge group in a fibre bundle. The last term, however, represents the drift, produced by the fact that the translation actually shifts the point in the manifold. This type of term is not present in a gauge transformation of the type (8), which means that the diffeomorphism group does not act as a local symmetry in a standard gauge theory, like Yang-Mills.

The best way to describe gauge transformations like (8) is in the language of fibre bundles. A fibre bundle is locally a direct product of a base manifold and a group, each fibre being a copy of the orbit of the group. In the case of the diffeomorphism group, the fibres lie along the base. Therefore this structure is not locally a product and the group of coordinate transformations on a manifold do not define a fibre bundle structure. Basically, the problem is that the translation group does not take a field at a given point into a different field at the same point, but changes the arguments of the fields, which is something gauge transformations never do, as can be seen in (8).

Apart from the obvious fact that the group of translations is a rather trivial group, whose gauging could hardly describe the richness of gravity, it is apparent that the translation symmetry is violated by the curvature of spacetime itself. This wouldn't happen in a genuine gauge theory, where gauge invariance is respected by all solutions of classical equations, and by all conceivable off-shell fields in the quantum theory. This is a key feature that makes gauge symmetries extremely useful in the quantum description: the invariance is an inherent property of the fields in the action and is not spoiled by dynamics, be it classical or quantum.

The generators of diffeomorphisms $\mathcal{H}_\mu$ form an algebra whose Poisson brackets are

$$\begin{aligned}
[\mathcal{H}_\perp(x), \mathcal{H}_\perp(y)] &= g^{ij}(x)\delta(x,y)_{,i}\,\mathcal{H}_j(y) - g^{ij}(y)\delta(y,x)_{,i}\,\mathcal{H}_j(x) \\
[\mathcal{H}_i(x), \mathcal{H}_j(y)] &= \delta(x,y)_{,i}\,\mathcal{H}_j(y) - \delta(x,y)_{,j}\,\mathcal{H}_i(y) \\
[\mathcal{H}_\perp(x), \mathcal{H}_i(y)] &= \delta(x,y)_{,i}\,\mathcal{H}_\perp(y)
\end{aligned} \quad , \qquad (18)$$

where $g^{ij}(x)$ is the inverse of the spatial metric at x, $\mathcal{H}_\perp$ generates diffeomorphisms along the normal direction $x^\perp$, $\mathcal{H}_i$ generates diffeomorphisms along the spatial direction x^i; $\delta(x,y)$ is the Dirac delta with support on $x{=}y$, and $\delta(x,y)_{,i} = (\partial/\partial y^i)\delta(x,y)$.

Note that this algebra is defined by a set of *structure functions* rather *structure constants*, as in ordinary Lie algebras. This type of structure is not a Lie algebra, but an *open algebra*,[10] and what is a more serious concern, in gravity, the structure functions involve the metric of the manifold, which is itself a dynamical variable. This represents a major drawback in a quantum theory, since the symmetry would depend on the local geometry of the manifold, which should be itself a state of the theory. A useful quantum symmetry, on the contrary, should not depend on the state of the system.

2.2.2. *Lorentz transformations*

Einstein's starting point in the construction of General Relativity was the observation that the effect of gravity can be neutralized by free fall. In a small freely falling laboratory, the effect of gravity can be eliminated so that the laws of physics there are indistinguishable from those observed in an inertial laboratory in Minkowski space. This trick is a local one: the lab has to be small enough and the time span of the experiments must be short enough. Under these conditions, the experiments will be indistinguishable from those performed in absence of gravity. In other words, in a local neighbourhood, spacetime possesses Lorentz invariance. In order to make this invariance manifest, it is necessary to perform an appropriate coordinate transformation to a particular reference system, viz., a freely-falling one. Conversely, Einstein argued, in the absence of gravity the gravitational field could be mocked by applying an acceleration to the laboratory.

This idea is known as the principle of equivalence meaning that, in a small spacetime region, gravitation and acceleration are equivalent effects. A freely falling observer defines a local inertial system. For a small enough region around him or her, the trajectories of projectiles (freely falling as well) are straight lines and the discrepancies with Euclidean geometry become negligible. Particle collisions mediated by short range forces, such as those between billiard balls, molecules or subnuclear particles, satisfy the conservation laws of energy and momentum that are valid in special relativity.

Since physical phenomena in a small neighbourhood of any spacetime should be invariant under Lorentz transformations, and since these transformations can be performed independently at each point, gravity must be endowed with local Lorentz symmetry. Hence, Einstein's observation that the principle of equivalence is a central feature of general relativity, makes gravitation a gauge theory for the group $SO(3,1)$, the first nonabelian gauge theory ever proposed.[11]

Note that while the Lorentz group can act independently at each spacetime point, the translations are a symmetry only in maximally symmetric spacetimes. The invariance of gravitation theory under $SO(3,1)$ is a minimal requirement, the complete group of invariance could be larger, $\mathbb{G} \supseteq SO(3,1)$. Natural options are the de Sitter ($SO(4,1)$), anti-de Sitter ($SO(3,2)$), conformal ($SO(4,2)$) and Poincaré $ISO(3,1)$) groups, or some of their supersymmetric extensions.

3. First order formalism

In order to implement the gauge symmetry, one should define an action principle the for the spacetime geometry, describing it in terms of fields that correspond to some nontrivial representation of $SO(D-1,1)$, where D is the spacetime dimension. This is most effectively done if the metric and affine features of the geometry are treated independently. This is achieved by the first-order formalism, introducing the notion of **tangent space** in an intuitively simple way.

The first order formalism uses the exterior calculus of differential forms, and since in exterior calculus, $d^2 \equiv 0$, the classical fields satisfy at most first order differential equations. It is sometimes held that the advantage of using differential forms is the compactness of the expressions as compared with the standard tensor calculus in coordinate bases.[16] Nevertheless, it is also true that coordinates are still necessary in order to solve Einstein's equations, so the advantage would be for the elegance in the formulation of the theory at best.

The real advantage of the first order formalism is in the decoupling between the gauge symmetry of gravity (local Lorentz invariance) and the particular configurations of the spacetime geometry. It is the fact that the symmetry generators form a Lie algebra –with structure constants–, and not an open algebra like (18), with structure functions that depend on the dynamical variables of the theory, that makes the first order formalism interesting and attractive.

The spacetime geometry can be captured by two fundamental fields, the vielbein $e^a_\mu(x)$ characterizing the metric structure, an the Lorentz connection, $\omega^a_{\ b\mu}(x)$, that codifies the affine features. Until further notice, these two fields are totally arbitrary and independent. [c]

[c]This section is a free interpretation of the ideas the author learned from lectures by B. Zumino[12] and T. Regge,[13] who in turn elaborated on earlier work by R. Utiyama[14] and T. W. Kibble.[15]

3.1. *The vielbein*

Spacetime is a smooth D-dimensional manifold M, of Lorentzian signature $(-1,1,1,\cdots,1)$. At every point on $x \in M$ there is a D-dimensional tangent space T_x, which is a good approximation of the manifold M in the neighbourhood of x. This tangent space corresponds to the reference frame of a freely falling observer mentioned in the Equivalence Principle. Every tangent space at one point of the manifold is a Minkowski space, identical to all the other tangent spaces, and each one invariant under the action of the Lorentz group. This endows the spacetime manifold with a fibre bundle structure, the *tangent bundle*, where the basis is the spacetime and the fibres are the tangent spaces on which the Lorentz group acts locally. This can also be regarded as a *collection of vector spaces parametrized by the manifold*, $\{T_x, x \in M\}$. Either way, the essential point is that the manifold M, labelled by the coordinates x^μ is the spacetime where we live, and the collection of tangent spaces over it is where the symmetry group acts.

The fact that any measurement carried out in any reference frame in spacetime can be translated to one in a freely falling frame, means that there is an isomorphism between tensors on M and tensors on T_x, represented by means of a linear mapping, also called *"soldering form"* or **vielbein**. It is sufficient to define this mapping on a complete set of vectors such as the coordinate separation dx^μ between two infinitesimally close points on M. The corresponding separation in T_x is defined to be

$$dz^a = e^a_\mu(x) dx^\mu, \tag{1}$$

where z^a represent an orthonormal coordinate basis in the tangent space. For this reason the vielbein is also viewed as a local orthonormal frame. Since T_x is a standard Minkowski space, it has a natural metric, η_{ab}, which defines a metric on M through the isomorphism e^a_μ. In fact,

$$\begin{aligned} ds^2 &= \eta_{ab} dz^a dz^b \\ &= \eta_{ab}\, e^a_\mu(x) dx^\mu\, e^b_\nu(x) dx^\nu \\ &= g_{\mu\nu}(x) dx^\mu dx^\nu, \end{aligned} \tag{2}$$

where

$$g_{\mu\nu}(x) \equiv \eta_{ab}\, e^a_\mu(x)\, e^b_\nu(x)\,, \tag{3}$$

is the metric on M, induced by the vielbein $e^a_\mu(x)$ and the tangent space metric η_{ab}.

This relation can be read as the vielbein being "the square root" of the metric. Given $e^a_\mu(x)$ one can find the metric and therefore, all the metric properties of spacetime are contained in the vielbein. The converse, however, is not true: given the metric, there exist infinitely many choices of vielbein that reproduce the same metric.

By the definition (1) the vielbein transforms as a covariant vector under diffeomorphisms on M and as a contravariant vector under local Lorentz rotations of T_x, $SO(D-1,1)$, as

$$e^a_\mu(x) \longrightarrow e'^a_\mu(x) = \Lambda^a_b(x) e^b_\mu(x), \tag{4}$$

where the matrix $\mathbf{\Lambda}(x)$ leaves the metric in the tangent space unchanged,

$$\Lambda^a_c(x)\Lambda^b_d(x)\eta_{ab} = \eta_{cd}, \tag{5}$$

and therefore, the metric $g_{\mu\nu}(x)$ is clearly unchanged under Lorentz transformations. The matrices of unit determinant that satisfy (5) form the Lorentz group $SO(D-1,1)$. The vielbein e^a_μ has D^2 independent components, whereas the metric has only $D(D+1)/2$. The mismatch is exactly $D(D-1)/2$, the number of independent rotations in D dimensions.

In this way, the **vielbein one-forms** $e^a(x) = e^a_\mu(x) dx^\mu$, define the metric structure of the manifold.

3.2. *The Lorentz Connection*

In order to define a derivative on the manifold, a connection is required so that the differential structure remains invariant under local Lorentz transformations $\Lambda(x)$, even if they act independently at each spacetime point. This is achieved by introducing the Lorentz connection,[d] which measures the rotation experienced by a vector relative to the local Lorentz frames when parallel transported between tangent spaces on neighbouring points, T_x and T_{x+dx}.

Consider a field $\phi^a(x)$ that transforms as a vector under Lorentz rotations defined on the tangent T_x. The parallel-transported field from $x+dx$ to x, is defined as

$$\begin{aligned} \phi^a_{||}(x) &\equiv \phi^a(x+dx) + dx^\mu \omega^a_{\ b\mu}(x)\phi^b(x) \\ &= \phi^a(x) + dx^\mu[\partial_\mu \phi^a(x) + \omega^a_{\ b\mu}(x)\phi^b(x)]. \end{aligned} \tag{6}$$

[d] In physics, this is often called the *spin connection*. The word "spin" is due to the fact that it arises naturally in the discussion of spinors, which carry a special representation of the group of rotations in the tangent space. For a more extended discussion, there are several texts such as those of Refs.[2,7,16] and[17]

In this case, both ϕ^a and $\phi^a_{||}$ are vectors defined at the same point, therefore their difference,

$$\begin{aligned} dx^\mu[\partial_\mu\phi^a(x) + \omega^a_{\ b\mu}(x)\phi^b(x)] &= dx^\mu D_\mu\phi^a(x) \\ &= D\phi^a(x). \end{aligned} \tag{7}$$

must also be a vector under Lorentz transformations at x. This new vector, which is defined as the covariant derivative, measures the difference in the components of the Lorentz vector produced by parallel transport between $x+dx$ and x. This expression is a Lorentz vector at x, provided ω transforms as a connection:

$$\omega^a_{\ b\mu}(x) \xrightarrow{\Lambda} \omega'^a_{\ b\mu}(x) = \Lambda^a_{\ c}(x)\Lambda_b^{\ d}(x)\omega^c_{\ d\mu}(x) + \Lambda^a_{\ c}(x)\partial_\mu\Lambda_b^{\ c}(x). \tag{8}$$

This notion of parallelism is analogous to the one defined for vectors whose components are referred to a coordinate basis,

$$\varphi^\mu_{||}(x) = \varphi^\mu(x) + dx^\lambda[\partial_\lambda\varphi^\mu(x) + \Gamma^\mu_{\lambda\rho}(x)\varphi^\rho(x)]. \tag{9}$$

These two definitions are independent as they refer to objects on different spaces, but it is always possible to express $\omega^a_{\ b\mu}$ as a function of e^a_μ and $\Gamma^\mu_{\lambda\rho}$.

It is apparent that (7-8) can be expressed in a more compact way using exterior forms, as

$$\begin{aligned} \phi^a_{||}(x) - \phi^a(x) = d\phi^a(x) + \omega^a_{\ b}(x)\phi^b(x) &= D\phi^a(x) \\ \omega'^a_{\ b}(x) = \Lambda^a_{\ c}(x)\Lambda_b^{\ d}(x)\omega^c_{\ d}(x) &+ \Lambda^a_{\ c}(x)d\Lambda_b^{\ c}(x), \end{aligned}$$

where $\omega^a_{\ b} = \omega^a_{\ b\mu}dx^\mu$ is the **connection one-form**.

3.3. *Lorentz invariant tensors*

The group $SO(D-1,1)$ has two invariant tensors, the Minkowski metric, η_{ab}, which we already mentioned, and the totally antisymmetric Levi-Civita tensor, $\epsilon_{a_1a_2\cdots a_D}$. Because they are the same in every tangent space, they are constant ($\partial_\mu\eta_{ab} = 0$, $\partial_\mu\epsilon_{a_1a_2\cdots a_D} = 0$) and since they are also invariant, they are covariantly constant,

$$d\eta_{ab} = D\eta_{ab} = 0, \tag{10}$$

$$d\epsilon_{a_1a_2\cdots s_D} = D\epsilon_{a_1a_2\cdots a_D} = 0. \tag{11}$$

This implies that the Lorentz connection satisfies two identities,

$$\eta_{ac}\omega^c_{\ b} = -\eta_{bc}\omega^c_{\ a}, \tag{12}$$

$$\epsilon_{b_1a_2\cdots a_D}\omega^{b_1}_{\ \ a_1} + \epsilon_{a_1b_2\cdots a_D}\omega^{b_2}_{\ \ a_2} + \cdots + \epsilon_{a_1a_2\cdots b_D}\omega^{b_D}_{\ \ a_D} = 0. \tag{13}$$

The first condition, the requirement that the Lorentz connection be compatible with the metric structure of the tangent space (12), implies antisymmetry of the connection $\omega^{ab} = -\omega^{ba}$. The second relation (13) does not impose further restrictions on the components of the Lorentz connection. Then, the number of independent components of $\omega^{a}_{\ b\mu}$ is $D^2(D-1)/2$, which is less than the number of independent components of the Christoffel symbol $(D^2(D+1)/2)$.

3.4. *Curvature*

The 1-form exterior derivative operator, $dx^\mu \partial_\mu \wedge$ is such that acting on a p-form, α_p, it yields a (p+1)-form, $d\alpha_p$. One of the fundamental properties of exterior calculus is that the second exterior derivative of a differential form vanishes identically on continuously differentiable forms,

$$d(d\alpha_p) = d^2\alpha_p = 0. \tag{14}$$

A consequence of this is that the square of the covariant derivative operator is not a differential operator, but an algebraic operator, the curvature two-form. For instance, applying twice the covariant derivative on a vector yields

$$\begin{aligned} D^2\phi^a &= D[d\phi^a + \omega^a_{\ b}\phi^b] \\ &= d[d\phi^a + \omega^a_{\ b}\phi^b] + \omega^a_{\ b}[d\phi^b + \omega^b_{\ c}\phi^c] \\ &= [d\omega^a_{\ b} + \omega^a_{\ c} \wedge \omega^c_{\ b}]\phi^b. \end{aligned} \tag{15}$$

The two-form within brackets in this last expression is a second rank Lorentz tensor known as the **curvature two-form** (see, e.g.,[16,17] for a formal definition of $R^a_{\ b}$),

$$\begin{aligned} R^a_{\ b} &= d\omega^a_{\ b} + \omega^a_{\ c} \wedge \omega^c_{\ b} \\ &= \frac{1}{2} R^a_{\ b\mu\nu} dx^\mu \wedge dx^\nu. \end{aligned} \tag{16}$$

The fact that $\omega^a_{\ b}(x)$ and the gauge potential in Yang-Mills theory, $A^a_{\ b} = A^a_{\ b\mu} dx^\mu$, are both 1-forms and have similar properties is not an accident since they are both connections of a gauge group. Their transformation laws have the same form, and the curvature $R^a_{\ b}$ is completely analogous to the field strength in Yang-Mills, $\mathbf{F} = d\mathbf{A} + \mathbf{A} \wedge \mathbf{A}$.

The curvature two form defined by (16) is a Lorentz tensor on the tangent space, related to the Riemann tensor $R^{\alpha\beta}_{\ \ \mu\nu}$ through

$$R^{ab} = \frac{1}{2} e^a_{\ \alpha} e^b_{\ \beta} R^{\alpha\beta}_{\ \ \mu\nu} dx^\mu \wedge dx^\nu. \tag{17}$$

This equivalence would not be true in a space with torsion, as we discuss in the next paragraph.

3.5. *Torsion*

The fact that the two independent geometrical ingredients, ω and e, play different roles is underscored by their different transformation properties under the Lorentz group. In gauge theories this is reflected by the fact that spinor fields in a vector representation play the role of matter, while the connections in an adjoint representation are the carriers of interactions.

Another important consequence of this asymmetry is the impossibility to construct a tensor two-form solely out of e^a and its exterior derivatives, in contrast with the curvature which is uniquely defined by the connection. The only tensor obtained by differentiation of e^a is its covariant derivative, also known the **torsion 2-form**,

$$T^a = de^a + \omega^a_{\ b} \wedge e^b, \tag{18}$$

which involves both the vielbein and the connection. In contrast with T^a, the curvature $R^a_{\ b}$ depends only on ω and is not the covariant derivative of something else. In a manifold with torsion, one can split the connection into a torsion-free part $\bar{\omega}$, and the so-called contorsion κ,

$$de^a + \bar{\omega}^a_{\ b} \wedge e^b \equiv 0, \quad \text{and} \quad T^a = \kappa^a_{\ b} \wedge e^b.$$

In this case, the curvature two-form reads

$$R^a_{\ b} = \bar{R}^a_{\ b} + \bar{D}\kappa^a_{\ b} + \kappa^a_{\ c}\kappa^c_{\ b}, \tag{19}$$

where $\bar{R}^a_{\ b}$ and $\bar{D}$ are the curvature and the covariant derivative constructed out of the torsion-free connection. It is the purely metric part of the curvature two-form, $\bar{R}^a_{\ b}$, that relates to the Riemann curvature through (17).

3.6. *Bianchi identity*

As we saw in **3.4**, taking the second covariant derivative of a vector amounts to multiplying by the curvature 2-form. A consequence of this is an important property known as Bianchi identity,

$$DR^a_{\ b} = dR^a_{\ b} + \omega^a_{\ c} \wedge R^c_{\ b} - \omega^c_{\ b} \wedge R^a_{\ c} \equiv 0\,. \tag{20}$$

This is an identity and not a set of equations because it is satisfied for *any* well defined connection 1-form whatsoever, and therefore it does not restrict the form of the field $\omega^a_{b\mu}$ in any way. This can be checked explicitly by substituting (19) in the second term of (20).

The Bianchi identity implies that the curvature R^{ab} is "transparent" for the exterior covariant derivative,

$$D(R^a_{\ b}\phi^b) = R^a_{\ b} \wedge D\phi^b\,. \tag{21}$$

An important direct consequence of this identity is that by taking successive exterior derivatives of e^a, ω^{ab} and T^a one does not generate new independent Lorentz tensors, in particular,

$$DT^a = R^a_{\ b} \wedge e^b. \tag{22}$$

The physical implication is that if no other fields are introduced, in the first order formulation there is a very limited number of possible Lagrangians that can be constructed out of these fields in any given dimension.[18]

3.7. *Building blocks*

The basic building blocks of first order gravity are the one forms e^a and $\omega^a_{\ b}$, and the two-forms $R^a_{\ b}$, T^a. In order to produce Lorentz invariants with them, the invariant tensors η_{ab} and $\epsilon_{a_1a_2\cdots a_D}$ can also be used to contract the indices. There are no more building blocks, and with them we must put together an action. We are interested in objects that transform in a controlled way under Lorentz rotations (scalars, vectors, tensors of various ranks). The existence of Bianchi identities implies that differentiating these fields, the only tensors that can be produced are combinations of the same objects. In the next sections we discuss the construction of the possible actions for gravity using these ingredients.

All the geometric properties features of the geometry of M are captured in the two fundamental fields

$$e^a \equiv e^a_\mu(x)dx^\mu \quad \text{and} \quad \omega^a_{\ b} \equiv \omega^a_{b\mu}(x)dx^\mu. \tag{23}$$

Hence, the action principle for a purely gravitational system could be expressed by an action functional of the form $I[e,\omega]$. Since e^a and $\omega^a_{\ b}$ describe independent features of the geometry, these two fields must be varied independently in the action. The metric is a derived expression given in (1) and not a fundamental field to be varied in the action .

Finally, since both e^a and $\omega^a_{\ b}$ carry only Lorentz indices but no coordinate indices (μ, ν, etc.), these 1-forms, like all exterior forms, are invariant under coordinate transformations of M. This is why a description of the geometry that only uses these forms and their exterior derivatives, is naturally coordinate-free and trivially coordinate invariant.

4. Gravity actions

We now turn to the construction of the action for gravitation. As mentioned above, we expect it to be a local functional of the one-forms e^a, ω^a_b and their exterior derivatives. In addition the two invariant tensors of the Lorentz group, η_{ab}, and $\epsilon_{a_1 \cdots a_D}$ can be used to raise, lower and contract indices. One needs not worry about invariance under general coordinate transformations as exterior forms are coordinate independent by construction.

The use of only exterior products of forms excludes the metric, its inverse and the Hodge $\star$-dual (see[12] and[13] for more on this). This postulate also excludes tensors like the Ricci tensor[e] $R_{\mu\nu} = E^\lambda_a \eta_{bc} e^c_\mu R^{ab}_{\ \ \lambda\nu}$, or $R_{\alpha\beta} R_{\mu\nu} R^{\alpha\mu\beta\nu}$, except in very special combinations like the Gauss-Bonnet form, that can be expressed as exterior products of forms.

The action principle cannot depend on the choice of basis in the tangent space and hence Lorentz invariance should be ensured. A sufficient condition to have Lorentz invariant field equations is to demand the Lagrangian itself to be Lorentz invariant, but this is not really necessary. If the Lagrangian is *quasi-invariant* so that it changes by a total derivative –and the action changes by a boundary term–, still gives rise to covariant field equations in the bulk, provided the fields satisfy appropriate boundary conditions.

4.1. *General Lorentz-invariant Lagrangians*

Let us consider first Lorentz invariant Lagrangians. By inspection, one concludes that they must be D-forms, consisting of linear combinations of products of e^a, R^a_b, T^a, contracted with η_{ab} and $\epsilon_{a_1 \cdots a_D}$, and no ω . Such invariants are:[18]

$$\mathbf{P}_{2k} =: R^{a_1}{}_{a_2} R^{a_2}{}_{a_3} \cdots R^{a_k}{}_{a_1} \tag{1}$$

$$\upsilon_k =: e_{a_1} R^{a_1}{}_{a_2} R^{a_2}{}_{a_3} \cdots R^{a_k}{}_b e^b, \quad \text{odd } k \tag{2}$$

$$\tau_k =: T_{a_1} R^{a_1}{}_{a_2} R^{a_2}{}_{a_3} \cdots R^{a_k}{}_b T^b, \quad \text{even } k \tag{3}$$

$$\zeta_k =: e_{a_1} R^{a_1}{}_{a_2} R^{a_2}{}_{a_3} \cdots R^{a_k}{}_b T^b \tag{4}$$

$$\mathbf{E}_D =: \epsilon_{a_1 a_2 \cdots a_D} R^{a_1 a_2} R^{a_3 a_4} \ldots R^{a_{D-1} a_D}, \quad \text{even } D \tag{5}$$

$$L_p =: \epsilon_{a_1 a_2 \cdots a_D} R^{a_1 a_2} R^{a_3 a_4} \ldots R^{a_{2p-1} a_{2p}} e^{a_{2p+1}} \ldots e^{a_D}. \tag{6}$$

Among these local Lorentz invariants, the Pontryagin forms $\mathbf{P}_{2k}$ and the Euler forms $\mathbf{E}_{2n}$ define **topological invariants** in $4k$ and $2n$ dimensions, respectively: Their integrals on compact manifolds without boundary have

[e] Here E^λ_a is the inverse vielbein, $E^\lambda_a e^b_{\ \lambda} = \delta^a_b$.

integral spectra,

$$\Omega_n \int_{M^{2n}} \mathbf{E}_{2n} \in \mathbb{Z}, \qquad \tilde{\Omega}_k \int_{M^{4k}} \mathbf{P}_{4k} \in \mathbb{Z}\,, \tag{7}$$

where Ω_n and $\tilde{\Omega}_k$ are normalization coefficients determined uniquely by the dimension of the manifold. Thus, in every even dimension $D = 2n$ there is a topological invariant of the Euler family. If the dimension is a multiple of four, there are invariants of the Pontryagin family as well, of the form

$$\mathbf{P}_D = \mathbf{P}_{4k_1}\mathbf{P}_{4k_2}\cdots\mathbf{P}_{4k_r}\,, \tag{8}$$

where $4(k_1 + k_2 + \cdots + k_r) = D$, so the number of Pontryagin invariants grows with the number of different ways to express the number $D/4$ as a sum of integers.[f]

4.2. *Lovelock theory*

If torsion is set to zero, the invariants τ_k, ζ_k clearly vanish, but also v_k must vanish, since the contraction $R^a{}_b e^b$ it equals the covariant derivative of the torsion. Hence, we are led to the following

Theorem [Lovelock, 1970[19] and Zumino, 1986[12]]: In the absence of torsion, the most general action for gravity $I[e,\omega]$ invariant, under Lorentz transformations that does not explicitly involve the metric, is of the form

$$I_D[e,\omega] = \kappa \int_M \sum_{p=0}^{[D/2]} \alpha_p L_p^D \tag{9}$$

where a_p are arbitrary constants, and L_p^D is given by

$$L_p^D = \epsilon_{a_1\cdots a_D} R^{a_1a_2}\cdots R^{a_{2p-1}a_{2p}} e^{a_{2p+1}}\cdots e^{a_D}\,. \;\blacksquare \tag{10}$$

The Lovelock series is an arbitrary linear combination where each term L_p^D is the continuation to dimension D of all the lower-dimensional Euler forms. In even dimensions, the last term in the series is the Euler form of the corresponding dimension, $L_{D/2}^D = \mathcal{E}_D$. Let us examine a few examples.

---

[f]As it is shown in,[18] the number of torsion-dependent terms grows as the partitions of $D/4$, which for large D is approximately given by the Hardy-Ramanujan formula, $p(D/4) \sim \frac{1}{\sqrt{3}D}\exp[\pi\sqrt{D/6}]$.

• $D=2$: The Lovelock Lagrangian reduces to 2 terms, the 2-dimensional Euler form and the spacetime volume (area),

$$\begin{aligned} I_2 &= \kappa \int_M \epsilon_{ab}[\alpha_1 R^{ab} + \alpha_0 e^a e^b] \\ &= \kappa \int_M \sqrt{|g|}\,(\alpha_1 R + 2\alpha_0)\, d^2x \\ &= \kappa\alpha_1 \cdot \mathbf{E}_2 + 2\kappa\alpha_0 \cdot V_2. \end{aligned} \tag{11}$$

This action has as a local extremum for $V_2 = 0$, which reflects the fact that, unless matter is included, I_2 does not produce a very interesting dynamical theory for the geometry. If the manifold M has Euclidean metric and a prescribed boundary, the first term picks up a boundary term and the action is extremized by a minimal surface, like a soap bubble, the famous Plateau problem.[20]

• $D=3$ and $D=4$: The action reduces to the Einstein-Hilbert and the cosmological constant terms. In four dimensions, the action also admits Euler invariant $\mathbf{E}_4$,

$$\begin{aligned} I_4 &= \kappa \int_M \epsilon_{abcd}\left[\alpha_2 R^{ab}R^{cd} + \alpha_1 R^{ab}e^c e^d + \alpha_0 e^a e^b e^c e^d\right] \\ &= -\kappa \int_M \sqrt{|g|}\left[\alpha_2\left(R^{\alpha\beta\gamma\delta}R_{\alpha\beta\gamma\delta} - 4R^{\alpha\beta}R_{\alpha\beta} + R^2\right) + 2\alpha_1 R + 24\alpha_0\right] d^4x \\ &= -\kappa\alpha_2 \cdot \mathbf{E}_4 - 2\alpha_1 \int_M \sqrt{|g|} R d^4x - 24\kappa\alpha_0 \cdot V_4. \end{aligned} \tag{12}$$

• $D=5$: The Euler form $\mathbf{E}_4$, also known as the Gauss-Bonnet density, provides the first nontrivial generalization of Einstein gravity occurring in five dimensions,

$$\epsilon_{abcde} R^{ab}R^{cd}e^e = \sqrt{|g|}\left[R^{\alpha\beta\gamma\delta}R_{\alpha\beta\gamma\delta} - 4R^{\alpha\beta}R_{\alpha\beta} + R^2\right] d^5x. \tag{13}$$

The fact that this term could be added to the Einstein-Hilbert action in five dimensions seems to have been known for many years, and is commonly attributed to Lanczos.[21]

4.2.1. *Dynamical content of Lovelock theory*

The Lovelock theory is the natural generalization of GR when the spacetime dimension is greater than four. In the absence of torsion this theory generically describes the same $D(D-3)/2$ degrees of freedom as the Einstein-Hilbert theory.[22] The action (9) has been identified as describing the only ghost-free[g] effective theory for a spin two field, generated from string theory at low energy.[12,23] The unexpected and nontrivial absence of ghosts seems to reflect the fact that, in the absence of torsion, the Lovelock action yields second order field equations for the metric, so that the propagators behave as k^{-2}, and not as $k^{-2}+k^{-4}$, as would be the case in a generic theory involving arbitrary combinations of the curvature tensor and higher derivatives.

Extremizing the action (9) with respect to e^a and ω^{ab}, yields

$$\delta I_D = \int [\delta e^a \mathcal{E}_a + \delta\omega^{ab}\mathcal{E}_{ab}] = 0, \tag{14}$$

modulo surface terms. The condition for I_D to have an extremum to first order under arbitrary infinitesimal variations is that $\mathcal{E}_a$ and $\mathcal{E}_{ab}$ vanish:

$$\mathcal{E}_a = \sum_{p=0}^{[\frac{D-1}{2}]} \alpha_p (D-2p)\mathcal{E}_a^{(p)} = 0, \tag{15}$$

and

$$\mathcal{E}_{ab} = \sum_{p=1}^{[\frac{D-1}{2}]} \alpha_p p(D-2p)\mathcal{E}_{ab}^{(p)} = 0, \tag{16}$$

where we have defined

$$\mathcal{E}_a^{(p)} := \epsilon_{ab_2\cdots b_D} R^{b_2b_3}\cdots R^{b_{2p}b_{2p+1}} e^{b_{2p+2}}\cdots e^{b_D}, \tag{17}$$

$$\mathcal{E}_{ab}^{(p)} := \epsilon_{aba_3\cdots a_D} R^{a_3a_4}\cdots R^{a_{2p-1}a_{2p}} T^{a_{2p+1}} e^{a_{2p+2}}\cdots e^{a_D}. \tag{18}$$

These equations cannot contain second or higher derivatives of e^a and ω^a_b, simply because $d^2=0$. Indeed, R^{ab} and T^a only involve first derivatives of e^a and ω^a_b, respectively. If one furthermore assumes –as is usually done– that the torsion vanishes identically,

$$T^a = de^a + \omega^a{}_b e^b = 0, \tag{19}$$

then Eq. (16) is automatically satisfied. Moreover, the torsion-free condition can be solved for ω as a function of the inverse vielbein (E^μ_a) and its derivatives,

$$\omega^a{}_{b\mu} = -E^\nu_b(\partial_\mu e^a_\nu - \Gamma^\lambda_{\mu\nu} e^a_\lambda), \tag{20}$$

where $\Gamma^\lambda_{\mu\nu}$ is symmetric in $\mu\nu$ and can be identified as the Christoffel symbol (torsion-free affine connection). Substituting this expression for the Lorentz connection back into (17) yields second order field equations for the metric.

[g]A Lagrangian containing arbitrarily high derivatives of fields generally leads to ghosts.

These equations are identical to the ones obtained from varying the Lovelock action written in terms of the Riemann tensor and the metric,

$$I_D[g] = \int_M d^D x\sqrt{g}\left[\alpha_0' + \alpha_1' R + \alpha_2'(R^{\alpha\beta\gamma\delta}R_{\alpha\beta\gamma\delta} - 4R^{\alpha\beta}R_{\alpha\beta} + R^2) + \cdots\right].$$

Now one can understand the remarkable feature of GR, where the field equations for the metric are second order and not fourth order, in spite of the fact that the Lagrangian involves the second derivatives of $g_{\mu\nu}$. This "miraculous accident" is a consequence of the fact that the action can be written using only wedge products and exterior derivatives of the fields, without using the $\star$-Hodge dual, and the fact that the torsion is assumed to vanish identically. In $f(R)$ theories, for example, this first condition is not respected and they generically give rise to higher order field equations. In the presence of fermionic matter, torsion does not vanish, hence the second condition would not hold and the use of a purely metric formulation would be unwarranted.

The standard, purely metric, form of the action is also called second order formalism, because it yields equations with up to second derivatives of the metric. The fact that the Lagrangian contains second derivatives of $g_{\mu\nu}$ has induced some authors to refer to the Lovelock actions as *higher derivative theories of gravity*, which is incorrect, as already mentioned.

One important feature that makes the behaviour of Lovelock theories very different for $D \leq 4$ and for $D > 4$ is that in the former case the field equations (15, 16) are linear in the curvature tensor, while in the latter case the equations are generically nonlinear in R^{ab}. For $D \leq 4$ the equations (18) imply the vanishing of torsion, which is no longer true for $D > 4$. In fact, the field equations evaluated in some configurations may leave some components of the curvature and torsion tensors completely undetermined. For example, Eq.(16) has the form of a polynomial in R^{ab} times T^a, and it is possible that the polynomial vanishes identically, imposing no conditions on the torsion tensor.

However, the configurations for which the equations leave some components of R^{ab} or T^a undetermined form sets of measure zero in the space of solutions. In a generic case, outside of these degenerate configurations, the Lovelock theory has the same number of degrees of freedom as ordinary gravity.[22] The problem of degeneracy, however, is a major issue in determining the time evolution of certain dynamical systems, usually associated with the splitting of the phase space into causally disconnected regions and irreversible loss of degrees of freedom.[24] These features might be associated to a dynamical dimensional reduction in gravitation theories,[25] and has been shown to survive even at the quantum level.[26]

4.3. *Torsional series*

Lovelock's theorem assumes equation (19) to be an identity. This means that e^a and $\omega^a_{\ b}$ are not independent fields, contradicting the assumption that these fields correspond to two independent features of the geometry on equal footing. For $D \leq 4$ (and in the absence of other fields), equations (19) and (18) coincide, so that imposing the torsion-free constraint may be seen as an unnecessary albeit harmless restriction. In fact, for 3 and 4 dimensions, the Lorentz connection can be algebraically solved from its own field equation and by the implicit function theorem, the first order and the second order actions have the same extrema and define equivalent theories, $I[\omega, e] = I[\omega(e, \partial e), e]$.

In general, although the torsion-free condition is always a solution of (16), (19) does not follow from the field equations. Since, (16) is not algebraic in ω, it is impossible to solve for $\omega(e, \partial e)$ globally, and therefore the second order action principle is not equivalent to the first order one.

Thus, it is reasonable to consider the generalization of the Lovelock action in which torsion *is not* assumed to vanish identically, adding all possible Lorentz invariants involving torsion that would vanish if $T^a = 0$.[18] This means allowing for combinations of terms included in the first four expressions (1-4). For example, a possible contribution to the Lagrangian in fifteen dimensions could be $\mathbf{P}_4 \upsilon_1 \tau_0 \zeta_0$. Let us examine other examples:

• For $D = 3$, there is one torsion term in addition to the Lovelock family,

$$\zeta_0 = e^a T_a, \tag{21}$$

• For $D = 4$, there are three such terms,

$$\upsilon_1 = e^a e^b R_{ab}, \quad \tau_0 = T^a T_a, \quad \mathbf{P}_4 = R^{ab} R_{ab}. \tag{22}$$

The integral of the last term in (22) is a topological invariant (the Pontryagin number), and the linear combination of the other two terms,

$$\mathbf{N}_4 = T^a T_a - e^a e^b R_{ab}, \tag{23}$$

known as the Nieh-Yan form, also yields a topological invariant.[27] The properly normalized integral of (23) over a 4-manifold is an integer, equal to the difference between the Pontryagin classes for $SO(5)$ and $SO(4)$ (or their related groups $SO(n, 5-n$ and $SO(m, 4-m)$).[28]

To make life even harder, there are some linear combinations of these products which are topological densities, as in (22). In 8 dimensions there are two Pontryagin forms

$$\mathbf{P}_8 = R^{a_1}_{\ a_2} R^{a_2}_{\ a_3} \cdots R^{a_4}_{\ a_1}, \tag{24}$$

$$(\mathbf{P}_4)^2 = (R^a_{\ b} R^b_{\ a})^2, \tag{25}$$

which also occur in the absence of torsion, and there are two generalizations of the Nieh-Yan form,

$$(\mathbf{N}_4)^2 = (T^a T_a - e^a e^b R_{ab})^2, \tag{26}$$

$$\mathbf{N}_4 \mathbf{P}_4 = (T^a T_a - e^a e^b R_{ab})(R^c_{\ d} R^d_{\ c}), \tag{27}$$

etc. (for details and extensive discussions, see Ref.[18]).

4.4. *Quasi-invariant Chern-Simons series*

The Pontryagin classes $\mathbf{P}_{2k}$ defined in (1), as well as those that involve torsion, like the Nieh-Yan forms (23), are all closed, which means that locally they can be expressed as the exterior derivative of some form (Poincaré Lemma, again). Therefore, following Section 2.1, one can look for locally defined CS forms whose exterior derivatives yield the corresponding closed forms. These CS forms can also be included as Lagrangian densities in the appropriate dimension.

The idea is best illustrated with examples. Consider the Pontryagin and the Nieh-Yan forms in four dimensions, $\mathbf{P}_4$ and $\mathbf{N}_4$, respectively. The corresponding CS three-forms are

$$C_3^{Lor} = \omega^a_{\ b} d\omega^b_{\ a} + \frac{2}{3}\omega^a_{\ b}\omega^b_{\ c}\omega^c_{\ a} \tag{28}$$

$$C_3^{Tor} = e^a T_a. \tag{29}$$

Both these terms are invariant under $SO(2,1)$ (Lorentz invariant in three dimensions), and are related to the four-dimensional Pontryagin and Nieh-Yan forms,

$$dC_3^{Lor} = R^{ab} R_{ab} = \mathbf{P}_4, \tag{30}$$

$$dC_3^{Tor} = T^a T_a - e^a e^b R_{ab} = \mathbf{N}_4. \tag{31}$$

The general recipe is simple. For each Pontryagin form in $4k$ dimensions, there is a CS form which provides a sensible action for gravity in $4k-1$ dimensions. For example, in $D = 7$, the Lorentz CS form is

$$C_7^{Lor} = Tr[\omega(d\omega)^3 + \frac{8}{5}\omega^3(d\omega)^2 + \frac{4}{5}\omega^2(d\omega)\omega(d\omega) + 2\omega^5(d\omega) + \frac{4}{7}\omega^7],$$

where the trace is over the suppressed the Lorentz indices.

Thus, the most general gravity Lagrangian in a given dimension would be a linear combination of Lorentz invariant and quasi-invariant D-forms of the three families described above: Lovelock, torsional and Lorentz Chern-Simons forms. While the Lovelock series has a simple systematic rule for any dimension (9), there is no simple recipe for the torsional Lagrangians. These look awkward, there is no systematic rule to even say how many terms appear in a given dimension and the number of elementary terms of the families υ, τ and ζ, grows wildly with the dimension. This proliferation problem is not purely aesthetic. It is like the cosmological constant problem but for a huge number of indeterminate parameters in the theory and not just one.

5. Selecting Sensible Theories

There is another serious aspect of the proliferation issue: the coefficients in front of each term in the Lagrangian are not only arbitrary but dimensionful. This problem already occurs in 4 dimensions, where Newton's constant and the cosmological constant have dimensions of $[\text{length}]^2$ and $[\text{length}]^{-2}$ respectively.

The presence of dimensionful parameters leaves little room for optimism in a quantum version of the theory. Dimensionful parameters in the action are potentially dangerous because they are likely to acquire uncontrolled quantum corrections. This is what makes ordinary gravity nonrenormalizable in perturbation theory: In 4 dimensions, Newton's constant has dimensions of $[\text{mass}]^{-2}$ in natural units. This means that as the order in perturbation series increases, more powers of momentum will occur in the Feynman graphs, making the ultraviolet divergences increasingly worse. Concurrently, the radiative corrections to these bare parameters require the introduction of infinitely many counterterms into the action to render them finite.[29] But an illness that requires infinite amount of medication is synonym of incurable.

The only safeguard against the threat of uncontrolled divergences in quantum theory is to have some symmetry principle that fixes the values of the parameters in the action, limiting the number of possible counterterms that could be added to the Lagrangian. Obviously, a symmetry endowed with such a high responsibility should be a *bona fide* quantum symmetry, and not just an approximate feature of its effective classical descendant. A symmetry that is only present in the classical limit but is not a feature of the quantum theory is said to be anomalous. This means that if one conceives the quantum theory as the result of successive quantum corrections to the classical theory, these corrections "break" the symmetry. An anomalous symmetry is an artifact of the classical limit, that does not correspond to a true symmetry of the microscopic world.

If a non anomalous symmetry fixes the values of the parameters in the action, this symmetry will protect those values under renormalization. A good indication that this might happen would be if all the coupling constants are dimensionless and could be absorbed in the fields, as in Yang-Mills theory. As shown below, in odd dimensions there is a unique choice of coefficients in the Lovelock action that yields a theory with an enlarged gauge symmetry. This action has no dimensionful parameters and can be seen to depend on a unique (dimensionless) coefficient (κ), analogous to Newton's constant. This coefficient can be shown to be quantized, following an argument similar to the one that yields Dirac's quantization of the product of magnetic and electric charge.[30] All these miraculous properties can be traced back to the fact that the particular choice of coefficients in that Lagrangian turns the Lovelock Lagrangian into a CS form for an enhanced gauge symmetry.

5.1. *Extending the Lorentz group*

The coefficients α_p in the Lovelock action (9) of section 4 have dimensions l^{D-2p}. This is because the canonical dimension of the vielbein is $[e^a] = l$, while the Lorentz connection has dimensions $[\omega^{ab}] = l^0$, as a true gauge field. This reflects the fact that gravity is naturally a gauge theory for the Lorentz group, where e^a plays the role of a matter field, *not* a connection field but a vector under Lorentz transformations.

5.1.1. *Poincaré group*

Three-dimensional gravity, where e^a can play the role of a connection, is an important exception to the previous statement. This is in part thanks to the coincidence in three dimensions that makes the Lorentz connection and the vielbein have the same number of components. Then, both ω^{ab} and $\hat{\omega}^{ab} = \omega^{ab} + \xi\epsilon^{abc}e_c$ define proper connections for the Lorentz group.

Consider the Einstein-Hilbert Lagrangian in three dimensions

$$L_3 = \epsilon_{abc}R^{ab}e^c. \tag{1}$$

Under an infinitesimal Lorentz transformation with parameter $\lambda^a{}_b$, the

Lorentz connection transforms as

$$\begin{aligned}\delta\omega^a{}_b &= d\lambda^a{}_b + \omega^a{}_c\lambda^c{}_b - \omega^c{}_b\lambda^a{}_c \\ &= D_\omega\lambda^a{}_b,\end{aligned} \tag{2}$$

while e^c, R^{ab} and ϵ_{abc} transform as tensors,

$$\begin{aligned}\delta e^a &= -\lambda^a{}_c e^c \\ \delta R^{ab} &= -(\lambda^a{}_c R^{cb} + \lambda^b{}_c R^{ac}), \\ \delta\epsilon_{abc} &= -(\lambda^d{}_a\epsilon_{dbc} + \lambda^d{}_b\epsilon_{adc} + \lambda^d{}_c\epsilon_{abd}) \equiv 0.\end{aligned}$$

Combining these relations, the Lorentz invariance of L_3 can be explicitly checked. What is unexpected is that the action defined by (1) is also invariant under the group of local translations in the three dimensional tangent space. For this additional symmetry e^a transforms as a gauge connection for the translation group.[h] In fact, if the vielbein transforms under "local translations" in tangent space, parametrized by λ^a as

$$\begin{aligned}\delta e^a &= d\lambda^a + \omega^a{}_b\lambda^b, \\ &= D_\omega\lambda^a\end{aligned} \tag{3}$$

while the Lorentz connection remains unchanged,

$$\delta\omega^{ab} = 0, \tag{4}$$

then, the Lagrangian L_3 changes by a total derivative,

$$\delta L_3 = d[\epsilon_{abc}R^{ab}\lambda^c], \tag{5}$$

which can be dropped from the action, under the assumption of standard boundary conditions. This means that in three dimensions ordinary gravity is gauge invariant under the whole Poincaré group. This can also be shown using the infinitesimal transformations δe and $\delta\omega$ to compute the commutators of the second variations, obtaining the Lie algebra of the Poincaré group.

5.1.2. *(Anti-)de Sitter group in 2+1 dimensions*

In the presence of a cosmological constant $\Lambda = \mp l^{-2}$ it is also possible to extend the local Lorentz symmetry. In this case, however, the invariance of the appropriate tangent space is not the local Poincaré symmetry, but

[h] This translational invariance in the tangent space is not to be confused with the local translations in the base manifold mentioned in Sect.1.3.1, Eq(15).

the local (anti)-de Sitter group. The point is that different spaces can be chosen as tangents to a given manifold M, provided they are diffeomorphic to the open neighbourhoods of M. However, a useful choice of tangent space corresponds to the covering space of a vacuum solution of the Einstein equations. In the previous case, flat space was singled out because it is the maximally symmetric solution of the Einstein equations. If $\Lambda \neq 0$, it is no longer flat spacetime, but the de Sitter or anti-de Sitter space, that solve the Einstein equations for $\Lambda > 0$ or $\Lambda < 0$, respectively.

The three-dimensional Lagrangian in (9) of section 4 reads

$$L_3^{AdS} = \epsilon_{abc}(R^{ab}e^c \pm \frac{1}{3l^2}e^a e^b e^c), \tag{6}$$

and the action is invariant, modulo surface terms, under infinitesimal (A)dS transformations

$$\delta\omega^{ab} = d\lambda^{ab} + \omega^a{}_c\lambda^{cb} + \omega^b{}_c\lambda^{ac} \pm [e^a\lambda^b - \lambda^a e^b]l^{-2} \tag{7}$$

$$\delta e^a = d\lambda^a + \omega^a{}_b\lambda^b - \lambda^a_b e^b. \tag{8}$$

These transformations can be cast in a more suggestive way as

$$\delta\begin{bmatrix} \omega^{ab} & e^a l^{-l} \\ -e^b l^{-l} & 0 \end{bmatrix} = d\begin{bmatrix} \lambda^{ab} & \lambda^a l^{-l} \\ -\lambda^b l^{-l} & 0 \end{bmatrix} + \begin{bmatrix} \omega^a_c & e^a l^{-l} \\ -e_c l^{-l} & 0 \end{bmatrix}\begin{bmatrix} \lambda^{cb} & \lambda^c l^{-1} \\ \pm\lambda^b l^{-1} & 0 \end{bmatrix} - \begin{bmatrix} \lambda^{ac} & \lambda^a l^{-1} \\ -\lambda^c l^{-1} & 0 \end{bmatrix}\begin{bmatrix} \omega_c^{\ b} & e_c l^{-1} \\ \pm e^b l^{-1} & 0 \end{bmatrix}. \tag{9}$$

This can also be written as

$$\delta W^{AB} = d\Lambda^{AB} + W^A_C\Lambda^{CB} - \Lambda^{AC}W_C^B,$$

where the 1-form W^{AB} and the 0-form Λ^{AB} stand for the combinations

$$W^{AB} = \begin{bmatrix} \omega^{ab} & e^a l^{-1} \\ -e^b l^{-1} & 0 \end{bmatrix} \tag{10}$$

$$\Lambda^{AB} = \begin{bmatrix} \lambda^{ab} & \lambda^a l^{-1} \\ -\lambda^b l^{-1} & 0 \end{bmatrix}, \tag{11}$$

(here $a, b, .. = 1, 2, ..D$, while $A, B, ... = 1, 2, .., D+1$). Clearly, W^{AB} transforms as a connection and Λ^{AB} can be identified as the infinitesimal transformation parameters, but for which group? Since $\Lambda^{AB} = -\Lambda^{BA}$, this indicates that the group is one that leaves invariant a symmetric, real bilinear form, so it must be a group in the $SO(r, s)$ family. The signs ($\pm$) in the transformation above can be traced back to the sign of the cosmological constant. It is easy to check that this structure fits well if indices are raised and lowered with the metric Υ^{AB} given by

$$\Upsilon^{AB} = \begin{bmatrix} \eta^{ab} & 0 \\ 0 & \mp 1 \end{bmatrix}, \tag{12}$$

so that, for example, $W^{AB} = \Upsilon^{BC} W^{A}_{\ C}$. Then, the covariant derivative in the connection W of this metric vanishes identically,

$$D_W \Upsilon^{AB} = d\Upsilon^{AB} + W^{A}_{\ C}\Upsilon^{CB} + W^{B}_{\ C}\Upsilon^{AC} = 0. \tag{13}$$

Since Υ^{AB} is constant, this last expression implies $W^{AB} + W^{BA} = 0$, in exact analogy with what happens with the Lorentz connection, $\omega^{ab} + \omega^{ba} = 0$, where $\omega^{ab} \equiv \eta^{bc}\omega^{a}_{\ c}$. Indeed, this is a way to discover in hindsight that the 1-form W^{AB} is actually a connection for the group which leaves invariant the metric Υ^{AB}. Here the two signs in Υ^{AB} correspond to the de Sitter (+) and anti-de Sitter (−) groups, respectively.

What we have found here is an explicit way to immerse the three-dimensional Lorentz group reflected here by the connection ω and the vielbein e, into a larger symmetry group in which the Lorentz connection and the vielbein have been incorporated on equal footing as components of a larger (A)dS connection. The Poincaré symmetry is obtained in the limit $l \to \infty$ ($\lambda \to 0$). In that case, instead of (7, 8) one has

$$\delta\omega^{ab} = d\lambda^{ab} + \omega^{a}_{\ c}\lambda^{cb} + \omega^{b}_{\ c}\lambda^{ac} = D_\omega\lambda^{ab} \tag{14}$$

$$\delta e^{a} = d\lambda^{a} + \omega^{a}_{\ b}\lambda^{b} - \lambda^{a}_{\ b}e^{b} = D_\omega\lambda^{a}. \tag{15}$$

The vanishing cosmological constant limit is actually a deformation of the (A)dS algebra analogous to the deformation that yields the Galileo group from the Poincaré symmetry in the limit of infinite speed of light ($c \to \infty$). These deformations are examples of what is known as a Inönü-Wigner contraction.[31,32] The procedure starts from a semisimple Lie algebra and some generators are rescaled by a parameter (l or λ in the above example). Then, in the limit where the parameter is taken to zero (or infinity), a new (not semisimple) algebra is obtained. For the Poincaré group which is the familiar symmetry of Minkowski space, the representation in terms of W becomes inadequate because, in the limit, the metric Υ^{AB} should be replaced by the degenerate (noninvertible) metric of the Poincaré group,

$$\Upsilon_0^{AB} = \begin{bmatrix} : \eta^{ab} & 0 \\ 0 & 0 \end{bmatrix}, \tag{16}$$

and is no longer clear how to raise and lower indices. Nevertheless, the Lagrangian (6) in the limit $l \to \infty$ takes the usual Einstein Hilbert form with vanishing cosmological constant,

$$L_3^{EH} = \epsilon_{abc} R^{ab} e^c, \tag{17}$$

which is still invariant under (15).

As Witten showed, General Relativity in three spacetime dimensions is a renormalizable quantum system.[33] It is strongly suggestive that precisely in 2+1 dimensions GR is also a gauge theory on a fibre bundle. It could be thought that the exact solvability miracle is due to the absence of propagating degrees of freedom in three-dimensional gravity, but the final power-counting argument of renormalizability rests on the fibre bundle structure of the Chern-Simons system and doesn't seem to depend on the absence of propagating degrees of freedom. In what follows we will generalize the gauge invariance of three-dimensional gravity to higher dimensions.

5.2. *More Dimensions*

Everything that has been said about embedding the Lorentz group into (A)dS and Poincaré groups for $D = 3$ in the previous section, can be generalized for higher D. In fact, it is always possible to embed the D-dimensional Lorentz group into the de-Sitter, or anti-de Sitter groups,

$$SO(D-1,1) \hookrightarrow \begin{cases} SO(D,1), & \Upsilon^{AB} = \mathrm{diag}(\eta^{\mathrm{ab}}, +1) \\ SO(D-1,2), & \Upsilon^{AB} = \mathrm{diag}(\eta^{\mathrm{ab}}, -1) \end{cases} \tag{18}$$

as well as into their Poincaré limit,

$$SO(D-1,1) \hookrightarrow ISO(D-1,1). \tag{19}$$

The question naturally arises, are there gravity actions in dimensions ≥ 3 which are also invariant, not just under the Lorentz group, but under some of its extensions, $SO(D,1)$, $SO(D-1,2)$, $ISO(D-1,1)$? As we will see now, the answer to this question is affirmative in odd dimensions: There exist gravity actions for every $D = 2n-1$, invariant under local $SO(2n-2,2)$, $SO(2n-1,1)$ or $ISO(2n-2,1)$ transformations, where the vielbein and the Lorentz connection combine to form the connection of the larger group. In even dimensions, in contrast, this cannot be done.

Why is it possible in three dimensions to enlarge the symmetry from local $SO(2,1)$ to local $SO(3,1)$, $SO(2,2)$ and $ISO(2,1)$? What happens if one tries to do this in four or more dimensions? Consider the Hilbert action for $D = 4$,

$$L_4 = \epsilon_{abcd} R^{ab} e^c e^d. \tag{20}$$

Let us examine how this transforms under the Poincaré group. Clearly, this is Lorentz invariant, so the question is what happens to this action under a local translation given by (15). A simple calculation yields

$$\begin{aligned}\delta L_4 &= 2\epsilon_{abcd} R^{ab} e^c \delta e^d \\ &= d(2\epsilon_{abcd} R^{ab} e^c \lambda^d) - 2\epsilon_{abcd} R^{ab} T^c \lambda^d. \end{aligned} \tag{21}$$

The first term in the r.h.s. of (21) is a total derivative and therefore gives a surface contribution to the action. The last term, however, need not vanish, unless one imposes the field equation $T^a = 0$. But this means that the invariance of the action only occurs *on shell.* Now, "on shell symmetries" are invariances of classical configurations and not symmetries of the action. This means that they are unlikely to survive quantization because quantum mechanics does not respect equations of motion.

On the other hand, the miracle in three dimensions was due to the Lagrangian (17) being linear in e. In fact, Lagrangians of the form

$$L_{2n+1} = \epsilon_{a_1 \cdots a_{2n+1}} R^{a_1 a_2} \cdots R^{a_{2n-1} a_{2n}} e^{a_{2n+1}}, \tag{22}$$

which are only defined in odd dimensions, can be easily checked to be also invariant under local translations (15).

Since the Poincaré group is a limit of (A)dS, it seems likely that there should exist Lagrangians in odd dimensions, invariant under local (A)dS transformations, whose limit for vanishing cosmological constant ($l \to \infty$) is (22). One way to find out what that Lagrangian might be, one could take the most general Lovelock Lagrangian and select the coefficients by requiring invariance under (7, 8). This route may be long, tedious and reliable. An alternative approach is to try to understand why it is that in three dimensions the gravitational Lagrangian with cosmological constant (6) is invariant under the (A)dS group.

Let us consider the three-dimensional case first. If we take seriously the notion that W^{AB} is a connection, then the associated curvature is

$$F^{AB} = dW^{AB} + W^A_{\ C} W^{CB},$$

where W^{AB} is defined in (10). Then, it is easy to prove that

$$F^{AB} = \begin{bmatrix} R^{ab} \pm l^{-2} e^a e^b & l^{-1} T^a \\ -l^{-1} T^b & 0 \end{bmatrix}. \tag{23}$$

where a, b run from 1 to 3 (or 0 to 2) and A, B from 1 to 4 (or 0 to 3). Since the (A)dS group has an invariant tensor ϵ_{ABCD}, one can construct the 4-form invariant

$$\mathbf{E}_4 = \epsilon_{ABCD} F^{AB} F^{CD}. \tag{24}$$

This invariant under the (A)dS group is readily recognized, up to a constant multiplicative factor, as the Euler form[i] for a four-dimensional manifold whose tangent space is not Minkowskian, but has the metric Υ^{AB} =diag $(\eta^{ab}, \mp 1)$. The form $\mathbf{E}_4$ can also be written explicitly in terms of R^{ab}, T^a, and e^a,

$$\begin{aligned}\mathbf{E}_4 &= 4\epsilon_{abc}(R^{ab} \pm l^{-2}e^a e^b)l^{-1}T^a \\ &= \frac{4}{l}d\left[\epsilon_{abc}\left(R^{ab} \pm \frac{1}{3l^2}e^a e^b\right)e^c\right],\end{aligned} \tag{25}$$

which is, up to constant factors, the exterior derivative of the Lagrangian (6),

$$\mathbf{E}_4 = dL_3^{AdS}. \tag{26}$$

This is the key point: the l.h.s. of (26) is invariant under local (A)dS$_3$ by construction. Therefore, the same must be true of the r.h.s.,

$$\delta\left(dL_3^{AdS}\right) = 0.$$

Since the variation is a linear operation, it commutes with the derivative,

$$d\left(\delta L_3^{AdS}\right) = 0,$$

which in turn means, by Poincaré's Lemma[34] that, locally, $\delta L_3^{AdS} = d\Omega$, for some two-form Ω. This explains why the action is (A)dS invariant up to surface terms, which is exactly what we found for the variation, [see, (5)]. The fact that three dimensional gravity can be written in this way was observed many years ago by Achúcarro and Townsend,[35] a fact that has been extensively used since then.[33]

The steps to generalize the (A)dS Lagrangian from 3 to $2n-1$ dimensions are now clear[j]:

- First, generalize the Euler density (24) to a $2n$-form,

$$\mathbf{E}_{2n} = \epsilon_{A_1 \cdots A_{2n}} F^{A_1 A_2} \cdots F^{A_{2n-1}A_{2n}}. \tag{27}$$

- Second, express $\mathbf{E}_{2n}$ explicitly in terms of R^{ab}, T^a and e^a using (23).
- Write this as the exterior derivative of a $(2n-1)$-form L_{2n-1}.
- L_{2n-1} can be used as a Lagrangian in $(2n-1)$ dimensions.

[i]This identification is formal, since the differential forms that appear here are still in three dimensions. However, they can be naturally defined in four dimensions by simply extending the range of coordinate indices. Thus, we assume the three-dimensional manifold as embedded in, or as a boundary of, a four-dimensional one.
[j]The construction we outline here was discussed in,[36,37] and also in.[38,39]

This procedure directly yields the $(2n-1)$-dimensional (A)dS-invariant Lagrangian as

$$L_{2n-1}^{(A)dS} = \sum_{p=0}^{n-1} \bar{\alpha}_p L_p^{2n-1}, \tag{28}$$

where L_p^D is given by (10) of section 4. This is a particular case of a Lovelock Lagrangian in which all the coefficients $\bar{\alpha}_p$ have been fixed to take the values

$$\bar{\alpha}_p = \kappa \cdot \frac{(\pm 1)^{p+1} l^{2p-D}}{(D-2p)} \binom{n-1}{p}, \; p = 1, 2, ..., n-1 = \frac{D-1}{2}, \tag{29}$$

where κ is an arbitrary dimensionless constant.

Another interesting exercise is to show that, for AdS, the action (28) can also be written as[40]

$$I_{2n-1} = \frac{\kappa}{l} \int_M \int_0^1 dt \, \epsilon_{a_1 \cdots a_{2n-1}} R_t^{a_1 a_2} \cdots R_t^{a_{2n-3} a_{2n-2}} e^{a_{2n-1}} , \tag{30}$$

where we have defined $R_t^{ab} := R^{ab} + (t^2/l^2) e^a e^b$.

Example: In five dimensions, the (A)dS Lagrangian reads

$$L_5^{(A)dS} = \frac{\kappa}{l} \epsilon_{abcde} \left[e^a R^{bc} R^{de} \pm \frac{2}{3l^2} e^a e^b e^c R^{de} + \frac{1}{5l^4} e^a e^b e^c e^d e^e \right]. \tag{31}$$

The parameter l is a length scale (Planck length) and cannot be fixed by theoretical considerations. Actually, l only appears in the combination

$$\tilde{e}^a = l^{-1} e^a,$$

that could be considered as the "true" dynamical field which is natural if W^{AB} is used instead of ω^{ab} and e^a separately. In fact, the Lagrangian (28) can also be written in terms of W^{AB} as

$$L_{2n-1}^{(A)dS} = \kappa \cdot \epsilon_{A_1 \cdots A_{2n}} \left[W (dW)^{n-1} + a_3 W^3 (dW)^{n-2} + \cdots a_{2n-1} W^{2n-1} \right],$$

where all indices are contracted appropriately and the coefficients a_3, $\cdots a_{2n-1}$ are rational numbers fixed by the condition $dL_{2n-1}^{(A)dS} = \mathbf{E}_{2n}$.

5.3. *Generic Chern-Simons forms*

The construction outlined above is not restricted to the Euler form, but applies to any gauge invariant of similar nature, generally known as characteristic classes, like the Pontryagin or Chern classes. The corresponding CS forms were studied first in the context of abelian and nonabelian gauge theories (see, e. g.,[2,41]). Tables 1 and 2 give examples of CS forms which define Lagrangians in three and seven dimensions, and their corresponding topological invariants,

$D = 3$ CS Lagrangians	Characteristic classes	Groups
$L_3^{(A)dS} = \epsilon_{abc}(R^{ab} \pm \frac{e^a e^b}{3l^2})e^c$	$\mathbf{E}_4 = \epsilon_{abc}(R^{ab} \pm \frac{e^a e^b}{l^2})T^c$	$SO(4)$, $SO(3,1)$ or $SO(2,2)$
$L_3^{Lor} = \omega^a{}_b d\omega^b{}_a + \frac{2}{3}\omega^a{}_b\omega^b{}_c\omega^c{}_a$	$\mathbf{P}_4^{Lor} = R^a{}_b R^b{}_a$	$SO(2,1)$
$L_3^{Tor} = e^a T_a$	$\mathbf{N}_4 = T^a T_a - e^a e^b R_{ab}$	$SO(2,1)$
$L_3^{U(1)} = AdA$	$\mathbf{P}_4^{U(1)} = FF$	$U(1)$
$L_3^{SU(N)} = Tr[\mathbf{A}d\mathbf{A} + \frac{2}{3}\mathbf{AAA}]$	$\mathbf{P}_4^{SU(4)} = Tr[\mathbf{FF}]$	$SU(N)$

Table 1: Three-dimensional gravitational CS Lagrangians, their related characteristic classes and the corresponding gauge groups.

Here R, F, and $\mathbf{F}$ are the curvatures for the connections $\omega^a_{\ b}$, A and $\mathbf{A}$, corresponding to of the Lorentz, the electromagnetic, and $SU(N)$ gauge groups respectively; T^a is the torsion; $\mathbf{E}_4$ and $\mathbf{P}_4$ are the Euler and the Pontryagin densities for the Lorentz group,[16] and $\mathbf{N}_4$ is the Nieh-Yan invariant.[27]

5.4. *Torsional Chern-Simons forms*

So far we have not included torsion in the CS Lagrangian, but as we see in the table above, some CS forms do include torsion. All the CS forms above are Lorentz invariant (up to an exact form), but there is a linear combination of the second and third which is invariant under the (A)dS group. This is the so-called exotic gravity,[33]

$$L_3^{Exotic} = L_3^{Lor} \pm \frac{2}{l^2} L_3^{Tor}. \tag{32}$$

As can be shown directly by taking its exterior derivative, this is invariant under (A)dS:

$$\begin{aligned} dL_3^{Exotic} &= R^a_{\ b} R^b_{\ a} \pm \frac{2}{l^2}\left(T^a T_a - e^a e^b R_{ab}\right) \\ &= F^A_{\ B} F^B_{\ A}. \end{aligned}$$

This exotic Lagrangian has the curious property of giving exactly the same field equations as the standard dL_3^{AdS}, but interchanged: varying dL_3^{AdS}

with respect to e^a gives the same equation as varying L_3^{Exotic} with respect to ω^{ab} and vice-versa.

In five dimensions, the only Lorentz invariant that can be formed using T^a is $R^{ab}T_a e_b$, which is a total derivative, and therefore there are no new Lagrangians involving torsion in this case. In seven dimensions there are two Lorentz-Pontryagin, and one torsional CS forms,

$D = 7$ CS Lagrangians	Characteristic classes
$L_7^{Lor} = \omega(d\omega)^3 + \frac{8}{5}\omega^3(d\omega)^2 + \cdots + \frac{4}{7}\omega^7$	$\mathbf{P}_8^{Lor} = R^a{}_b R^b{}_c R^c{}_d R^d{}_a$
$L_7^I = (\omega^a{}_b d\omega^b{}_a + \frac{2}{3}\omega^a{}_b\omega^b{}_c\omega^c{}_a)R^a{}_b R^b{}_a$	$(\mathbf{P}_4^{Lor})^2 = (R^a{}_b R^b{}_a)^2$
$L_7^{II} = (e^a T_a)(R^a{}_b R^b{}_a)$	$\mathbf{N}_4\mathbf{P}_4^{Lor} =$ $(T^a T_a - e^a e^b R_{ab})R^c{}_d R^d{}_c$

Table 2: Seven-dimensional gravitational CS Lagrangians, their related characteristic classes and the corresponding gauge groups.

There exist no CS forms in even dimensions for the simple reason that there are no characteristic classes in odd dimensions. The characteristic forms, like the Euler and the Pontryagin classes are exterior products of curvature two-forms and therefore are forms of even degree. The idea of characteristic class is one of the unifying concepts in mathematics connecting algebraic topology, differential geometry and analysis. The theory of characteristic classes explains mathematically why it is not always possible to perform a gauge transformation that makes the connection vanish everywhere even if it is locally pure gauge. The nonvanishing value of a topological invariant signals an obstruction to the existence of a gauge transformation that trivializes the connection globally.

The two basic invariants relevant for a Lorentz gauge theory in $2n$ dimensions are:

• The Euler class, associated with the $O(D-n,n)$ groups. For $D = 2$, the Euler number is related to the genus (g) of the surface, $\chi = 2 - 2g$.

• The Pontryagin class, associated with any classical semisimple group G. It counts the difference between self dual and anti-self dual gauge connections that are admitted in a given manifold.

The Nieh-Yan invariants correspond to the difference between Pontryagin classes for $SO(D+1)$ and $SO(D)$ in D dimensions.[28]

No similar invariants exist in odd dimensions. Hence, there are no CS actions for gravity for even D, invariant under the (anti-) de Sitter or Poincaré groups. In this light, it is fairly obvious that although ordinary Einstein-Hilbert gravity can be given a fibre bundle structure for the Lorentz group, this structure cannot be extended to include local translational invariance.

5.5. *Quantization of the gravitation constant*

The only free parameter in a CS action is κ, the multiplicative coefficient in front. If this action is to describe a quantum system, it could only take discrete values, i.e., it is quantized. To see this, consider a simply connected, compact $2n-1$ dimensional manifold M whose geometry is determined by an Euler-CS Lagrangian. Suppose M to be the boundary of a $2n$-dimensional compact orientable manifold Ω. Then the action for the geometry of M can be expressed as the integral of the Euler density $\mathbf{E}_{2n}$ over Ω, multiplied by κ. But since there can be many different manifolds with the same boundary M, the integral over Ω should give the same physical predictions as that over a different manifold, Ω'. In order for this change to leave the path integral unchanged, a minimal requirement would be

$$\kappa\left[\int_{\Omega}\mathbf{E}_{2n}-\int_{\Omega'}\mathbf{E}_{2n}\right]=2n\pi\hbar. \qquad (33)$$

The sign in front of the second integral can be changed by reversing the orientation of the second manifold $\Omega'\to\widetilde{\Omega}'=-\Omega'$, and $\int_{\Omega'}\to-\int_{\widetilde{\Omega}'}$, so that

$$\kappa\left[\int_{\Omega}\mathbf{E}_{2n}+\int_{\widetilde{\Omega}'}\mathbf{E}_{2n}\right]=2n\pi\hbar. \qquad (34)$$

With this orientation, the quantity in brackets –with the appropriate normalization– is the Euler number of the manifold obtained by gluing Ω and $\widetilde{\Omega}'$ along M in the right way to produce an orientable manifold, $\chi[\Omega\cup\widetilde{\Omega}']$. This integral can take an arbitrary integer value and hence κ must be quantized,

$$\kappa=nh,$$

where h is Planck's constant.[30]

5.6. *CS and Born-Infeld degenerate vacua*

If the Lovelock coefficients α_p are chosen according to the CS prescription (alphasCS), the field equations (15, 16) of section 4 adopt a particularly simple form (we take $D=2n+1$ and the sign corresponding to anti-de Sitter)

$$\epsilon_{ab_1b_2\cdots b_{2n}}\left[R^{b_1b_2}+\frac{1}{l^2}e^{b_1}e^{b_2}\right]\cdots\left[R^{b_{2n-1}b_{2n}}+\frac{1}{l^2}e^{b_{2n-1}}e^{b_{2n}}\right]=0 \qquad (35)$$

$$\epsilon_{abb_2\cdots b_{2n}}\left[R^{b_2b_3}+\frac{1}{l^2}e^{b_2}e^{b_3}\right]\cdots\left[R^{b_{2n-2}b_{2n-1}}+\frac{1}{l^2}e^{b_{2n-2}}e^{b_{2n-1}}\right]T^{b_{2n}}=0 \tag{36}$$

These equations are very special polynomials in R^{ab} and T^a. The first one has a n-fold degenerate root $R^{ab} = \frac{1}{l^2}e^a e^b$, which corresponds to a locally constant negative curvature geometry (locally AdS). The globally AdS solution[k] is a maximally symmetric geometry, a configuration with no features, where every point is identical to every other point and where all directions look the same. Such a boring state is naturally recognized as the vacuum.

This vacuum is an extreme state, not only because of its absence of features, but it is also an extremely dead place, around which the perturbations of the geometry are pure gauge modes. The reason is that a small perturbation of (35) around this geometry is always multiplied by zero. That means that the perturbations are not dynamically determined, as is the case of a propagating wave. Degrees of freedom that are not governed by dynamical equations are arbitrary functions, unphysical degrees of freedom also called gauge degrees of freedom or gauge modes. In this AdS vacuum, moreover, the torsion tensor is also completely indeterminate.

The only nontrivial features that are allowed by these equations are topologically distinct spaces of the form AdS/K, obtained by identification by a set of Killing vectors. In this sense, this vacuum is very similar to the vacuum of 2+1 gravity, which has no propagating degrees of freedom and all configurations are expected to be topological excitations.

Although as already mentioned, in even dimensions there are no CS theories, there exists a choice of Lovelock coefficients that yields equations that look very similar to (35 and 36). The corresponding theories are defined by a choice of α_ps such that the Lagrangian reads

$$\begin{aligned} L_{2n}^{BI} &= \epsilon_{a_1b_1a_2b_2\cdots a_nb_n}\left[R^{a_1b_1}+\frac{1}{l^2}e^{a_1}e^{b_1}\right]\cdots\left[R^{a_nb_n}+\frac{1}{l^2}e^{a_n}e^{b_n}\right] \\ &= \text{Pfaff}\left[R^{ab}+\frac{1}{l^2}e^a e^b\right]. \end{aligned} \tag{37}$$

This Lagrangian is the so-called Born-Infeld (**BI**) form and is the closest to a CS theory in even dimensions.[38,39,42] The name originates from the fact that the Pfaffian of an antisymmetric matrix M^{ab} is related to its

[k]Global AdS is denoted as $SO(D,2)/SO(D-1,1)$, which means a D-dimensional space of Lorentzian signature and isometry group $SO(D,2)$.

determinant as

$$\mathrm{Pfaff}[M] = \epsilon_{a_1 b_1 a_2 b_2 \cdots a_n b_n} M^{a_1 b_1} \cdots M^{a_n b_n} = \sqrt{det[M]}.$$

Hence, one can understand L_{2n}^{BI} as $\sqrt{det[R^{ab} + \frac{1}{l^2} e^a e^b]}$, which is reminiscent of the electromagnetic Born-Infeld Lagrangian, $L_{EM}^{BI} = \sqrt{det[F_{\mu\nu} - \lambda \eta_{\mu\nu}]}$.

A bizarre feature of all Lovelock theories for $D > 4$, including the special cases of CS and BI, is the fact that the dynamical properties are not homogeneous throughout the phase space. The number of degrees of freedom, and correspondingly, the amount of gauge invariance, depend on the location in phase space. As we saw above, CS and BI theories for $D > 4$ posses no propagating degrees of freedom around the maximally symmetric vacuum configuration. There are other less symmetrical classical solutions around which there are propagating degrees of freedom. In fact the number of degrees of freedom is almost everywhere maximal and it is only on sets of measure zero in phase space that some of those degrees of freedom become non propagating gauge modes.[43,44] This phenomenon can be seen to result from Lagrangians that are polynomials in the velocities possessing several nontrivial critical points,[45,46] or more generally, systems whose symplectic form becomes degenerate on some singular surfaces in phase space. An interesting phenomenon occurs if a system presenting these features starts from a generic configuration with maximum number of degrees of freedom, and its evolution takes it into a degenerate surface. It can be shown that in some cases this literally means that the dynamical system changes abruptly and irreversibly, losing degrees o freedom and even losing information about the initial state.[47]

Apart from those uncommon dynamical features, CS and BI Lagrangians have rather simple field equations which admit black hole solutions,[4] cosmological models, etc.. The simplification comes about because the equations admit a unique maximally symmetric configuration given by $R^{ab} + \frac{1}{l^2} e^a e^b = 0$, in contrast with the situation when all α_p's are arbitrary.

5.7. *Finite Action and the Beauty of Gauge Invariance*

Classical symmetries are defined as transformations that only change the action at most by a surface term. This is because, under appropriate boundary conditions, the boundary terms do not contribute to the variation of the action and therefore they do not alter the conditions for extrema, and the equation of motion remain the same. An example of "appropriate" boundary conditions could be those that keep the values of the fields fixed at the boundary: Dirichlet conditions. In a gauge theory, however, it may be more relevant to fix gauge invariant properties at the boundary –like the curvature. These boundary conditions are not of the Dirichlet type, but Neumann or some mixed Dirichlet-Neumann types.

To illustrate the problem consider the variation of an action $I[\phi^r]$ with respect to its arguments, defined on a manifold M with boundary ∂M,

$$\delta I[\phi] = \int_M \delta\phi^r \frac{\delta L}{\delta\phi^r} + \int_{\partial M} \Theta(\phi, \delta\phi). \tag{38}$$

The first integral vanishes identically on the classical orbits. The problem arises when the boundary conditions under which one varies the action are not sufficient to ensure that the second integral vanishes. A cynic would consider changing the boundary conditions in order to make sure $\Theta = 0$, but this is not always physically acceptable. The boundary conditions under which the action is varied reflect actual experimentally feasible situations, like the presence of a conductor fixing the electric field to be normal to the surface and freely moving charges to comply with this condition, or fixing the temperature and pressure in a fluid, or requiring the spacetime to be asymptotically flat, etc. Moreover, those boundary conditions are the same under which the classical equations admit a unique solution.

If the boundary conditions are insufficient to ensure $\Theta = 0$, before declaring the problem intractable, an option is to supplement the action by adding a suitably defined function of the fields at the boundary, $B(\phi^r|_{\partial M})$, so that

$$\delta B(\phi^r|_{\partial M}) + \int_{\partial M} \Theta(\phi, \delta\phi) = 0.$$

Changing the Lagrangian by a boundary term may seem innocuous because the Euler-Lagrange equations are not changed, but it is a delicate operation. The empirical fact is that adding a total derivative to a Lagrangian in general changes the expression for the conserved Noether charges, and again, possibly by an infinite amount.

On the other hand, it is desirable to have an action which has a finite value when evaluated on a physically observable configuration –e.g., on a classical solution. This is not just for the sake of elegance, it is a requirement when studying the semiclassical thermodynamic properties of the theory. This is particularly true for a theory possessing black holes, which exhibit interesting thermodynamic features. Moreover, quasi-invariant actions like those of CS systems, if defined on infinitely extended spacetimes are potentially ill defined because, under gauge transformations, the boundary terms could give infinite contributions to the action integral. This would not only cast doubt on the meaning of the action itself, but it would violently contradict the wish to have a gauge invariant action principle. The way to prevent this problem is to supplement the action principle with some boundary terms such that the variation of the action

The conclusion from this discussion is that a regularization principle must be in place in order for the action to be finite on physically interesting configurations, and that assures it remains finite under gauge transformations, and yields well defined conserved charges. So, the optimum would be to have the action principle, supplemented with boundary terms so that for a given set of boundary conditions:

- The action attains an extremum when the equations of motion are satisfied;
- The action changes by a finite boundary term under a gauge transformations;
- The Noether charges are finite.

In[40] it is shown that the CS action given by (28) with asymptotically AdS boundary conditions has an extremum when the field equations hold and is finite on classically interesting configurations if the action is supplemented with a boundary term of the form

$$B_{2n} = -\kappa n \int_0^1 dt \int_0^t ds \; \epsilon\theta e \left(\widetilde{R} + t^2\theta^2 + s^2 e^2\right)^{n-1}, \tag{39}$$

where we have suppressed all Lorentz indices ($\epsilon_{abc...} \to \epsilon$, etc.). Here $\widetilde{R}$ and θ are the intrinsic and extrinsic curvatures of the boundary. The resulting action

$$I_{2n-1} = \int_M L^{(A)dS}_{2n-1} + \int_{\partial M} B_{2n}, \tag{40}$$

attains an extremum for boundary conditions that fix the extrinsic curvature of the boundary. In[40] it is also shown that this action principle yields finite charges (mass, angular momentum) without resorting to ad-hoc subtractions or prescribed backgrounds, and the change in the action under gauge transformations is a finite boundary term. It can be asserted that in this case –as in many others–, the demand of gauge invariance is sufficient to cure other seemingly unrelated problems.

5.7.1. *Transgressions*

The boundary term (39) that ensures convergence of the action (40) and finite charges, turns out to have other remarkable properties. It makes the action gauge invariant –and not just quasi-invariant– under gauge transformations that keep both, the intrinsic AdS geometry, and the extrinsic curvature, fixed at the boundary. The condition of having a fixed asymptotic AdS geometry is natural for localized matter distributions such as black holes. Fixing the extrinsic curvature, on the other hand, implies that the connection approaches a fixed reference connection at infinity in a prescribed manner.

On closer examination, this boundary term can be seen to convert the action into a *transgression*, a mathematically well-defined object. A transgression form is a gauge invariant expression whose exterior derivative yields the difference of two Chern classes,[2]

$$d\mathcal{T}_{2n-1}(A,\bar{A}) = \mathcal{P}_{2n}(A) - \mathcal{P}_{2n}(\bar{A}), \tag{41}$$

where A and $\bar{A}$ are two connections in the same Lie algebra. There is an explicit expression for the transgression form in terms of the Chern-Simons forms for A and $\bar{A}$,

$$\mathcal{T}_{2n+1}(A,\bar{A}) = \mathcal{C}_{2n+1}(A) - \mathcal{C}_{2n+1}(\bar{A}) + d\mathcal{B}_{2n}(A,\bar{A}). \tag{42}$$

The last term in the R.H.S. is uniquely determined by the condition that the transgression form be invariant under a simultaneous gauge transformation of both connections throughout the manifold M

$$A \to A' = \Lambda^{-1}A\Lambda + \Lambda^{-1}d\Lambda \tag{43}$$

$$\bar{A} \to \bar{A}' = \bar{\Lambda}^{-1}\bar{A}\bar{\Lambda} + \bar{\Lambda}^{-1}d\bar{\Lambda} \tag{44}$$

with the condition that at the boundary the two gauge transformations match,

$$\bar{\Lambda}(x) - \Lambda(x) = 0, \quad \text{for} \quad x \in \partial M. \tag{45}$$

It can be seen that the boundary term in (39) is precisely the boundary term $\mathcal{B}_{2n}$ in the transgression form. The interpretation now presents some subtleties. Clearly one is not inclined to duplicate the fields by introducing a second dynamically independent set of fields ($\bar{A}$), having exactly the same couplings, gauge symmetry and quantum numbers.

One possible interpretation is to view the second connection as a nondynamical reference field. This goes against the principle according to which every quantity that occurs in the action that is not a coupling constant, a mass parameter, or a numerical coefficient like the spacetime dimension or a combinatorial factor, should be a dynamical quantum variable.[48] Even if one accepts the existence of this unwelcome guest, an explanation would be needed to justify its not being seen in nature. However, other possibilities exists, as we discuss next.

5.7.2. *Cobordism*

An alternative interpretation could be to assume that the spacetime is duplicated and we happen to live in one of the two parallel words where A is present, while $\bar{A}$ lives in the other. An obvious drawback of this interpretation is that the action for $\bar{A}$ has the wrong sign, which would lead to ghosts or rather unphysical negative energy states. We could ignore this fact because the ghosts would live in the "parallel universe" that we don't see, but this is not completely true because, at least at the boundary, the two universes meet.

Interestingly, A and $\bar{A}$ only couple at the boundary through the term $\mathcal{B}_{2n}(A, \bar{A})$, and therefore the bulk where A is defined need not be the same one where $\bar{A}$ lives. These two worlds must only share the same boundary, where condition (45) makes sense; they are independent but *cobordant* manifolds.

The negative sign in front of $\mathcal{C}(\bar{A})$ is an indication that the orientation of the parallel universe must be reversed. Then, the action can be written as

$$I[A, \bar{A}] = \int_{M} \mathcal{C}(A) + \int_{\bar{M}} \mathcal{C}(\bar{A}) + \int_{\partial M} \mathcal{B}(A, \bar{A}), \tag{46}$$

where the orientation of $\bar{M}$ has been reversed so that at the common boundary, $\partial M = -\partial \bar{M}$. In other words, the two cobordant manifolds M and $\bar{M}$, with the new orientation, define a single, uniformly oriented surface sewn at the common boundary which is also correctly oriented.

The picture that emerges in this interpretation is this: we live in a region of spacetime (M) characterized by the dynamical field A. At he boundary of our region, ∂M, there exists another field with identical properties as A and matching gauge symmetry. This second field $\bar{A}$ extends onto a cobordant manifold $\bar{M}$, to which we have no direct access except through the interaction of $\bar{A}$ with our A.[49] If the spacetime we live in is asymptotically AdS, this could be a reasonable scenario since the boundary is then causally connected to the bulk and can be easily viewed as the common boundary of two –or more– asymptotically AdS spacetimes.[50]

6. CS as brane couplings

Although CS forms appeared in high energy physics more than 30 years ago, in order to achieve local supersymmetry[51] and as Lagrangians in quantum field theory,[9,52] recently a different use has been identified. As noted in section 1, CS forms change under gauge transformation essentially like an abelian connection. Hence, they can be used like the electromagnetic vector potential A, to couple to a conserved current. However, since the CS forms have support on a $(2n+1)$-dimensional manifold, they couple gauge fields to extended sources ($2n$-dimensional membranes), charged with respect to some gauge color.[53–55]

The CS form provides a gauge-invariant coupling between the gauge potential and an extended source in a consistent manner, something that could be achieved by the "more natural" minimal couplings $A_{\mu_1\mu_2\cdots\mu_p}j^{\mu_1\mu_2\cdots\mu_p}$, only for an abelian potential.[56]

6.1. *Minimal coupling and geometry*[l]

As mentioned in the introduction, the other common situation in which total derivatives play a fundamental role is in the coupling between a field and a conserved current. The epitome of such coupling is the interaction between the electromagnetic field and an electric current,

$$I_{\text{Int}} = \int_{\Gamma} A_{\mu} j^{\mu} d^4x \,, \tag{1}$$

where A_{μ} is the vector potential and j^{μ} is the electromagnetic 4-vector current density. This coupling has two nontrivial properties besides its obvious Lorentz invariance, gauge invariance, and metric independence, both of which are common to all CS forms.

The invariance of I_{Int} under gauge transformations is a consequence of two properties of the current: its gauge invariance, $j'^{\mu} = j^{\mu}$ –as expected for any physical observable–, and its conservation, $\partial_{\mu}j^{\mu} = 0$. Conversely, the minimal coupling means that the vector potential *can only couple consistently* (in a gauge invariant way) to a conserved gauge-invariant current. This statement makes no reference to the equations of motion of the charges or to Maxwell's equations. Additionally, gauge invariance of the action implies the conservation of electric charge. The conserved charge is precisely the generator of the symmetry that implies its conservation.

---

[l]This discussion is based on.[57]

The metric independence of the minimal coupling has a more subtle meaning. It can be trivially verified from the fact that under a change of coordinates, A_μ transforms as a covariant vector while j^μ is a vector density and, therefore, the integrand in (1) is coordinate invariant. No "$\sqrt{|g|}$" is required in the integration measure, which implies that the same coupling can be used in a curved background or in flat space and in any coordinate frame. This ultimately means that the integrand in (1) is an intrinsic property of the field A defined over the worldlines swept by the point charges in their evolution, but it is independent of the geometry of those worldlines.

In the case of one point charge, the current is best understood as the dual the three-form Dirac delta (a density) supported on the worldline of the particle, $j = q * \delta(\Gamma)$ where e is the electric charge. Hence, $A_\mu j^\mu d^4x = qA \wedge \delta(\Gamma)$, and (1) reads

$$I_{\text{Int}} = q \int_\Gamma A_\mu(z) dz^\mu = q \int_\Gamma A \,, \tag{2}$$

where the coordinate z^μ is any convenient parametrization of the worldline Γ. This is correct since distributions are elements in the dual of a space of test functions, that upon integration yield numbers; in the case of the Dirac delta, it yields the value of the test function at the support, which is exactly the content of the equivalence between (1) and (2). In this case, the current merely projects the one-form A defined everywhere in spacetime, onto the worldline. The result is clearly independent of the metric of the ambient spacetime and of the metric on the worldline, which in this simple case corresponds to the choice of coordinate z^μ.

It is reassuring that not only the coupling, but also the conservation law $\partial_\mu j^\mu = 0$ does not require a metric, since ∂ is the ordinary derivative, and j is a contravariant vector density, which makes the conservation equation valid in any coordinate basis and for any metric. Metric independence ultimately means that the coupling is insensitive to deformations of the worldline Γ, and of the spacetime metric. Thus, regardless of how the particle twists and turns in its evolution, or what are the metric properties of spacetime where the interaction takes place, the coupling remains consistently gauge invariant. This fact is crucial for the dynamical consistency of the coupling to membranes or other extended objects.

6.2. *Extended sources and CS couplings*

A $(2p+1)$-CS form describes the coupling between a connection $\mathbf{A}$ and a membrane whose time evolution sweeps a $(2p+1)$-dimensional volume. The consistency of this scheme follows from the precise form of the coupling,[58]

$$I[\mathbf{A};\mathbf{j}] = \int \langle \mathbf{j}_{2p} \wedge \mathfrak{C}_{2p+1}(\mathbf{A}) \rangle, \tag{3}$$

where $\mathfrak{C}_{2p+1}$ is the algebra-valued form whose trace is the $(2p+1)$-CS form living on the brane history, $C_{2p+1}(\mathbf{A}) = \langle \mathfrak{C}_{2p+1}(\mathbf{A}) \rangle$. The current generated by the $2p$-brane is represented by the $(D-2p-1)$-form $\mathbf{j}$ supported on the worldvolume of the brane,

$$\mathbf{j}_{2p} = qj^{a_1 a_2 \cdots a_s} \mathbf{K}_{a_1} \mathbf{K}_{a_2} \cdots \mathbf{K}_{a_s} \delta(\Gamma) dx^{\alpha_1} \wedge dx^{\alpha_2} \wedge \cdots dx^{\alpha_{D-2p-1}}, \tag{4}$$

where dx^{α_i} are transverse directions to Γ. The integration over the $D-2p-1$ transverse directions yields a $(2p+1)$-CS form integrated over the worldvolume of the $2p$-brane. The invariance under the gauge transformations can be checked directly by noting that, $I[\mathbf{A};\mathbf{j}]$ changes by a (locally) exact form, provided the current is covariantly conserved,

$$D\mathbf{j}_{2p} = d\mathbf{j}_{2p} + [\mathbf{A}, \mathbf{j}_{2p}] = 0, \tag{5}$$

which can be independently checked for (4). Moreover, if the current $\mathbf{j}_{2p}$ results from particles or fields whose dynamics is governed by an action invariant under the same gauge group $\mathbf{G}$, then its conservation is guaranteed by consistency. On the worldvolume, however, the gauge symmetry is reduced to the subgroup that commutes with $j^{a_1 a_2 \cdots a_s} \mathbf{K}_{a_1} \mathbf{K}_{a_2} \cdots \mathbf{K}_{a_s}$.

Possibly the simplest example of such embedded brane occurs when an identification is made in the spatial slice of AdS_3, using a rotational Killing vector with a fixed point. In that case, a deficit angle is produced giving rise to a conical geometry around the singularity, which can be identified with a point particle;[59] the singularity is the worldline of the particle where the curvature behaves as a Dirac delta. The geometry is analogous to that of the 2+1 black hole,[60] but the naked singularity results from a wrong sign in the mass parameter of the solution.[61] A similar situation arises also with a co-dimension 2 brane in higher dimensions.[55] In all these cases it is confirmed that the coupling between this 0-brane and the (nonabelian) connection is indeed of the form (3). These branes only affect the topological structure of the geometry, but the local geometry outside the worldline of the source remains unchanged.

6.3. *3D-CS systems and condensed matter*

There may be realistic situations where a three-dimensional CS theory could give rise to interesting effects. For example, materials in ordinary four-dimensional spacetime whose excitations propagate on two-dimensional layers, display a quantum Hall effect responsible for superconductivity of high critical temperature.[62,63]

A (2+1)-dimensional CS system is naturally generated at the interface separating two regions of three-dimensional space in which a Yang-Mills theory has different vacua. Consider an action of the form

$$I[A] = \frac{1}{2}\int_M \left(Tr[F \wedge *F] - \Theta(x) Tr[F \wedge F]\right), \tag{6}$$

where the last term has the form of a topological invariant, but it fails to be topological precisely because it is multiplied by a function. If this function Θ takes a constant value θ_1 in the region $\tilde{M} \subset M$ and θ_2 elsewhere, then this term can also be written as a coupling between the Chern-Simons and a surface current,

$$\int_M \Theta Tr[F \wedge F] = \int_{\partial \tilde{M}} j \wedge Tr[A \wedge dA], \tag{7}$$

where the surface current is the one-form $j = d\Theta = (\theta_1 - \theta_2)\delta(\Sigma)dz$, and z is the coordinate along outward normal to the surface of $\tilde{M}$, $\Sigma = \partial\tilde{M}$. Since the θ-term is locally exact, the field equations, both inside and outside $\tilde{M}$ are the same as in vacuum. However, this term modifies the behavior of the field *at* the surface Σ.

This coupling between the CS form and the spacetime boundary of the region $\tilde{M}$ has physical consequences even in the simple case of the abelian theory (electromagnetism). Although Maxwell's equations are not affected by the θ-term, the interface rotates the polarization plane of an electromagnetic wave that crosses it.[64] Another effect of such coupling is a modification of the Casimir energy inside an empty region surrounded by a "material" in which $\theta \neq 0$.[65] It turns out that the so-called topological insulators are materials that produce the effect of modifying the "θ vacuum" of electromagnetic theory, a phenomenon that might be relevant for condensed matter applications.[66]

7. Summary

The relevance of gauge symmetry in physics cannot be overemphasized. One of the great achievements of physics in the last century was to establish that all interactions in nature are based on gauge invariance. The fact that nature possesses this fundamental symmetry explains the binding forces in the atomic nucleus, the functioning of stars, the chemistry that supports life, and the geometry of the universe. This unifying principle is comparable to the invention of mechanics in the XVII century, or electrodynamics and statistical mechanics of the XIX century.

It is a remarkable feature rooted in the equivalence principle, that gravity is a gauge theory for the Lorentz group. The spacetime geometry can be described by two independent notions, metricity and affinity, each one described by a fundamental field that transforms under the local symmetry in a definite representation. It is an even more remarkable feature that, in odd dimensions, these two fundamental objects can combine to become a connection for an enlarged gauge symmetry, the (anti-) de Sitter or the Poincaré groups. The resulting theory is described by a CS form that has no arbitrary free parameters, no dimensionful couplings, and whose gauge invariance is independent of the spacetime geometry.

None of this seems random. One cannot help feeling that something profound and beautiful lies in these structures. Whether the CS theories of gravity, or their more ambitious supersymmetric extensions turn out to be the way to understand the connection between gravitation and quantum mechanics, remains to be seen. However, the fact that CS forms are singled out in gravity, the fact that they play such an important role in the couplings between gauge fields and sources, their deep relation with quantum mechanics, strongly suggests that there is some meaning to it. This does not look like a contingent result of natural chaos.

Ivar Ekeland reflects on the sense of nature as is revealed to us: *But is contingency complete or is there room for meaning? Must we be content to merely note the facts, or should we look for reasons? De events follow one another randomly, or does the world function according to certain rules that we can reveal and make use of? We often don't like the way things are, and some people go so far as to give their lives to change them; the quest for meaning must therefore be part of human existence.*[67]

We may not go as far as to give our lives to convince anyone of the virtues of Chern-Simons theories. But CS forms surely make our quest for meaning a more aesthetic endeavour.

Acknowledgments: Many interesting discussions with P.Alvarez, E.Ayón-Beato, A.P.Balachandran, C.Bunster, F.Canfora, G.Giribet, J.D.Edelstein, L.Huerta, O.Mišković, R.Troncoso, M.Valenzuela, are warmly acknowledged. Special thanks are due to Sylvie Paycha, for her patient and careful reading of the manuscript that led to this much improved version. Needless to say, any remaining errors are my own responsibility.

This work has been partially supported by FONDECYT grants 1100755, 1085322, 1100328, 1110102, and by the Southern Theoretical Physics Laboratory ACT- 91 grants from CONICYT. The Centro de Estudios Científicos (CECs) is funded by the Chilean Government through the Centers of Excellence Base Financing Program of CONICYT.

References

1. S. S. Chern and J. Simons, *Characteristic forms and geometric invariants*, Annals Math. **99** (1974) 48.
2. M. Nakahara, *Geometry, Topology and Physics*, Graduate Student Series in Physics, Taylor & Francis; 2nd edition (2003).
3. J. Baez, and J. P. Munian, *Gauge fields, knots and gravity*, World Scientific, Singapore (1994).
4. J. Crisóstomo, R. Troncoso and J. Zanelli, *Black Hole Scan*, Phys. Rev. **D62** (2000) 084013; [hep-th/0003271].
5. C. Charmousis, *Higher order gravity theories and their black hole solutions*, Lect. Notes Phys. **769** (2009) 299. [arXiv:0805.0568/gr-qc].
6. C. Garraffo and G. Giribet, *The Lovelock Black Holes*, Mod. Phys. Lett. A **23** (2008) 1801; [arXiv:0805.3575/gr-qc].
7. B. Schutz, *Geometrical methods of mathematical physics* Cambridge U. Press, UK (1980).
8. See, e.g., http://math.ucr.edu/home/baez/lie/node1.html. This is also quoted in Lewis Ryder *Introduction to General Relativity*, Cambridge University Press, Cambriddge, UK (2009).
9. S. Deser, R. Jackiw and S. Templeton, *Three-Dimensional Massive Gauge Theories*, Phys. Rev. Lett. **48** (1982) 975. *Topologically massive gauge theories* Annals Phys. **140** (1982) 372; Errata-ibid.:**185** (1988) 406, **281** (2000) 409. J. Schonfeld, *A mass term for three dimensional gauge fields*, Nucl. Phys. **B185** (1981) 157.
10. M. Henneaux, *Hamiltonian Form of the Path Integral for Theories with a Gauge Freedom*, Phys. Rept. **126** (1985) 1.
11. L. O'Raifeartaigh and N. Straumann, *Gauge theory: Historical origins and some modern developments*, Rev. Mod. Phys. **72** (2000) 1.
12. B. Zumino, *Gravity theories in more than four dimensions*, Phys. Rep. **137** (1986) 109.
13. T. Regge, *On Broken Symmetries and Gravity*, Phys. Rep. **137** (1986) 31.
14. R. Utiyama, *Invariant Theoretical Interpretation of Interaction*, Phys. Rev. **101** (1955) 1597.
15. T. W. B. Kibble, *Lorentz Invariance and the Gravitational Field* J. Math. Phys. **2** (1961) 212.
16. T. Eguchi, P. Gilkey and A. J. Hanson, *Gravitation, Gauge Theories and*

Differential Geometry, Phys. Rept. **66** (1980) 213.
17. M. Goeckeler and T. Schuecker, *Differential Geometry, Gauge Theories and Gravity*, Cambridge University Press (1987).
18. A. Mardones and J. Zanelli, *Lovelock-Cartan theory of gravity*, Class. Quant. Grav. **8** (1991) 1545.
19. D. Lovelock, *The Einstein tensor and its generalizations*, J. Math. Phys. **12** (1971) 498.
20. Jesse Douglas, *Solution of the problem of Plateau*, Trans. Amer. Math. Soc. **33** (1931), 263.
21. C. Lanczos, *A remarkable property of the Riemann-Christoffel tensor in four dimensions*, Ann. Math. **39** (1938) 842.
22. C. Teitelboim and J. Zanelli, *Dimensionally continued topological gravitation theory in hamiltonian form*, Class. and Quantum Grav. **4** (1987) L125; and in *Constraint Theory and Relativistic Dynamics*, G. Longhi and L. Lussana, eds., World Scientific, Singapore (1987).
23. B. Zwiebach, *Curvature squared terms and string theories*, Phys. Lett. **B156** (1985) 315.
24. J. Saavedra, R. Troncoso and J. Zanelli, *Degenerate dynamical systems*, J. Math. Phys **42** (2001) 4383; [hep-th/0011231].
25. M. Hassaïne, R. Troncoso and J. Zanelli, *Poincaré-invariant gravity with local supersymmetry as a gauge theory for the M-algebra*, Phys. Lett. **B 596** (2004) 132; [hep-th/0306258].
26. F. de Micheli and J. Zanelli, *Quantum Degenerate Systems*, J. Math. Phys. **53** (2012) 102112. [arXiv:1203.0022 [hep-th]].
27. H. T. Nieh and M. L. Yan, *An identity in Riemann-Cartan geometry*, J. Math. Phys. **23**, 373 (1982).
28. O. Chandía and J. Zanelli, *Topological invariants, instantons and chiral anomaly on spaces with torsion*, Phys. Rev. **D55**, 7580-7585 (1997), [hep-th/9702025]; *Supersymmetric particle in a spacetime with torsion and the index theorem*, **D58**, 045014 (1998); [hep-th/9803034].
29. G. t Hooft and M. Veltman, *One loop divergences in the theory of gravitation*, Ann. Inst. H. Poincaré (Phys. Theor.) **A20** (1974) 69-94. G. t Hooft, *Quantum Gravity: A Fundamental Problem And Some Radical Ideas*, 1978 Cargese Summer School, and also in NATO Adv. Study Inst. Ser. B Phys.**44**, 323 (1979).
30. J. Zanelli, *Quantization of the gravitational constant in odd dimensions*, Phys. Rev. **D51** (1995) 490; [hep-th/9406202].
31. R. Gilmore, *Lie Groups Lie Algebras and Some of Their Applications*, Wiley, New York (1974).
32. E. Inönü and E.P. Wigner, On the contraction of groups and their representations, Proc. Nat. Acad. Sci. U.S.A. 39, 510-524 (1953).
33. E. Witten, *(2+1)-Dimensional gravity as an exactly soluble system*, Nucl.

Phys. **B311** (1988) 46.

34. Michael Spivak, *Calculus on Manifolds: A Modern Approach to Classical Theorems of Advanced Calculus*, Perseus Pub. Co., Boulder, (1965).
35. A. Achúcarro and P. K. Townsend, *A Chern-Simons action for three-dimensional anti-de Sitter supergravity theories*, Phys. Lett. **B180** (1986) 89.
36. A. H. Chamseddine, *Topological gravity and supergravity in various dimensions*, Nucl. Phys. **B346** (1990) 213.
37. F. Müller-Hoissen, *From Chern-Simons to Gauss-Bonnet*, Nucl. Phys. **B346** (1990) 235.
38. M. Bañados, C. Teitelboim and J. Zanelli, *Lovelock-Born-Infeld theory of gravity*, in J. J. Giambiagi Festschrift, La Plata, May 1990, World Scientific, Singapore (1991).
39. M. Bañados, C. Teitelboim and J. Zanelli, *Dimensionally continued black holes*, Phys. Rev. **D49**, (1994) 975; [gr-qc/9307033].
40. P. Mora, R. Olea, R. Troncoso and J. Zanelli, *Finite action principle for Chern-Simons adS gravity*, JHEP **0406** (2004) 036; [hep-th/0405267].
41. R. Jackiw, *Diverse Topics in Theoretical and Mathematical Physics*, World Scientific, Singapore (1995).
42. R. Troncoso and J. Zanelli, *Higher-dimensional gravity, propagating torsion and adS gauge invariance*, Class. and Quantum Grav. **17** (2000) 4451; [hep-th/9907109].
43. M. Banados, L. J. Garay and M. Henneaux, *The Dynamical structure of higher dimensional Chern-Simons theory*, Nucl. Phys. B **476** (1996) 611. [hep-th/9605159].
44. O. Miskovic, R. Troncoso and J. Zanelli, *Canonical sectors of five-dimensional Chern-Simons theories*, Phys. Lett. B **615** (2005) 277. [hep-th/0504055].
45. C. Teitelboim and J. Zanelli, *Dimensionally continued topological gravitation theory in Hamiltonian form*, Class. Quant. Grav. **4** (1987) L125.
46. M. Henneaux, C. Teitelboim, and J. Zanelli, *Gravity In Higher Dimensions*, in Proceedings of SILARG VI Conference, Rio de Janeiro 1987, M. Novello, editor, World Scientific, Singapore, 1988; *Quantum Mechanics of Multivalued Hamiltonians*, Phys. Rev. **A36** (1987) 4417.
47. J. Saavedra, R. Troncoso and J. Zanelli, Degenerate dynamical systems, J. Math. Phys. **42** (2001) 4383 [hep-th/0011231].
48. C. N. Yang, *Integral formalism for gauge fields* Phys. Rev. Lett. **33** (1974) 445.
49. P. Mora, R. Olea, R. Troncoso and J. Zanelli, *Transgression forms and extensions of Chern-Simons gauge theories*, JHEP **0601** (2006) 067; [hep-th/0601081].
50. J. Zanelli, *Chern-Simons forms and transgression actions, or the universe as a subsystem*, J. Phys. Conf. Ser. **68** (2007) 012002.

51. E. Cremmer, B. Julia and J. Scherk, *Supergravity theory in 11 dimensions*, Phys. Lett. **B76** (1978) 409.
52. J. Schonfeld, *A mass term for three dimensional gauge fields*, Nucl. Phys. **B185** (1981) 157.
53. J. Zanelli, *Uses of Chern-Simons actions*, in "Ten Years of AdS/CFT", AIP Conf. Proc. **1031** (2008) 115; [arXiv:0805.1778/hep-th].
54. J. D. Edelstein and J. Zanelli, *Sources for Chern-Simons theories*, in "Quantum Mechanics of Fundamental SystemsVII", Springer-Verlag, N.Y. (2009); [arXiv:0807.4217/hep-th].
55. O. Mišković and J. Zanelli, *Couplings between Chern-Simons gravities and 2p-branes*, Phys. Rev. D **80** (2009) 044003; [arXiv:0901.0737/hep-th].
56. C. Teitelboim, *Gauge invariance for extended objects*, Phys. Lett. B **167**, 63 (1986). M. Henneaux and C. Teitelboim, *P-Form Electrodynamics*, Found. Phys. **16** (1986) 593.
57. J. Zanelli, *Introductory Lectures on Chern-Simons Theories*, Advanced Summer School in Physics 2011. AIP Conf. Proc. 1420 (2012) 11; doi: 10.1063/1.3678608.
58. J. D. Edelstein, A. Garbarz, O. Mišković and J. Zanelli, *Naked singularities, topological defects and brane couplings*, Int. J. Mod. Phys. **D20** (2011) 839; [arXiv:1009.4418/hep-th].
59. S. Deser, R. Jackiw and G. 't Hooft, *Three-Dimensional Einstein Gravity: Dynamics Of Flat Space*, Annals Phys. **152** (1984) 220. S. Deser and R. Jackiw, *Three-Dimensional Cosmological Gravity: Dynamics Of Constant Curvature*, Annals Phys. **153** (1984) 405; *String sources in (2+1)-dimensional gravity*, Annals Phys. **192** (1989) 352.
60. M. Bañados, C. Teitelboim, J. Zanelli, *The Black hole in three-dimensional space-time*, Phys. Rev. Lett. **69** (1992) 1849; [hep-th/9204099]. M. Bañados, M. Henneaux, C. Teitelboim, J. Zanelli, *Geometry of the (2+1) black hole*, Phys. Rev. **D48** (1993) 1506; [arXiv:gr-qc/9302012].
61. O. Mišković and J. Zanelli, *On the negative spectrum of the 2+1 black hole*, Phys. Rev.**D79** (2009) 105011; [arXiv:0904.0475/hep-th].
62. S. C. Zhang, T. H. Hansson, and S. Kivelson, *An effective field theory model for the fractional quantum Hall effect*, Phys. Rev. Lett. **62** (1989) 82; S. C. Zhang, *The Chern-Simons-Landau-Ginzburg theory of the fractional quantum Hall effect*, Int. J. Mod. Phys. **B6** (1992) 25.
63. A. P. Balachandran, L. Chandar, and B. Sathiapalan, *Chern-Simons duality and the quantum Hall effect*, Int. J. Mod. Phys. **A 11** (1996) 3587.
64. L. Huerta and J. Zanelli, *Introduction to Chern-Simons theories*, PoS **ICFI2010** (2010) 004; *Optical Properties of a θ-Vacuum*, Phys. Rev. **D85** (2012) 085024; [arXiv:1202.2374/hep-th]
65. F. Canfora, L. Rosa and J. Zanelli, *Theta term in a bounded region*, Phys. Rev. **D84** (2011) 105008; [arXiv:1105.2490/hep-th].

66. M. A. H. Vozmediano, M. I. Katsnelson and F. Guinea, *Gauge fields in graphene*, Phys. Rept. **496** (2010) 109.
67. I. Ekeland, *The Broken Dice and Other Mathematical Tales of Chance*, University of Chicago Press, Chicago (1993).

NONCOMMUTATIVE GEOMETRY MODELS FOR PARTICLE PHYSICS

Matilde Marcolli

Department of Mathematics
California Institute of Technology
1200 E California Boulevard, Pasadena, CA 91125, USA
E-mail: matilde@caltech.edu

Keywords: Noncommutative geometry, particle models.

1. Introduction

This is a writeup of part of a series of lectures delivered at the Villa de Leyva summer school "Geometric, Algebraic and Topological Methods for Quantum Field Theory". I am very grateful to Sylvie Paycha for the invitation and to the organizers and the participants for the very friendly and nice atmosphere and the very lively and useful discussions. This paper follows the same format of the lectures and the same very informal style, but we refer the reader, wherever needed, to more detailed references for additional material and a more rigorous treatment. The lectures originally covered both the particle physics applications and some recent applications to cosmology, along the lines of the work.[12,33,35,36,46] This writeup only deals with the particle physics part, while the cosmology part of the lectures will be written up elsewhere.[34]

1.1. *Geometrization of physics*

A very general philosophical standpoint behind the material I am going to present in these lectures stems from the question of whether, and to what extent, geometry can be used to model physics. It is well known that Riemannian gemetry was extremely successful in providing the mathematical formalism necessary to the formulation of General Relativity, which is the classical theory of gravity, and that differential geometry (in particular differential form, de Rham cohomology and Hodge theory) evolved as a mathematical language that was originally very closely tied up with the formulation of Maxwell's equations and the classical theory of electromagnetism.

An interesting combination of these two approaches can be seen in the Kaluza-Klein theory, where electromagnetism is given a geometric formulation in terms of a circle bundle over a spacetime manifold, with the electromagnetic potential described by a bundle connection.

As our understanding of modern physics progressed over the span of the 20th century, more elaborate geometries came to play a role in the formulation of physical models. Most prominent among them is the case of Yang–Mills gauge theories, developed to describe the weak and strong interactions in the setting of particle physics. The classical theory can be described geometrically in terms of bundles (typically $SU(2)$ or $SU(3)$-bundles) over 4-dimensional spacetime manifolds, where the gauge potentials are again described by connections and the fermions they interact with are described by sections of associated vector bundles.

String theory provided another sort of geometric modeling of physics, where 6 extra dimensions, in the form of a Calabi-Yau variety, are added to the extended 4 dimensions of spacetime, and particle types are described in terms of string vibrations on the resulting geometry.

The kind of models we will be discussing here are a further type of geometry, which was studied since the mid '90s as models of gravity coupled to matter. In these models again one enriches a 4-dimensional spacetime with extra dimensions, but these are neither the internal bundle geometry of classical gauge theories, nor the extra dimensions of string theories. In these models, over each point of space time there is a "small" (in a sense that will be made precise later) *noncommutative space*. The type of noncommutative geometry involved determines the type of fields (fermions and bosons) that will be present in the theory. The main idea behind these models is that an action functional for gravity coupled to matter on an ordinary spacetime can be seen as pure gravity on a noncommutative product geometry.

One can ask the question of why do we want or need such a geometrization of physics: after all one can explicitly write the Lagrangian of the Standard Model (SM) of elementary particle physics without resorting to the use of noncommutative geometry, so what does one gain by this mathematical abstraction? One main observation is that, explicit as it may be, the full SM Lagrangian is extremely complicated (it is written out in full in[24] for whoever wants to admire it in its entirety). It is one of the fundamental principles of both mathematics and theoretical physics that complicated phenomena should *follow* from simple laws and simple principles. An advantage of the noncommutative geometry approach I will be describing in these lectures is that the Lagrangian is no longer seen as the primary object of the physical theory but it is completely *computed* (using the asymptotic expansion of the spectral action) from simpler and more basic data. Another reason to study the noncommutative geometry models of gravity coupled to matter is that the coupling to gravity is interesting and non-trivial: it can be seen as a sort of *modified gravity* model that allows for predictions that are not exactly the same that one would obtain from the usual SM of elementary particles minimally coupled to Einstein–Hilbert gravity. We will explain some of the possible interesting differences in what follows. For cosmological implications we refer the reader to the upcoming survey,[34] or to a shorter review.[32]

1.2. *Noncommutative geometry models*

The main features of the type of noncommutative geometry models we will be discussing are summarized as follows.

The model contains the usual Einstein–Hilbert action for gravity,

$$S_{EH}(g_{\mu\nu}) = \frac{1}{16\pi G}\int_M R\sqrt{g}\, d^4x,$$

which gives General Relativity as we know it.

Gravity is coupled to matter in the sense that the action functional computed from the asymptotic expansion of the spectral action has terms of the form $S = S_{EH} + S_{SM}$ with S_{SM} the action functional for a particle physics model, which includes the complete *Standard Model Lagrangian.*

The particle physics content is an extension of the Minimal Standard Model (MSM) by the addition of right handed neutrinos with Majorana masses.

The gravity sector includes a *modified gravity model* with additional terms $f(R, R^{\mu\nu}, C_{\lambda\mu\nu\kappa})$ that include conformal gravity. This means that there is a term in the action functional involving the Weyl curvature tensor

$$C_{\lambda\mu\nu\kappa} = R_{\lambda\mu\nu\kappa} - \frac{1}{2}(g_{\lambda\nu}R_{\mu\kappa} - g_{\lambda\kappa}R_{\mu\nu} - g_{\mu\nu}R_{\lambda\kappa} + g_{\mu\kappa}R_{\lambda\nu})$$

$$+\frac{1}{6}R^{\alpha}_{\alpha}(g_{\lambda\nu}g_{\mu\kappa} - g_{\lambda\kappa}g_{\mu\nu}).$$

The coupling of gravity to matter also includes a *non-minimal coupling* of the Higgs field to gravity of the form

$$\int_M \left(\frac{1}{2}|D_\mu \mathbf{H}|^2 - \mu_0^2|\mathbf{H}|^2 - \xi_0\, R\, |\mathbf{H}|^2 + \lambda_0|\mathbf{H}|^4\right) \sqrt{g}\, d^4x.$$

On the good side, these models present a very natural geometric setup for the coupling of gravity to matter, where "all forces become gravity" (on a noncommutative space); on the bad side, there isn't a well developed quantum field theory for the spectral action, hence there are serious difficulties in trying to implement these models as a true quantum field theory (as is to be expected in the presence of gravity).

2. Spectral triples and particle physics

We recall in this section some basic facts about noncommutative geometry and how these are employed in the construction of the particle physics content of the model.

It is well known that, by the Gelfand–Naimark theorem, assigning a compact Hausdorff topological space X is equivalent to assigning its algebra of continuous function $C(X)$, which is an abelian unital C^*-algebra. Out of this correspondence comes the idea that, if one considers more general types of C^*-algebras, that are not necessarily commutative, one can still view them as being algebras of "continuous functions on a space", except that, this time, the space does not exist as an ordinary space, but only through its algebra of functions. Such "spaces" are called *noncommutative spaces*, see[19] for an in depth view of this whole philosophy. What this really means, of course, is that one uses the commutative case of ordinary spaces as a guideline to reformulate as much as possible the geometric properties of the space in terms of algebraic notions that only require knowledge of the algebra of functions. Thus reformulated, these in turn give a good way of defining analogs of the classical notions of geometry for noncommutative spaces, that is, after dropping the commutativity hypothesis on the algebra. This way of describing geometry in terms of the algebra $C(X)$, and a dense subalgebras like $C^\infty(X)$ if X is a smooth manifold, leads to good algebraic notions of differential forms, vector bundles, connections, cohomology: all of these then continue to make sense without assuming commutativity, hence they define the same geometric notions for noncommutative geometry, see.[19]

One immediate advantage of this generalization is that noncommutative geometry provides a very efficient way of describing "bad quotients" (of equivalence relations, such as group actions) as if they were still good spaces (manifolds). The main idea here consists of replacing the usual description of functions on the quotients as functions invariant under the equivalence relation with functions defined on the graph of the equivalence relation with the commutative pointwise product replaced by a noncommutative convolution product related to the associativity property of the equivalence relation. We will not explore this particular aspect of noncommutative geometry in these notes, but the interested reader can consult,[19,24]

Our main interest here is models of gravity coupled to matter, and to this purpose one needs to deal with a particular aspect of geometry, namely *Riemannian geometry*, which one knows is the basic mathematical theory underlying the classical theory of gravity, general relativity.

2.1. *Spectral triples*

The formalism in noncommutative geometry that best encodes a generalization of Riemannian geometry is the theory of *spectral triples*,[20] The basic data of a spectral triple $(\mathcal{A}, \mathcal{H}, D)$ consist of the following:

- an involutive algebra $\mathcal{A}$;
- a representation $\pi : \mathcal{A} \to \mathcal{L}(\mathcal{H})$ in the algebra $\mathcal{L}(\mathcal{H})$ of bounded linear operators on a separable Hilbert space;
- a self adjoint operator $D = D^*$ acting on $\mathcal{H}$ with a dense domain;
- the compact resolvent condition that $(1+D^2)^{-1/2}$ belongs to the algebra $\mathcal{K}$ of compact operators;
- the condition that commutators $[a, D]$ of elements in the algebra with the Dirac operator are bounded for all $a \in \mathcal{A}$.
- The spectral triples is *even* if there is also a $\mathbb{Z}/2$- grading γ on $\mathcal{H}$ such that

$$[\gamma, a] = 0, \ \forall a \in \mathcal{A}, \qquad \text{and} \qquad D\gamma = -\gamma D.$$

The main example of a classical space that gives rise to data of a spectral triple is a compact smooth spin Riemannian manifold M, for which the canonical spectral triple has the form $(C^\infty(M), L^2(M, S), \partial\!\!\!/_M)$, with S the spinor bundle and $L^2(M, S)$ the Hilbert space of square integrable sections, on which the algebra of smooth functions $C^\infty(M)$ acts as multiplication operators. The Dirac operator is in this case the usual Dirac operator $D = \partial\!\!\!/_M$. In the 4-dimensional case we will be mostly interested in here, the even structure is given by the chirality operator γ_5, the product of the gamma functions. Even in the commutative case there are several interesting generalizations of this basic case, see.[22,31,49]

One sees already from this formulation one evident difficulty in constructing physical models using the formalism of spectral triples, namely the fact that one is forced to work with 4-manifolds that are compact and Riemannian. The compactness property can be circumvented more easily using a "local" version of the notion of spectral triples (for simplicity we will not get into it in these notes). However, the Riemannian property is more difficult to get rid of, as there does not exist at present a fully satisfactory theory of Lorentzian spectral triples. Thus, this will lead to working with Euclidean gravity, and only after the computation based on the spectral action have been carried out, one can check which of the obtained results still makes sense after a Wick rotation to Minkowskian signature. One hopes that a more satisfactory version of the theory will be available in the future, but for the time being one should get the best possible results out of the existing theory.

2.2. *Real structures*

An additional property of spectral triples that plays an important role in the particle physics models is the notion of a *real structure*. This depends on a further notion of dimension, the *KO-dimension*, which is defined as $n \in \mathbb{Z}/8\mathbb{Z}$, an integer modulo 8.

A real structure on a spectral triple consists of an antilinear isometry $J : \mathcal{H} \to \mathcal{H}$ satisfying

$$J^2 = \varepsilon, \qquad JD = \varepsilon' DJ, \text{ and } J\gamma = \varepsilon'' \gamma J,$$

where $\epsilon, \epsilon', \epsilon'' \in \{\pm 1\}$ are signed satisfying one of the possibilities listed in the table below (which correspond to the different possible KO-dimensions:

n	0	1	2	3	4	5	6	7
ε	1	1	-1	-1	-1	-1	1	1
ε'	1	-1	1	1	1	-1	1	1
ε''	1		-1		1		-1	

The antilinear involution J is also required to satisfy the commutation relation $[a, b^0] = 0$, for all $a, b \in \mathcal{A}$, where the notation b^0 stands for $b^0 = Jb^* J^{-1}$ for $b \in \mathcal{A}$, and the *order one condition*:

$$[[D, a], b^0] = 0 \qquad \forall\, a, b \in \mathcal{A}.$$

The table of signs listed above is modeled on the case of ordinary manifolds, where it reflects the properties of the mod eight periodicity of KO-homology and real K-theory. However, imposing that real structures on noncommutative spectral triples have signs $\epsilon, \epsilon', \epsilon''$ restricted to the cases in the table creates problems with having a good category of spectral triples with products. This was observed in Dabrowsky–Dossena.[25] A more detailed study of the notion of real structure that identifies a good generalization of the data listed above, and which solves this problem, was recently developed by Ćaćić in.[11]

2.3. *Finite real spectral triples*

Noncommutative spectral triples can be very far from ordinary manifolds: they include quantum groups, fractals, noncommutative tori, and other such unusual spaces that one cannot identify in any way with ordinary smooth geometries. However, the models of direct interest to particle physics and cosmology only use noncommutative geometry in a very limited form, which is, in a suitable sense, close enough to an ordinary smooth manifolds. Namely, for models matter coupled to gravity one considers noncommutative spaces that are (locally) a product $M \times F$ of a 4-dimensional spacetime manifold M by a "finite noncommutative space" F, which represents the extra dimensions of the model.

This leads to an interesting class of spaces, called *almost commutative geometries*. In their more general form, they consist of fibrations (not necessarily products) over a manifold, with fiber a finite noncommutative space. They have been studied in this general form in.[5,9]

A *finite noncommutative space* means a spectral triple $F = (\mathcal{A}_F, \mathcal{H}_F, D_F)$ where A_F is a *finite dimensional* algebra. This means that, in a sense, F should be thought of as a zero-dimensional noncommutative space, although we will see that, due to the existence of different notions of dimension in noncommutative geometry, these can for instance have positive KO-dimension.

The data of $F = (\mathcal{A}_F, \mathcal{H}_F, D_F)$ are given by

- a finite dimensional (real) C^*-algebra $\mathcal{A}$, which by Wedderburn theorem can always be written as

$$\mathcal{A} = \oplus_{i=1}^{N} M_{n_i}(\mathbb{K}_i)$$

  where $\mathbb{K}_i = \mathbb{R}$ or $\mathbb{C}$ or the quaternions $\mathbb{H}$;
- a representation on a finite dimensional Hilbert space $\mathcal{H}_F$;
- an antilinear involution J on $\mathcal{H}_F$ with signs $\epsilon, \epsilon', \epsilon''$ (see[11] for details on the real structure);

- a bimodule structure given by the condition $[a, b^0] = 0$ for the real structure J;
- a self adjoint liner operator $D^* = D$ satisfying the order one condition $[[D, a], b^0] = 0$.

Since the algebra and Hilbert space are finite dimensional the bounded commutator condition is automatically satisfied, and all the data reduce to linear algebra conditions.

One can consider the *moduli spaces* of all Dirac operators D up to unitary equivalence on a given $(\mathcal{A}_F, \mathcal{H}_F)$. These moduli spaces will provide a geometric interpretation for the physical parameters of particle physics models, such as masses and mixing angles. For a general study of moduli spaces of Dirac operators on finite noncommutative geometries see.[10]

2.4. *Building a particle physics model*

We will give here one explicit example of particle physics model constructed with the spectral triples method, taken from.[17] Earlier constructions that played an important role in the development of the field and are still in use for some purposes include the Connes–Lott model,[23] the treatment of the Minimal Standard Model in.[21] A useful reference is also the treatment of the low energy regime in.[37] More recently, very interesting models that include supersymmetry were developed by Thijs van den Broek, Walter D. van Suijlekom,,[78] The model of[17] gives rise to an extension of the Minimal Standard Model with right handed neutrinos with Majorana mass terms. Supersymmetric extensions of the Standard Model have been obtained in,[7] a model that only captures the electrodynamics is constructed in,[28] while the Minimal Standard Model was considered in.[21] For the model of[17] the minimal input is given by choosing an *ansatz* for the algebra, in the form of the *left-right symmetric algebra*

$$\mathcal{A}_{LR} = \mathbb{C} \oplus \mathbb{H}_L \oplus \mathbb{H}_R \oplus M_3(\mathbb{C}).$$

This has an involution $(\lambda, q_L, q_R, m) \mapsto (\bar{\lambda}, \bar{q}_L, \bar{q}_R, m^*)$ and two subalgebras: $\mathbb{C} \oplus M_3(\mathbb{C})$, regarded as the integer spin $\mathbb{C}$-algebra, and $\mathbb{H}_L \oplus \mathbb{H}_R$, the half-integer spin $\mathbb{R}$-algebra. More general choices of an initial ansatz leading to the same model were considered in.[14]

An important conceptual point in these constructions is the difference between working with associative algebras rather than Lie algebras, more familiar to particle physicists. The reason why associative algebras are better for model building than Lie algebras is that there are much more restrictive constraints on the representation theory of associative algebras. More precisely, in the classical geometry of gauge theories one considers a bundle over spacetime, with connections and sections, and with automorphisms gauge group. This naturally leads to Lie groups and their representations. This approach has been extremely successful, in the history of theoretical particle physics, in explaining the decomposition of composite particles into elementary particles in terms Lie group representations and their decomposition into irreducible representations. This is the mathematical basis for the theory of hadrons and quarks. However, if one is only interested in identifying the elementary particles themselves, then associative algebras have very few representations, which leads to a very constrained choice that suffices to pin down directly the elementary particles in the model. One still obtains the gauge groups, in the form of inner automorphisms of the algebra.

Given a bimodule $\mathcal{M}$ over $\mathcal{A}$, and a unitary $u \in \mathcal{U}(\mathcal{A})$, one has the adjoint action

$$\mathrm{Ad}(u)\xi = u\xi u^* \quad \forall \xi \in \mathcal{M}.$$

For the left-right symmetric algebra, one can then say that a bimodule is odd of $s = (1,-1,-1,1)$ acts by $Ad(s) = -1$. These bimodules can equivalently be described as representations of the complex algebra $\mathcal{B} = (\mathcal{A}_{LR} \otimes_{\mathbb{R}} \mathcal{A}_{LR}^{op})_p$, where $p = \frac{1}{2}(1 - s \otimes s^0)$, with $\mathcal{A}^0 = \mathcal{A}^{op}$, see.[17,24] This algebra is

$$\mathcal{B} = \oplus^{4-times} M_2(\mathbb{C}) \oplus M_6(\mathbb{C}).$$

The contragredient bimodule of $\mathcal{M}$ is then defined as

$$\mathcal{M}^0 = \{\bar{\xi}\,;\, \xi \in \mathcal{M}\}, \quad a\,\bar{\xi}\,b = \overline{b^* \xi\, a^*}$$

and one can consider the bimodule $\mathcal{M}_F$ which is the sum of all the inequivalent irreducible odd $\mathcal{A}_{LR}$-bimodules. This has the following properties: as a vector space $\dim_{\mathbb{C}} \mathcal{M}_F = 32$; it decomposes as $\mathcal{M}_F = \mathcal{E} \oplus \mathcal{E}^0$, with

$$\mathcal{E} = \mathbf{2}_L \otimes \mathbf{1}^0 \oplus \mathbf{2}_R \otimes \mathbf{1}^0 \oplus \mathbf{2}_L \otimes \mathbf{3}^0 \oplus \mathbf{2}_R \otimes \mathbf{3}^0,$$

where $\mathbf{1}$, $\mathbf{2}$, $\mathbf{3}$ denote the standard representations of $\mathbb{C}$, $\mathbb{H}$ and $M_3(\mathbb{C})$ of dimensions one, two, and three, respectively. This is also a sum of irreducible representations of $\mathcal{B}$,

$$\mathbf{2}_L \otimes \mathbf{1}^0 \oplus \mathbf{2}_R \otimes \mathbf{1}^0 \oplus \mathbf{2}_L \otimes \mathbf{3}^0 \oplus \mathbf{2}_R \otimes \mathbf{3}^0$$

$$\oplus \mathbf{1} \otimes \mathbf{2}_L^0 \oplus \mathbf{1} \otimes \mathbf{2}_R^0 \oplus \mathbf{3} \otimes \mathbf{2}_L^0 \oplus \mathbf{3} \otimes \mathbf{2}_R^0$$

There is an antilinear isomorphism $\mathcal{M}_F \cong \mathcal{M}_F^0$ given by $J_F(\xi, \bar{\eta}) = (\eta, \bar{\xi})$ for $\xi\,,\, \eta \in \mathcal{E}$

$$J_F^2 = 1\,, \quad \xi\, b = J_F b^* J_F\, \xi \qquad \xi \in \mathcal{M}_F\,,\; b \in \mathcal{A}_{LR}.$$

There is a $\mathbb{Z}/2\mathbb{Z}$-grading $\gamma_F = c - J_F\, c\, J_F$ with $c = (0, 1, -1, 0) \in \mathcal{A}_{LR}$, satisfying

$$J_F^2 = 1\,, \quad J_F\, \gamma_F = -\,\gamma_F\, J_F.$$

These properties of the grading show that the KO-dimension is equal to 6 mod 8, by the table of signs for real structures.

A basis for $\mathcal{M}_F$ has a natural interpretation as the list of elementary fermions for the model. We will see later that this identification is justified by the fact that it gives the correct quantum numbers (hypercharges) to the particles. One fixes a notation for a basis $|\uparrow\rangle$ and $|\downarrow\rangle$ of $\mathbf{2}$ determined by

$$q(\lambda) = \begin{pmatrix} \lambda & 0 \\ 0 & \bar{\lambda} \end{pmatrix} \qquad q(\lambda)|\uparrow\rangle = \lambda|\uparrow\rangle, \qquad q(\lambda)|\downarrow\rangle = \bar{\lambda}|\downarrow\rangle.$$

One then uses the following identification of the particles:

- the subspace $\mathbf{2}_L \otimes \mathbf{1}^0$: gives the neutrinos $\nu_L \in |\uparrow\rangle_L \otimes \mathbf{1}^0$ and the charged leptons $e_L \in |\downarrow\rangle_L \otimes \mathbf{1}^0$;
- the subspace $\mathbf{2}_R \otimes \mathbf{1}^0$ gives the right-handed neutrinos $\nu_R \in |\uparrow\rangle_R \otimes \mathbf{1}^0$ and the charged leptons $e_R \in |\downarrow\rangle_R \otimes \mathbf{1}^0$;
- the subspace $\mathbf{2}_L \otimes \mathbf{3}^0$, where $\mathbf{3}$ represents the color indices gives the u/c/t quarks $u_L \in |\uparrow\rangle_L \otimes \mathbf{3}^0$ and the d/s/b quarks $d_L \in |\downarrow\rangle_L \otimes \mathbf{3}^0$;
- the subspace $\mathbf{2}_R \otimes \mathbf{3}^0$ (with color indices) gives the u/c/t quarks $u_R \in |\uparrow\rangle_R \otimes \mathbf{3}^0$ and the d/s/b quarks $d_R \in |\downarrow\rangle_R \otimes \mathbf{3}^0$;
- the subspace $\mathbf{1} \otimes \mathbf{2}^0_{L,R}$ gives the antineutrinos $\bar{\nu}_{L,R} \in \mathbf{1} \otimes \uparrow\rangle^0_{L,R}$ and the charged antileptons $\bar{e}_{L,R} \in \mathbf{1} \otimes |\downarrow\rangle^0_{L,R}$;
- the subspace $\mathbf{3} \otimes \mathbf{2}^0_{L,R}$ (with color indices) gives the antiquarks $\bar{u}_{L,R} \in \mathbf{3} \otimes |\uparrow\rangle^0_{L,R}$ and $\bar{d}_{L,R} \in \mathbf{3} \otimes |\downarrow\rangle^0_{L,R}$.

2.5. *Subalgebras and the order one condition*

One needs to input by hand in the model the number of particle generations $N = 3$. This means taking as the Hilbert space of the finite geometry $\mathcal{H}_F = \mathcal{M}_F \oplus \mathcal{M}_F \oplus \mathcal{M}_F$. The left action of $\mathcal{A}_{LR}$ is a sum of representations $\pi|_{\mathcal{H}_f} \oplus \pi'|_{\mathcal{H}_{\bar{f}}}$ with $\mathcal{H}_f = \mathcal{E} \oplus \mathcal{E} \oplus \mathcal{E}$ and $\mathcal{H}_{\bar{f}} = \mathcal{E}^0 \oplus \mathcal{E}^0 \oplus \mathcal{E}^0$ and with no equivalent subrepresentations. This implies that any linear map D that mixes $\mathcal{H}_f$ and $\mathcal{H}_{\bar{f}}$ would not satisfy the order one condition for $\mathcal{A}_{LR}$. Thus, the question of finding a finite spectral triple $(\mathcal{A}_F, \mathcal{H}_F, D_F)$ can be phrased as a coupled problem for the pair $(\mathcal{A}_F, D_F)$, for fixed $\mathcal{H}_F$ as above, where $\mathcal{A}_F \subset \mathcal{A}_{LR}$ is a subalgebra and D_F has off diagonal terms. The algebra for the finite spectral triple will be the maximal $\mathcal{A}_F$ (up to isomorphism) that admits a D_F with order one condition. This gives the solution

$$\mathcal{A}_F = \{(\lambda, q_L, \lambda, m) \mid \lambda \in \mathbb{C}\,,\ q_L \in \mathbb{H}\,,\ m \in M_3(\mathbb{C})\}$$

$$\sim \mathbb{C} \oplus \mathbb{H} \oplus M_3(\mathbb{C}),$$

unique up to $\mathrm{Aut}(\mathcal{A}_{LR})$, which represents a spontaneous breaking of the left-right symmetry of $\mathcal{A}_{LR}$ imposed by the order one condition of the Dirac operator.

2.6. *Symmetries*

Up to a finite abelian group one has symmetry groups (inner automorphisms with trivial determinant in the given representation) given by

$$\mathrm{SU}(\mathcal{A}_F) \sim \mathrm{U}(1) \times \mathrm{SU}(2) \times \mathrm{SU}(3),$$

as expected for the Standard Model. The adjoint action of U(1), expressed in powers of $\lambda \in \mathrm{U}(1)$ gives

	$\uparrow \otimes \mathbf{1}^0$	$\downarrow \otimes \mathbf{1}^0$	$\uparrow \otimes \mathbf{3}^0$	$\downarrow \otimes \mathbf{3}^0$
$\mathbf{2}_L$	-1	-1	$\frac{1}{3}$	$\frac{1}{3}$
$\mathbf{2}_R$	0	-2	$\frac{4}{3}$	$-\frac{2}{3}$

These numbers are in fact the correct hypercharges of fermions and this confirms the identification of the chosen basis of $\mathcal{H}_F$ with the fermions of the particle model.

2.7. *Classifying Dirac operators*

As mentioned above, one can classify up to unitary equivalence all the possible Dirac operators for $(\mathcal{A}_F, \mathcal{H}_F, \gamma_F, J_F)$, namely all the self-adjoint linear maps D_F on $\mathcal{H}_F$, commuting with J_F, anticommuting with γ_F and satisfying $[[D, a], b^0] = 0$, for all $a, b \in \mathcal{A}_F$. Here one also imposes an additional physical condition on the solution, which is that the photon remains massless. This is expressed by a commutativity condition of D_F with the subalgebra

$$\mathbb{C}_F \subset \mathcal{A}_F\,, \quad \mathbb{C}_F = \{(\lambda, \lambda, 0)\,, \lambda \in \mathbb{C}\}.$$

Then one finds that the possible Dirac operators have to be of the form

$$D(Y) = \begin{pmatrix} S & T^* \\ T & \bar{S} \end{pmatrix} \quad \text{with} \quad S = S_1 \oplus (S_3 \otimes 1_3),$$

where

$$S_1 = \begin{pmatrix} 0 & 0 & Y^*_{(\uparrow 1)} & 0 \\ 0 & 0 & 0 & Y^*_{(\downarrow 1)} \\ Y_{(\uparrow 1)} & 0 & 0 & 0 \\ 0 & Y_{(\downarrow 1)} & 0 & 0 \end{pmatrix}$$

same for S_3, with

$$S_3 = \begin{pmatrix} 0 & 0 & Y^*_{(\uparrow 3)} & 0 \\ 0 & 0 & 0 & Y^*_{(\downarrow 3)} \\ Y_{(\uparrow 3)} & 0 & 0 & 0 \\ 0 & Y_{(\downarrow 3)} & 0 & 0 \end{pmatrix}$$

$Y_{(\downarrow 1)}, Y_{(\uparrow 1)}, Y_{(\downarrow 3)}, Y_{(\uparrow 3)} \in \mathrm{GL}_3(\mathbb{C})$ and Y_R symmetric:

$$T : E_R = \uparrow_R \otimes \mathbf{1}^0 \to J_F\, E_R.$$

The unitary equivalence classes of such operators are then given by the moduli space $\mathcal{C}_3 \times \mathcal{C}_1$, where $\mathcal{C}_3$ is the set of pairs $(Y_{(\downarrow 3)}, Y_{(\uparrow 3)})$ modulo the equivalence

$$Y'_{(\downarrow 3)} = W_1\, Y_{(\downarrow 3)}\, W_3^*, \qquad Y'_{(\uparrow 3)} = W_2\, Y_{(\uparrow 3)}\, W_3^*$$

where the W_j are unitary matrices, so that

$$\mathcal{C}_3 = (K \times K)\backslash(G \times G)/K,$$

with $G = \mathrm{GL}_3(\mathbb{C})$ and $K = U(3)$. The real dimension is $\dim_{\mathbb{R}} \mathcal{C}_3 = 10 = 3 + 3 + 4$. These correspond to $3 + 3$ eigenvalues, 3 angles and 1 phase. Similarly, $\mathcal{C}_1$ is the set of triplets $(Y_{(\downarrow 1)}, Y_{(\uparrow 1)}, Y_R)$ with Y_R a symmetric matrix, modulo the equivalence

$$Y'_{(\downarrow 1)} = V_1\, Y_{(\downarrow 1)} V_3^*, \quad Y'_{(\uparrow 1)} = V_2\, Y_{(\uparrow 1)}\, V_3^*, \quad Y'_R = V_2\, Y_R\, \bar{V}_2^*,$$

by unitary matrices. There is a projection $\pi : \mathcal{C}_1 \to \mathcal{C}_3$ that forgets Y_R, so the fiber is given by symmetric matrices modulo $Y_R \mapsto \lambda^2 Y_R$ and the real dimension is therefore $\dim_{\mathbb{R}}(\mathcal{C}_3 \times \mathcal{C}_1) = 31$, with the dimension of the fiber $12 - 1 = 11$.

The coordinates on this moduli space have an important physical interpretation: they give the Yukawa parameters and Majorana masses. In fact, representatives in $\mathcal{C}_3 \times \mathcal{C}_1$ can be written in the form

$$Y_{(\uparrow 3)} = \delta_{(\uparrow 3)} \qquad Y_{(\downarrow 3)} = U_{CKM}\, \delta_{(\downarrow 3)}\, U^*_{CKM}$$

$$Y_{(\uparrow 1)} = U^*_{PMNS}\, \delta_{(\uparrow 1)}\, U_{PMNS} \qquad Y_{(\downarrow 1)} = \delta_{(\downarrow 1)}$$

with $\delta_\uparrow$, $\delta_\downarrow$ diagonal matrices specifying the Dirac masses and

$$U = \begin{pmatrix} c_1 & -s_1 c_3 & -s_1 s_3 \\ s_1 c_2 & c_1 c_2 c_3 - s_2 s_3 e_\delta & c_1 c_2 s_3 + s_2 c_3 e_\delta \\ s_1 s_2 & c_1 s_2 c_3 + c_2 s_3 e_\delta & c_1 s_2 s_3 - c_2 c_3 e_\delta \end{pmatrix}$$

specifying the angles and phase, $c_i = \cos\theta_i$, $s_i = \sin\theta_i$, $e_\delta = \exp(i\delta)$. For the quark and lepton sector ($\mathcal{C}_3$ and $\mathcal{C}_1$, respectively) these are the Cabibbo–Kobayashi–Maskawa mixing matrix U_{CKM} and the Pontecorvo–Maki–Nakagawa–Sakata mixing matrix U_{PMNS}. In particular, because of the presence in the model of the latter matrix, this model can account for the experimental phenomenon of neutrino mixing. The matrix Y_R contains the Majorana mass terms for the right-handed neutrinos. The CKM matrix is subject to very strict experimental constraints, such as the unitarity triangle, determined by the condition that the off diagonal elements of V^*V add up to 0. There are experimental constraints also on the neutrino mixing matrix.

Summarizing this geometric point of view: the data of the CKM and PMNS matrices are coordinates on the moduli space of Dirac operators of the finite geometry; experimental constraints on these parameters define subvarieties in the moduli space; the moduli spaces are described by symmetric spaces given by double quotients of Lie groups $(K \times K)\backslash(G \times G)/K$. These are interesting geometries extensively studied for their mathematical properties. Thus, one expects to get interesting parameter relations from "interesting geometric subvarieties".

Moreover, the matter content of the model obtained in this way agrees with what is known as the νMSM: an extension of the Minimal Standard Model with additional right handed neutrinos with Majorana mass terms. The free parameters in the model are:

- 3 coupling constants
- 6 quark masses, 3 mixing angles, 1 complex phase
- 3 charged lepton masses, 3 lepton mixing angles, 1 complex phase
- 3 neutrino masses

- 11 Majorana mass matrix parameters
- 1 QCD vacuum angle

The moduli space of Dirac operators on the finite geometry F, as we have seen, accounts for all the masses, mixing angles, phases, and Majorana mass terms. Among the remaining parameters, the coupling constants will appear in the product geometry, when one considers the spectral action functional, while the vacuum angle is not there in the model (though this may reflect the fact that there is still no completely satisfactory way to accommodate quantum corrections).

3. Spectral action and Lagrangians

We have described above the finite geometry that determines the particle content of the model and the Yukawa parameters. We now introduce the action functional and the coupling of gravity to matter through the *spectral action* of.[13]

3.1. *Product geometries and almost commutative geometries*

While the finite geometry F is pure matter, the coupling to gravity happens on the product $M \times F$ with a 4-dimensional spacetime manifold M. It is important to keep in mind here that the setting of spectral triples, at present, works very well for the case of Riemannian manifolds (Euclidean signature), but the encoding of geometry via spectral triples and the spectral action functional do not extend well to Lorentzian manifolds (Minkowskian signature), though some partial results in that direction exist, based on the theory of Krein spaces with indefinite bilinear forms replacing the positive definite inner products of Hilbert spaces. Thus, for the purposes of these models of matter coupled to gravity one considers *Euclidean gravity* with compact Riemannian manifolds M.

Given two spectral triples $(\mathcal{A}_i, \mathcal{H}_i, D_i, \gamma_i, J_i)$ of KO-dim 4 and 6, their product is the spectral triple with data

$$\mathcal{A} = \mathcal{A}_1 \otimes \mathcal{A}_2 \qquad \mathcal{H} = \mathcal{H}_1 \otimes \mathcal{H}_2$$

$$D = D_1 \otimes 1 + \gamma_1 \otimes D_2$$

$$\gamma = \gamma_1 \otimes \gamma_2 \qquad J = J_1 \otimes J_2.$$

In the case of a 4-dimensional compact spin manifold M and the finite noncommutative geometry F described above, one has

$$\mathcal{A} = C^\infty(M) \otimes \mathcal{A}_F = C^\infty(M, \mathcal{A}_F)$$

$$\mathcal{H} = L^2(M, S) \otimes \mathcal{H}_F = L^2(M, S \otimes \mathcal{H}_F)$$

$$D = \partial\!\!\!/_M \otimes 1 + \gamma_5 \otimes D_F$$

where D_F is chosen in the moduli space described above.

There are different notions of dimension for a spectral triple $(\mathcal{A}, \mathcal{H}, D)$:

- the *metric dimension* is determined by the growth of the eigenvalues of the Dirac operator;
- the *KO-dimension* (an integer modulo 8) is determined by the signs of the commutation relations of J, γ, and D;
- The *dimension spectrum* is the set of poles of the zeta functions $\zeta_{a,D}(s) = \mathrm{Tr}(a|D|^{-s})$ associated to the spectral triple.

In the case of ordinary manifolds the first two notions agree (the KO-dimension is the metric dimension modulo eight) and the dimension spectrum contains the usual dimension as one of its points. However, for more general noncommutative spaces there is no reason why these different notions of dimension would continue to be so related. The dimension spectrum, as a subset of $\mathbb{C}$, can contain non-integer as well as non-real points; the KO-dimension is in general not equal to the metric dimension modulo 8, etc. In our setting $X = M \times F$ is metrically four dimensional; its KO-dimension is $10 = 4 + 6$ (equal 2 mod 8) and the dimension spectrum is given by the set $\{k \in \mathbb{Z}_{\geq 0},\ k \leq 4\}$.

The more general setting one should consider here is the *almost commutative geometries* of[5] and,[9] of the form

$$(C^\infty(M, \mathcal{E}), L^2(M, \mathcal{E} \otimes S), \mathcal{D}_\mathcal{E}),$$

where

- M is a smooth manifold, $\mathcal{E}$ is an algebra bundle with fiber $\mathcal{E}_x$ a finite dimensional algebra $\mathcal{A}_F$
- $C^\infty(M, \mathcal{E})$ are the smooth sections of the algebra bundle $\mathcal{E}$
- The Dirac operator is of the form $\mathcal{D}_\mathcal{E} = c \circ (\nabla^\mathcal{E} \otimes 1 + 1 \otimes \nabla^S)$ with a spin connection ∇^S and a hermitian connection on the bundle
- The triple has a compatible grading and real structure (in the more general sense discussed in[11]).

An equivalent intrinsic (abstract) characterization of almost commutative geometry is given in.[9] For simplicity, in the following we assume that the almost-commutative geometry is a product $M \times F$ as above.

3.2. *Inner fluctuations and gauge fields*

While the fermions of the particle physics model are determined by the representation $\mathcal{H}_F$ of the finite geometry, the bosons are determined by inner fluctuations of the Dirac operator on the product geometry. The setup is as follows:

- one has a right $\mathcal{A}$-module structure on $\mathcal{H}$

$$\xi\, b = b^0\, \xi, \qquad \xi \in \mathcal{H}\,, \quad b \in \mathcal{A}$$

- The unitary group acts in the adjoint representation as

$$\xi \in \mathcal{H} \to \mathrm{Ad}(u)\, \xi = u\, \xi\, u^* \qquad \xi \in \mathcal{H}$$

The *inner fluctuations* are then given by

$$D \to D_A = D + A + \varepsilon'\, J\, A\, J^{-1}$$

with $A = A^*$ a self-adjoint operator of the form

$$A = \sum a_j [D, b_j]\,, \quad a_j, b_j \in \mathcal{A}.$$

Notice that this is not an equivalence relation because, for a finite geometry, the Dirac operator D can fluctuate to zero, but the zero operator cannot be further deformed by inner fluctuations (which are themselves all zero in that case). Since inner fluctuations can be seen as "self Morita equivalences" of a spectral triple, this shows that there are problems in extending the notion of Morita equivalence from algebras to spectral triples in such as way as to still have an equivalence relation. This issue was discussed more in depth in the work of D. Zhang,[49] which proposes a different form of Morita equivalence that restores the symmetric property, and in the work of Venselaar,[47] which shows that the usual notion is an equivalence relation in the case of noncommutative tori.

3.3. *Gauge bosons and Higgs boson*

Given an algebra $\mathcal{A}$, its unitary group is given by $\mathrm{U}(\mathcal{A}) = \{u \in \mathcal{A} \mid uu^* = u^*u = 1\}$. In the case of a finite dimensional algebra $\mathcal{A}_F$ represented on a finite dimensional Hilbert space $\mathcal{H}_F$, one also has a notion of special unitary group

$$\mathrm{SU}(\mathcal{A}_F) = \{u \in \mathrm{U}(\mathcal{A}_F) \,|\, \det(\mathrm{u}) = 1\},$$

where the determinant is taken in the representation of u acting on $\mathcal{H}_F$. In the case of the finite geometry of[17] described above, up to a finite abelian group one has

$$\mathrm{SU}(\mathcal{A}_F) \sim \mathrm{U}(1) \times \mathrm{SU}(2) \times \mathrm{SU}(3),$$

which is the correct gauge group of the standard model, or more precisely, the unimodular subgroup of $\mathcal{U}(\mathcal{A})$ in the adjoint representation $\mathrm{Ad}(u)$ on $\mathcal{H}$ is the gauge group of the standard model. The unimodular inner fluctuations of the Dirac operator (in the M directions) give the gauge bosons of SM: $U(1)$, $SU(2)$ and $SU(3)$ gauge bosons. On the other hand, the inner fluctuations of the Dirac operator in the noncommutative F direction provide the Higgs field of the particle physics model.

More precisely, the inner fluctuations of the form $A^{(1,0)} = \sum_i a_i[\partial\!\!\!/_M \otimes 1, a_i']$, with $a_i = (\lambda_i, q_i, m_i)$ and $a_i' = (\lambda_i', q_i', m_i')$ in $\mathcal{A} = C^\infty(M, \mathcal{A}_F)$ give rise to terms that can be collected into the following cases:

- a $U(1)$ gauge field $\Lambda = \sum_i \lambda_i\, d\lambda_i' = \sum_i \lambda_i[\partial\!\!\!/_M \otimes 1, \lambda_i']$;
- an $SU(2)$ gauge field $Q = \sum_i q_i\, dq_i'$, with $q = f_0 + \sum_\alpha i f_\alpha \sigma^\alpha$ and $Q = \sum_\alpha f_\alpha[\partial\!\!\!/_M \otimes 1, i f_\alpha' \sigma^\alpha]$;
- a $U(3)$ gauge field $V' = \sum_i m_i\, dm_i' = \sum_i m_i[\partial\!\!\!/_M \otimes 1, m_i']$;
- after reducing the gauge field V' to $SU(3)$, passing to the unimodular subgroup $\mathrm{SU}(\mathcal{A}_F)$ and unimodular gauge potential with $\mathrm{Tr}(A) = 0$, one gets an $SU(3)$-gauge field

$$V' = -V - \frac{1}{3}\begin{pmatrix} \Lambda & 0 & 0 \\ 0 & \Lambda & 0 \\ 0 & 0 & \Lambda \end{pmatrix} = -V - \frac{1}{3}\Lambda 1_3.$$

As in the case of the fermions, the identification with the physical particles is justified by the fact that they have the correct hypercharges for gauge bosons. This can be seen from the fact that the $(1,0)$ part of $A + JAJ^{-1}$ acts on quarks and leptons by

$$\begin{pmatrix} \frac{4}{3}\Lambda + V & 0 & 0 & 0 \\ 0 & -\frac{2}{3}\Lambda + V & 0 & 0 \\ 0 & 0 & Q_{11} + \frac{1}{3}\Lambda + V & Q_{12} \\ 0 & 0 & Q_{21} & Q_{22} + \frac{1}{3}\Lambda + V \end{pmatrix}$$

$$\begin{pmatrix} 0 & 0 & 0 & 0 \\ 0 & -2\Lambda & 0 & 0 \\ 0 & 0 & Q_{11}-\Lambda & Q_{12} \\ 0 & 0 & Q_{21} & Q_{22}-\Lambda \end{pmatrix}$$

and the coefficients that appear in this expression are indeed the correct hypercharges.

The Higgs boson corresponds instead to the inner fluctuations of the form $A^{(0,1)}$, in the F-space direction. These are given by

$$\sum_i a_i[\gamma_5 \otimes D_F, a_i'](x)|_{\mathcal{H}_f} = \gamma_5 \otimes (A_q^{(0,1)} + A_\ell^{(0,1)})$$

where

$$A_q^{(0,1)} = \begin{pmatrix} 0 & X \\ X' & 0 \end{pmatrix} \otimes 1_3 \qquad A_1^{(0,1)} = \begin{pmatrix} 0 & Y \\ Y' & 0 \end{pmatrix}$$

$$X = \begin{pmatrix} \Upsilon_u^* \varphi_1 & \Upsilon_u^* \varphi_2 \\ -\Upsilon_d^* \bar\varphi_2 & \Upsilon_d^* \bar\varphi_1 \end{pmatrix} \quad \text{and} \quad X' = \begin{pmatrix} \Upsilon_u \varphi_1' & \Upsilon_d \varphi_2' \\ -\Upsilon_u \bar\varphi_2' & \Upsilon_d \bar\varphi_1' \end{pmatrix}$$

$$Y = \begin{pmatrix} \Upsilon_\nu^* \varphi_1 & \Upsilon_\nu^* \varphi_2 \\ -\Upsilon_e^* \bar\varphi_2 & \Upsilon_e^* \bar\varphi_1 \end{pmatrix} \quad \text{and} \quad Y' = \begin{pmatrix} \Upsilon_\nu \varphi_1' & \Upsilon_e \varphi_2' \\ -\Upsilon_\nu \bar\varphi_2' & \Upsilon_e \bar\varphi_1' \end{pmatrix}$$

with $\varphi_1 = \sum \lambda_i(\alpha_i' - \lambda_i')$, $\varphi_2 = \sum \lambda_i \beta_i'$ $\varphi_1' = \sum \alpha_i(\lambda_i' - \alpha_i') + \beta_i \bar\beta_i'$ and $\varphi_2' = \sum(-\alpha_i\beta_i' + \beta_i(\bar\lambda_i' - \bar\alpha_i'))$, for $a_i(x) = (\lambda_i, q_i, m_i)$ and $a_i'(x) = (\lambda_i', q_i', m_i')$ and

$$q = \begin{pmatrix} \alpha & \beta \\ -\bar\beta & \bar\alpha \end{pmatrix}.$$

Thus, this discrete part of the inner fluctuations gives rise to a quaternion valued function $H = \varphi_1 + \varphi_2 j$ or $\varphi = (\varphi_1, \varphi_2)$, which is the Higgs field of the model.

One has

$$D_A^2 = (D^{1,0})^2 + 1_4 \otimes (D^{0,1})^2 - \gamma_5 \, [D^{1,0}, 1_4 \otimes D^{0,1}]$$

$$[D^{1,0}, 1_4 \otimes D^{0,1}] = \sqrt{-1}\, \gamma^\mu \, [(\nabla_\mu^s + \mathbb{A}_\mu), 1_4 \otimes D^{0,1}]$$

and this gives $D_A^2 = \nabla^*\nabla - E$ where $\nabla^*\nabla$ is the Laplacian of $\nabla = \nabla^s + \mathbb{A}$ and

$$-E = \frac{1}{4}\, s \otimes \mathrm{id} + \sum_{\mu<\nu} \gamma^\mu \gamma^\nu \otimes \mathbb{F}_{\mu\nu} - i\, \gamma_5\, \gamma^\mu \otimes \mathbb{M}(D_\mu \varphi) + 1_4 \otimes (D^{0,1})^2,$$

with $s = -R$ the scalar curvature and $\mathbb{F}_{\mu\nu}$ the curvature of $\mathbb{A}$, with

$$D_\mu \varphi = \partial_\mu \varphi + \frac{i}{2} g_2 W_\mu^\alpha \varphi \, \sigma^\alpha - \frac{i}{2} g_1 B_\mu \, \varphi$$

involving the $SU(2)$ and $U(1)$ gauge potentials.

3.4. *The spectral action functional*

A good action functional for noncommutative geometries[13] is defined in terms of the spectral decomposition of the Dirac operator as

$$\mathrm{Tr}(f(D/\Lambda)),$$

where D is the Dirac operator of the spectral triple, Λ is a mass scale that makes D/Λ dimensionless, and $f > 0$ is an even smooth function, which is usually taken to be a smooth approximation of a cutoff function.

In the case where the spectral triple has simple dimension spectrum, it is shown in[13] that there is an asymptotic expansion for $\Lambda \to \infty$ of the spectral action functional of the form

$$\mathrm{Tr}(f(D/\Lambda)) \sim \sum_k f_k \, \Lambda^k \fint |D|^{-k} + f(0) \, \zeta_D(0) + o(1),$$

where $f_k = \int_0^\infty f(v) \, v^{k-1} \, dv$ are the momenta of f and where the summation runs over the part of the dimension spectrum $\mathrm{DimSp}(\mathcal{A}, \mathcal{H}, D)$, namely the poles of $\zeta_{b,D}(s) = \mathrm{Tr}(b|D|^{-s})$, contained in the positive axis.

This asymptotic expansion of the spectral action is closely related to the heat kernel expansion

$$\mathrm{Tr}(e^{-t\Delta}) \sim \sum a_\alpha \, t^\alpha \qquad (t \to 0)$$

and its relation to the the ζ function

$$\zeta_D(s) = \mathrm{Tr}(\Delta^{-s/2}).$$

More precisely, a non-zero terms a_α with $\alpha < 0$ corresponds to a *pole* of ζ_D at -2α with

$$\mathrm{Res}_{s=-2\alpha} \, \zeta_D(s) = \frac{2 \, a_\alpha}{\Gamma(-\alpha)}.$$

This follows from

$$|D|^{-s} = \Delta^{-s/2} = \frac{1}{\Gamma\left(\frac{s}{2}\right)} \int_0^\infty e^{-t\Delta} \, t^{s/2-1} \, dt$$

with $\int_0^1 t^{\alpha+s/2-1}\,dt = (\alpha+s/2)^{-1}$. Moreover, the fact that no $\log t$ terms are present in the expansion implies regularity at 0 for ζ_D with $\zeta_D(0) = a_0$. This, in turn, follows from

$$\frac{1}{\Gamma\left(\frac{s}{2}\right)} \sim \frac{s}{2} \quad \text{as} \quad s \to 0.$$

The contribution to the zeta function $\zeta_D(0)$ from pole part at $s = 0$ of

$$\int_0^\infty \mathrm{Tr}(e^{-t\Delta})\, t^{s/2-1}\,dt$$

is given by $a_0 \int_0^1 t^{s/2-1}\,dt = a_0 \frac{2}{s}$.

3.5. *Spectral action with fermionic terms*

The spectral action functional and its asymptotic expansion, as we discuss more in detail below, only account for the purely bosonic terms of the action functional. To have also the correct terms involving fermions or fermion-boson interaction, one needs additional fermionic terms, which can be expressed as a usual Dirac action, so that the resulting action functional will be of the form

$$S = \mathrm{Tr}(f(D_A/\Lambda)) + \frac{1}{2}\langle J\tilde{\xi}, D_A\tilde{\xi}\rangle, \quad \tilde{\xi} \in \mathcal{H}_{cl}^+,$$

with D_A the Dirac operator with unimodular inner fluctuations, J the real structure, and $\mathcal{H}_{cl}^+$ the space of classical spinors, seen as Grassmann variables. In fact, the Dirac part of the action functional has to take into account the fact that the KO-dimension of the finite geometry is six, so that the bilinear form

$$\frac{1}{2}\langle J\tilde{\xi}, D_A\tilde{\xi}\rangle$$

is an antisymmetric form $\mathfrak{A}(\tilde{\xi})$ on

$$\mathcal{H}_{cl}^+ = \{\xi \in \mathcal{H}_{cl} \,|\, \gamma\xi = \xi\}$$

which is nonzero when considered as acting on Grassmann variables. The Euclidean functional integral for this action is then given by a Pfaffian

$$Pf(\mathfrak{A}) = \int e^{-\frac{1}{2}\mathfrak{A}(\tilde{\xi})} D[\tilde{\xi}],$$

which avoids the Fermion doubling problem of previous models based on a symmetric $\langle \xi, D_A\xi\rangle$ for a finite noncommutative space with KO-dimension equal to zero. Grassman variables are a standard method to treat Majorana fermions in the Euclidean signature setting.

The explicit computation of

$$\frac{1}{2}\langle J\tilde{\xi}, D_A\tilde{\xi}\rangle$$

for the finite geometry of[17] described above gives those parts of the Standard Model Larangian involving the coupling of the Higgs to fermions, the coupling of gauge bosons to fermions and all the purely fermionic terms.

3.6. *Bosonic part and the spectral action*

The asymptotic expansion of the spectral action (without fermionic terms) gives the following terms:[17]

$$\begin{aligned}
S = & \frac{1}{\pi^2}(48\, f_4\, \Lambda^4 - f_2\, \Lambda^2\, \mathfrak{c} + \frac{f_0}{4}\, \mathfrak{d}) \int \sqrt{g}\, d^4x \\
+ & \frac{96\, f_2\, \Lambda^2 - f_0\, \mathfrak{c}}{24\pi^2} \int R\, \sqrt{g}\, d^4x \\
+ & \frac{f_0}{10\,\pi^2} \int (\frac{11}{6}\, R^* R^* - 3\, C_{\mu\nu\rho\sigma}\, C^{\mu\nu\rho\sigma})\, \sqrt{g}\, d^4x \\
+ & \frac{(-2\, \mathfrak{a}\, f_2\, \Lambda^2\, +\, \mathfrak{e}\, f_0)}{\pi^2} \int |\varphi|^2\, \sqrt{g}\, d^4x \\
+ & \frac{f_0 \mathfrak{a}}{2\,\pi^2} \int |D_\mu \varphi|^2\, \sqrt{g}\, d^4x \\
- & \frac{f_0 \mathfrak{a}}{12\,\pi^2} \int R\, |\varphi|^2\, \sqrt{g}\, d^4x \\
+ & \frac{f_0 \mathfrak{b}}{2\,\pi^2} \int |\varphi|^4\, \sqrt{g}\, d^4x \\
+ & \frac{f_0}{2\,\pi^2} \int (g_3^2\, G_{\mu\nu}^i\, G^{\mu\nu i} + g_2^2\, F_{\mu\nu}^\alpha\, F^{\mu\nu\alpha} + \frac{5}{3}\, g_1^2\, B_{\mu\nu}\, B^{\mu\nu})\, \sqrt{g}\, d^4x,
\end{aligned}$$

where the parameters are the following:

- f_0, f_2, f_4 are three free parameters, which depend only on the test function f, through $f_0 = f(0)$ and, for $k > 0$,

$$f_k = \int_0^\infty f(v) v^{k-1} dv.$$

- the parameters $\mathfrak{a}, \mathfrak{b}, \mathfrak{c}, \mathfrak{d}, \mathfrak{e}$ are functions of Yukawa parameters of νMSM (that is, of the coordinates on the moduli space of Dirac operators D_F

of the finite geometry) and they are given explicitly by

$$\begin{aligned}
\mathfrak{a} &= \mathrm{Tr}(Y_\nu^\dagger Y_\nu + Y_e^\dagger Y_e + 3(Y_u^\dagger Y_u + Y_d^\dagger Y_d)) \\
\mathfrak{b} &= \mathrm{Tr}((Y_\nu^\dagger Y_\nu)^2 + (Y_e^\dagger Y_e)^2 + 3(Y_u^\dagger Y_u)^2 + 3(Y_d^\dagger Y_d)^2) \\
\mathfrak{c} &= \mathrm{Tr}(MM^\dagger) \\
\mathfrak{d} &= \mathrm{Tr}((MM^\dagger)^2) \\
\mathfrak{e} &= \mathrm{Tr}(MM^\dagger Y_\nu^\dagger Y_\nu).
\end{aligned}$$

The above is obtained using $D_A^2 = \nabla^*\nabla - E$ and applying *Gilkey's theorem* for a differential operator $P = -\left(g^{\mu\nu} I\, \partial_\mu \partial_\nu + A^\mu \partial_\mu + B\right)$ with A, B bundle endomorphisms and $m = \dim M$ (see a detailed account in §1 of[24]). One has

$$\mathrm{Tr}\, e^{-tP} \sim \sum_{n \geq 0} t^{\frac{n-m}{2}} \int_M a_n(x, P)\, dv(x)$$

where $P = \nabla^* \nabla - E$ and $E_{;\mu}{}^{\mu} := \nabla_\mu \nabla^\mu E$, with

$$\nabla_\mu = \partial_\mu + \omega'_\mu, \qquad \omega'_\mu = \frac{1}{2}\, g_{\mu\nu}(A^\nu + \Gamma^\nu \cdot \mathrm{id})$$

$$E = B - g^{\mu\nu}(\partial_\mu\, \omega'_\nu + \omega'_\mu\, \omega'_\nu - \Gamma^\rho_{\mu\nu}\, \omega'_\rho)$$

$$\Omega_{\mu\nu} = \partial_\mu\, \omega'_\nu - \partial_\nu\, \omega'_\mu + [\omega'_\mu, \omega'_\nu].$$

The Seeley-DeWitt coefficients are then given by

$$\begin{aligned}
a_0(x,P) &= (4\pi)^{-m/2} \mathrm{Tr}(\mathrm{id}) \\
a_2(x,P) &= (4\pi)^{-m/2} \mathrm{Tr}\left(-\tfrac{R}{6}\, \mathrm{id} + E\right) \\
a_4(x,P) &= (4\pi)^{-m/2} \tfrac{1}{360} \mathrm{Tr}(-12 R_{;\mu}{}^{\mu} + 5R^2 - 2R_{\mu\nu}\, R^{\mu\nu} \\
&\quad + 2R_{\mu\nu\rho\sigma}\, R^{\mu\nu\rho\sigma} - 60\, R\, E + 180\, E^2 + 60\, E_{;\mu}{}^{\mu} \\
&\quad + 30\, \Omega_{\mu\nu}\, \Omega^{\mu\nu}).
\end{aligned}$$

3.7. *Normalization and coefficients*

Two standard choices of normalization are made in the expression above:

- Rescale Higgs field by $H = \frac{\sqrt{\mathfrak{a}}\, f_0}{\pi} \varphi$ to normalize the kinetic term to

$$\int \frac{1}{2} |D_\mu \mathbf{H}|^2 \sqrt{g}\, d^4x;$$

- Normalize the Yang-Mills terms to

$$\frac{1}{4} G^i_{\mu\nu} \overline{G}^{\mu\nu i} + \frac{1}{4} F^\alpha_{\mu\nu} \overline{F}^{\mu\nu\alpha} + \frac{1}{4} B_{\mu\nu} \overline{B}^{\mu\nu}.$$

In this normalized form the terms in the asymptotic expansion of the spectral action can be written as

$$\begin{aligned} S = &\frac{1}{2\kappa_0^2}\int R\,\sqrt{g}\,d^4x + \gamma_0\int \sqrt{g}\,d^4x \\ &+ \alpha_0\int C_{\mu\nu\rho\sigma}\,C^{\mu\nu\rho\sigma}\sqrt{g}\,d^4x + \tau_0\int R^*R^*\sqrt{g}\,d^4x \\ &+ \frac{1}{2}\int |DH|^2\sqrt{g}\,d^4x - \mu_0^2\int |H|^2\sqrt{g}\,d^4x \\ &- \xi_0\int R\,|H|^2\sqrt{g}\,d^4x + \lambda_0\int |H|^4\sqrt{g}\,d^4x \\ &+ \frac{1}{4}\int (G^i_{\mu\nu}\,G^{\mu\nu i} + F^{\alpha}_{\mu\nu}\,F^{\mu\nu\alpha} + B_{\mu\nu}\,B^{\mu\nu})\sqrt{g}\,d^4x \end{aligned}$$

where $R^*R^* = \frac{1}{4}\epsilon^{\mu\nu\rho\sigma}\epsilon_{\alpha\beta\gamma\delta}R^{\alpha\beta}_{\mu\nu}R^{\gamma\delta}_{\rho\sigma}$ integrates to the Euler characteristic $\chi(M)$ and $C^{\mu\nu\rho\sigma}$ is the Weyl curvature tensor. The coefficients are given by:

$$\begin{aligned} \tfrac{1}{2\kappa_0^2} &= \frac{96 f_2\Lambda^2 - f_0\mathfrak{c}}{24\pi^2} & \gamma_0 &= \frac{1}{\pi^2}(48 f_4\Lambda^4 - f_2\Lambda^2\mathfrak{c} + \frac{f_0}{4}\mathfrak{d}) \\ \alpha_0 &= -\frac{3f_0}{10\pi^2} & \tau_0 &= \frac{11 f_0}{60\pi^2} \\ \mu_0^2 &= 2\frac{f_2\Lambda^2}{f_0} - \frac{\mathfrak{e}}{\mathfrak{a}} & \xi_0 &= \tfrac{1}{12} \\ \lambda_0 &= \frac{\pi^2\mathfrak{b}}{2f_0\mathfrak{a}^2} \end{aligned}$$

The normalization of the Yang–Mills terms fixes a preferred energy scale, the *unification energy*, which in particle physics is assumed to be between 10^{15} and 10^{17} GeV. The common value of the coupling constants at unification is related to the first parameter f_0 of the model by the relation

$$\frac{g^2 f_0}{2\pi^2} = \frac{1}{4}.$$

4. Renormalization group analysis

We have seen that the asymptotic expansion of the spectral action depends on parameters $\mathfrak{a}, \mathfrak{b}, \mathfrak{c}, \mathfrak{d}, \mathfrak{e}$, which in turn are functions of the Yukawa parameters. The latter run with the renormalization group flow, according to the renormalization group equations (RGE). Moreover, the fact that the noncommutative geometry model lives naturally at unification scale means that there are some constraints dictated by the geometry on the possible values of the Yukawa parameters at unification energy and these initial conditions for the renormalization group flow should be compatible with observed physics at low energies.

We will discuss here the renormalization group analysis of the NCG models following.[17,29,30,33] A renormalization group analysis of NCG models, in the Minimal Standard Model setting, was also performed in.[37]

4.1. *Renormalization group equations for the νMSM*

We have seen above that the particle physics content of the NCG model of[17] is given by the νMSM, namely an extension of the minimal standard model with right handed neutrinos and Majorana mass terms. Thus, one can, in first approximation, use the renormalization group equations of this model to study the low energy behavior of the NCG model, although we will see below that, in order to have a realistic Higgs mass estimate, one needs to correct the model and the corresponding RGEs. The equations for the νMSM model were extensively studied by particle physicists, see for instance.[1]

At one 1-loop, the RGE equations take the form $\Lambda \frac{df}{d\Lambda} = \beta_f(\Lambda)$ with

$$16\pi^2\ \beta_{g_i} = b_i\, g_i^3 \quad \text{with } (b_{SU(3)}, b_{SU(2)}, b_{U(1)}) = (-7, -\frac{19}{6}, \frac{41}{10})$$

$$16\pi^2\ \beta_{Y_u} = Y_u(\frac{3}{2}Y_u^\dagger Y_u - \frac{3}{2}Y_d^\dagger Y_d + \mathfrak{a} - \frac{17}{20}g_1^2 - \frac{9}{4}g_2^2 - 8g_3^2)$$

$$16\pi^2\ \beta_{Y_d} = Y_d(\frac{3}{2}Y_d^\dagger Y_d - \frac{3}{2}Y_u^\dagger Y_u + \mathfrak{a} - \frac{1}{4}g_1^2 - \frac{9}{4}g_2^2 - 8g_3^2)$$

$$16\pi^2\ \beta_{Y_\nu} = Y_\nu(\frac{3}{2}Y_\nu^\dagger Y_\nu - \frac{3}{2}Y_e^\dagger Y_e + \mathfrak{a} - \frac{9}{20}g_1^2 - \frac{9}{4}g_2^2)$$

$$16\pi^2\ \beta_{Y_e} = Y_e(\frac{3}{2}Y_e^\dagger Y_e - \frac{3}{2}Y_\nu^\dagger Y_\nu + \mathfrak{a} - \frac{9}{4}g_1^2 - \frac{9}{4}g_2^2)$$

$$16\pi^2\ \beta_M = Y_\nu Y_\nu^\dagger M + M(Y_\nu Y_\nu^\dagger)^T$$

$$16\pi^2\ \beta_\lambda = 6\lambda^2 - 3\lambda(3g_2^2 + \frac{3}{5}g_1^2) + 3g_2^4 + \frac{3}{2}(\frac{3}{5}g_1^2 + g_2^2)^2 + 4\lambda\mathfrak{a} - 8\mathfrak{b}.$$

These differ from the equations for the minimal standard model in the coupled equations for the Dirac and the Majorana masses of neutrinos. Because the presence of the Majorana mass terms introduces a non-renormalizable interaction, the usual way to deal with these equations (see[1]) is to consider the case where one has a non-degenerate spectrum of Majorana masses, and different effective field theories in between the three see-saw scales that each time integrate out the heavier degrees of freedom. More precisely, the procedure to run down the renormalization group flow from unification energy consists of the following steps:

- Run the RGE flow from unification energy Λ_{unif} down to first see-saw scale (largest eigenvalue of M), starting with boundary conditions at unification that are compatible with the geometry of the model.
- Pass to an effective field theory with $Y_\nu^{(3)}$ obtained by removing the last row of Y_ν in a basis in which M is diagonal and with $M^{(3)}$ obtained by removing the last row and column.
- Restart the new RGE flow down to second see-saw scale, with the boundary conditions at the first see-saw scale that match the value coming from the previous running.
- Introduce a second effective field theory with $Y_\nu^{(2)}$ and $M^{(2)}$ again obtained as in the previous step, and with matching boundary conditions.
- Run the induced RGE flow down to first see-saw scale.
- Introduce a new effective field theory with $Y_\nu^{(1)}$ and $M^{(1)}$ again obtained by the same procedure, and with matching boundary conditions.
- Run the induced RGE down to the electoweak energy scale Λ_{ew}.

Using the effective field theories $Y_\nu^{(N)}$ and $M^{(N)}$ between see-saw scales gives rise to a renormalization group flow that has singularities at the see-saw scales. The effects on the gravity model of running the coefficients $\mathfrak{a}, \mathfrak{b}, \mathfrak{c}, \mathfrak{d}, \mathfrak{e}$ according to this flow was analyzed in,[33] in terms of possible models of the very early universe and the inflation epoch between the unification and the electroweak epoch in the cosmological timeline. We will discuss here a simplified model, where one only keeps some dominant terms in the renormalization group equations and sets the three seesaw scales all concentrated near unification energy as in,[17] so that one has a single equation without singularities all the way down to the electroweak scale.

4.2. *Equation for the coupling constants*

At 1-loop the RG equations for the three coupling constants uncouple from the equations for the Yukawa parameters and can be solved separately. This is no longer true at 2-loops.

The running of the coupling constants at one loop is given by equations of the form

$$\partial_t x_i(t) = \beta_{x_i}(x(t)),$$

in the variable $t = \log(\Lambda/M_Z)$, where the beta functions for the three coupling constants are given by

$$\beta_1 = \frac{41}{96\pi^2}\, g_1^3, \qquad \beta_2 = -\frac{19}{96\pi^2}\, g_2^3, \qquad \beta_3 = -\frac{7}{16\pi^2} g_3^3. \tag{1}$$

These equations can be solved exactly. The values of the coupling constants at $\Lambda = M_Z$, determined in agreement with experiments, are

$$g_1(0) = 0.3575, \qquad g_2(0) = 0.6514, \qquad g_3(0) = 1.221 \tag{2}$$

When one plots solutions with these boundary conditions, one sees clearly the appearance of a well known problem: if the low energy limits are taken in agreement with experimentally determined values, at unification energy the coupling constants do *not* meet in the expected relation $g_3^2 = g_2^2 = 5g_1^2/3$, but they form a triangle: this fact is considered as evidence suggesting the existence of new physics beyond the Standard Model. It is known, for instance, that in supersymmetric extensions of the Standard Model, the size of the triangle becomes smaller.

4.3. *A geometric point of view*

The fact that at 1-loop the equations for the coupling constant completely decouple from the rest of the renormalization group equations means that we can consider the solutions of the RGE for the coupling constants as known functions and input them in the other equations. The remaining part of the renormalization group equations then determine a flow (a vector field) on the moduli space $\mathcal{C}_3 \times \mathcal{C}_1$ of Dirac operators on the finite noncommutative space F. Subvarieties that are invariant under the flow are physical relations between the Standard Model parameters that hold at all energies. This is a very important problem in high energy physics, to determine whether there are relations between the parameters of the Standard Model, and it seems natural to use the geometry of the moduli space $\mathcal{C}_3 \times \mathcal{C}_1$ described above and a better geometric understanding of the flow determined by the RGEs to look for nontrivial relations. At two loops or higher, the equations for the coupling constants no longer decouples. Apart from the extraordinary technical difficulty in obtaining the full renormalization group equations at more than one loop (for the two loop case for the minimal standard model the reader can consult[2]). However, the geometric point of view described here can still be applied, by viewing the RGEs as a flow on a rank three vector bundle, whose fiber has coordinates given by the coupling constants, over the moduli space $\mathcal{C}_3 \times \mathcal{C}_1$.

4.4. *Constraints at unification*

The geometry of the model imposes several conditions at unification energy. These conditions are specific to this NCG model, see[17] and §1 of[24] for a more detailed description. They are summarized as follows.

- λ parameter constraint

$$\lambda(\Lambda_{unif}) = \frac{\pi^2}{2f_0} \frac{\mathfrak{b}(\Lambda_{unif})}{\mathfrak{a}(\Lambda_{unif})^2}$$

- Higgs vacuum constraint

$$\frac{\sqrt{\mathfrak{a} f_0}}{\pi} = \frac{2M_W}{g}$$

- See-saw mechanism and $\mathfrak{c}$ constraint

$$\frac{2f_2\Lambda_{unif}^2}{f_0} \leq \mathfrak{c}(\Lambda_{unif}) \leq \frac{6f_2\Lambda_{unif}^2}{f_0}$$

- Mass relation at unification

$$\sum_{generations} (m_\nu^2 + m_e^2 + 3m_u^2 + 3m_d^2)|_{\Lambda=\Lambda_{unif}} = 8M_W^2|_{\Lambda=\Lambda_{unif}}$$

The presence of these geometric constraints implies that one needs to have boundary conditions for the model at unification energy that are at the same time compatible with these requirements and also that give rise to low energy limits of the solutions of the RGE that are compatible with known experimental values.

4.5. *Boundary conditions at unification*

A set of initial conditions at unification for the renormalization group flow of[1] that is compatible with all the geometric constraints described above and with experimental values at low energy was identified explicitly in,[30] using the maximal mixing conditions

$$\zeta = \exp(2\pi i/3)$$

$$U_{PMNS}(\Lambda_{unif}) = \frac{1}{3}\begin{pmatrix} 1 & \zeta & \zeta^2 \\ \zeta & 1 & \zeta \\ \zeta^2 & \zeta & 1 \end{pmatrix}$$

$$\delta_{(\uparrow 1)} = \frac{1}{246}\begin{pmatrix} 12.2\times 10^{-9} & 0 & 0 \\ 0 & 170\times 10^{-6} & 0 \\ 0 & 0 & 15.5\times 10^{-3} \end{pmatrix}$$

$$Y_\nu = U^\dagger_{PMNS}\delta_{(\uparrow 1)}U_{PMNS}$$

It was also shown in[30] that the RGE of[1] presents a sensitive dependence on the initial conditions at unification, which was shown explicitly by changing only one parameter in the diagonal matrix Y_ν and following its effect on the running of the top quark term. This sensitive dependence problem creates a possible fine tuning problem in the model. We will see below how the choice of initial conditions at unification and the form of the RGE affect the Higgs mass estimate.

4.6. *Low energy limits*

When one extracts from the model estimates on the low energy behavior, through the running of the renormalization group flow, one usually makes some drastic simplifications on the form of the renormalization group equations, so that only the "dominant" terms are retained. Within the Minimal Standard Model, one usually assumes that the top quark Yukawa parameter is the dominant term in the RGE equations. This means that one can simplify the rest of the RGE, to just two equations with

$$\beta_y = \frac{1}{16\pi^2}\left(\frac{9}{2}y^3 - 8g_3^2 y - \frac{9}{4}g_2^2 y - \frac{17}{12}g_1^2 y\right). \tag{3}$$

$$\beta_\lambda = \frac{1}{16\pi^2}\left(24\lambda^2 + 12\lambda y^2 - 9\lambda(g_2^2 + \frac{1}{3}g_1^2) - 6y^4 + \frac{9}{8}g_2^4 + \frac{3}{8}g_1^4 + \frac{3}{4}g_2^2 g_1^2\right), \tag{4}$$

where g_1, g_2, g_3 are the solutions for the equations for the coupling constants discussed above.

In the case of the νMSM model considered in,[17] where the equations for the neutrino terms are coupled to large Majorana masses, the Yukawa coupling for the τ neutrino can also be of comparable magnitude. This issue was addressed in[17] by modifying the boundary condition of the top Yukawa parameter, consistently with the quadratic mass relation at unification mentioned above as one of the geometric constraints.

4.7. *From heavy to light Higgs mass*

The earlier results on Higgs mass estimates derived from NCG models were all based on the technique described above, namely fixing initial conditions at unification compatible with the geometric constraints and with known experimental low energy results. On then considers the Higgs scattering parameter

$$\frac{f_0}{2\pi^2}\int b\,|\varphi|^4\,\sqrt{g}\,d^4x = \frac{\pi^2}{2\,f_0}\frac{\mathfrak{b}}{\mathfrak{a}^2}\int |\mathbf{H}|^4\,\sqrt{g}\,d^4x$$

and uses the relation at unification, where $\tilde{\lambda}$ is the $|\mathbf{H}|^4$ coupling,

$$\tilde{\lambda}(\Lambda) = g_3^2\,\frac{\mathfrak{b}}{\mathfrak{a}^2}.$$

Running of Higgs scattering parameter according to the equation

$$\frac{d\lambda}{dt} = \lambda\gamma + \frac{1}{8\pi^2}(12\lambda^2 + B)$$

where

$$\gamma = \frac{1}{16\pi^2}(12y_t^2 - 9g_2^2 - 3g_1^2) \quad B = \frac{3}{16}(3g_2^4 + 2g_1^2g_2^2 + g_1^4) - 3y_t^4$$

produces an estimate for the Higgs mass,

$$m_H^2 = 8\lambda\,\frac{M^2}{g^2}, \quad m_H = \sqrt{2\lambda}\,\frac{2M}{g}.$$

With $\lambda(M_Z) \sim 0.241$ one obtains a heavy Higgs mass ~ 170 GeV, which can be lowered slightly as in[17] with corrections from the τ-neutrino term to ~ 168 GeV. This is still too heavy a Higgs mass to be compatible with the observed experimental value found by the LHC to be around 125 GeV.

This means that some more substantial correction to the model is needed, in order to obtain compatibility with the experimental Higgs mass. There are at present two different methods of correcting the NCG model so that it gives a realistic Higgs mass estimate. They were announced at around the same time by Chamseddine–Connes[16] and Estrada–Marcolli.[29] The method of[16] is based on introducing an additional scalar field in the model, which arises as a scalar perturbation of the Dirac operator, and which couples non-trivially to the Higgs field, thus modifying its running. The method of[29] is based on the coupling of matter to gravity that is a fundamental feature of the NCG models. Without introducing any additional field, but taking into account the effect of the gravitational terms on the RGEs for matter, along the lines of Weinberg's asymptotic safety scenario, one obtains modified RGEs that allow for a realistic Higgs mass. We will discuss more in detail this second method in the following section.

4.8. *Running of the gravitational terms*

In NCG models of matter coupled to gravity, there are different ways of treating the running of the gravitational terms that appear in the asymptotic expansion of the spectral action.

The method followed in[17] assumes that the dependence of the coefficients κ_0, γ_0, α_0, τ_0, μ_0, ξ_0, λ_0 in the asymptotic expansion of the spectral action on the parameters $\mathfrak{a}, \mathfrak{b}, \mathfrak{c}, \mathfrak{d}, \mathfrak{e}$, which in turn are functions of the Yukawa parameters of the matter sector of the model, *only* holds at unification energy. Thus, these relations that express the coefficients of the gravitational terms in terms of the matter parameters serve the purpose of imposing constraints (and therefore establish exclusion regions) in the space of all possible initial conditions at unification. Once the initial conditions are fixed according to these relations, the matter parameters are expected to run according to the RG flow for the corresponding particle physics model (as discussed above), while the running of the gravitational parameters is analyzed independently of the matter part of the model. In[17] the running of the gravitational parameters is based on the renormalization group analysis of modified gravity models of,[3],[18],[27] for a gravity action functional of the form

$$\int \left(\frac{1}{2\eta} C_{\mu\nu\rho\sigma} C^{\mu\nu\rho\sigma} - \frac{\omega}{3\eta} R^2 + \frac{\theta}{\eta} R^* R^* \right) \sqrt{g}\, d^4x$$

and the corresponding beta functions of the form

$$\begin{aligned} \beta_\eta &= -\frac{1}{(4\pi)^2} \frac{133}{10} \eta^2 \\ \beta_\omega &= -\frac{1}{(4\pi)^2} \frac{25 + 1098\,\omega + 200\,\omega^2}{60} \eta \\ \beta_\theta &= \frac{1}{(4\pi)^2} \frac{7(56 - 171\,\theta)}{90} \eta. \end{aligned}$$

It turns out that, in this approach, the running of the gravitational terms does not alter the RGEs for the matter sector, hence one does not obtain any significant correction to the Higgs mass. The approach of[16] keeps this same running of the gravitational terms, but achieves the desired lowering of the Higgs mass through the coupling to an additional scalar field.

A different point of view on the running of the gravitational terms in the NCG model of[17] was taken in[33] in order to study early universe models with variable gravitational and cosmological constants. In this approach, one assumes that the dependence of the coefficients of the asymptotic expansion of the spectral action on the Yukawa parameters holds, not only at unification energy, but across the part of the cosmological timeline between unification and electroweak epochs. While the electroweak scale corresponds in the cosmological timeline to only at most 10^{-12} seconds after the Big Bang, this very early universe period is especially significant cosmologically, because it includes the inflationary epoch. Thus, in this approach, one runs the gravitational terms according to the relations that express them in terms of the parameters $\mathfrak{a}, \mathfrak{b}, \mathfrak{c}, \mathfrak{d}, \mathfrak{e}$. The cosmological implications are described in.[33] In terms of the effect of this approach on the Higgs mass estimate, what changes with respect to the RGE analysis of[17] is that instead of applying the relation $\lambda_0 = \lambda \cdot (\pi^2 \mathfrak{b})/(f_0 \mathfrak{a}^2)$ only at the initial condition at unification one imposes it all the way down to the electroweak scale. This leads to a lower Higgs mass prediction of around 158 GeV, which is still too large to be compatible with the experimental value found by the LHC.

Yet another possible way of dealing with the running of the gravitational terms in the NCG model is the asymptotic safety scenario of gravity, originally proposed by Weinberg[48] and further studied in relation to the RGEs of the Standard Model coupled to gravity and the Higgs mass by Shaposhnikov and Wetterich.[45] This is the approach that leads to the Higgs mass estimate of[29] described below.

4.9. *Correcting the Higgs mass prediction*

In the asymptotic safety scenario of[48] the RGEs for the matter sector acquire correction terms coming from the gravitational parameters in the form

$$\partial_t x_j = \beta_j^{\mathrm{SM}} + \beta_j^{\mathrm{grav}}, \tag{5}$$

where x_j are, as before, the running parameters with β_j^{SM} is the Standard Model beta function for x_j described previously, and where β_j^{grav} are the gravitational correction terms, which are of the form

$$\beta_j^{\mathrm{grav}} = \frac{a_j}{8\pi} \frac{\Lambda^2}{M_P^2(\Lambda)} x_j, \tag{6}$$

The coefficients a_j are called *anomalous dimensions.* The scale dependence of the Newton constant can be estimated as in[41] (see also,[4038]) to be

$$M_P^2(\Lambda) = M_P^2 + 2\rho_0 \Lambda^2, \tag{7}$$

with $\rho_0 \sim 0.024$. The anomalous dimensions are usually taken to be $a_1 = a_2 = a_3 = a_g$, with $|a_g| \sim 1$ and a negative sign for the coupling constants and $a_g \leq a_y < 0$, while a_λ is expected to be positive (see[45]).

The main effect of the correction terms in the RGEs is that the presence of the gravitational terms with negative anomalous dimensions makes the Yukawa couplings and the forces asymptotically free. This is also known as "Gaussian matter fixed point". In this scenario, the solutions of the decoupled equations for the coupling constants decay rapidly to zero near unification energy and the coupled equations for the top Yukawa parameter and the Higgs self coupling further simplify by neglecting the coupling constants terms. The resulting equation for the Higgs self-coupling can be solved explicitly in terms of hypergeometric functions,[29] and one can show that it admits solutions that give a realistic Higgs mass $m_H \sim 125.4$, compatible with a realistic top quark mass of $m_t \sim 171.3$ GeV.

The Gaussian matter fixed point can be made compatible with the geometric constraints of the NCG model at unification,,[29] with one important change in the role of the Majorana mass terms. In the original setting of[17] the eigenvalues of the Majorana mass terms matrix M are assumed to be near unification scale. This leads to a nice geometric interpretation of the see-saw mechanism for neutrino masses, as described in.[17] In the setting with anomalous dimensions and Gaussian matter fixed point, the Majorana mass terms become much smaller (below the electroweak scale), and consistent with a different physical model, namely the dark matter models of Shaposhnikov and Tkachev.[42–44] For a recent and extensive reference on Majorana mass terms below the electroweak scale the reader may also consult.[6]

As a final consideration, while the construction of particle physics models via noncommutative geometry has had so far, as main goals, reducing the complexity of the Standard Model Lagrangian to something that can be derived from simple mathematical principle, as well as obtaining realistic estimates of particles in the low energy regime, there are many mysterious aspects of the Standard Model on which the NCG approach has not yet been tested, including charge quantization, the presence of "accidental" symmetries, the strong CP problem, and the hierarchy problem. In addition to exploring possible implications of the NCG formulation on some of these important problems, the coupling of matter to gravity that is a crucial part of these models allow for an extensive investigation of the cosmological implications of the model. This latter aspect has already been more extensively developed and, as announced, will be treated in detail elsewhere.[34]

References

1. S. Antusch, J. Kersten, M. Lindner, M. Ratz, M.A. Schmidt *Running neutrino mass parameters in see-saw scenarios*, JHEP 03 (2005) 024.
2. H. Arason, D.J. Castano, B. Kesthlyi, E.J. Piard, P. Ramond, B.D. Wright, *Renormalization group study of the standard model and its extensions: the standard model*, Phys. Rev. D, 46 (1992) N.9, 3945–3965.
3. I.G. Avramidi, *Covariant methods for the calculation of the effective action in quantum field theory and investigation of higher- derivative quantum gravity*, PhD Thesis, Moscow University, 1986, hep-th/9510140.
4. J.D. Barrow, *Gravitational memory?* Phys. Rev. D Vol.46 (1992) N.8 R3227, 4pp.
5. Jord Boeijink, Walter D. van Suijlekom, *The noncommutative geometry of Yang-Mills fields*, arXiv:1008.5101.
6. A. Boyarsky, O. Ruchayskiy, M. Shaposhnikhov, *The role of sterile neutrinos in cosmology and astrophysics*, Ann. Rev. Nucl. Part. Sci. Vol.59 (2009) 191–214.
7. Thijs van den Broek, Walter D. van Suijlekom, *Supersymmetric QCD and noncommutative geometry*, Comm. Math. Phys. 303 (2011), no. 1, 149–173.
8. Thijs van den Broek, Walter D. van Suijlekom, *Going beyond the Standard Model with noncommutative geometry*, arXiv:1211.0825.
9. Branimir Ćaćić, *A reconstruction theorem for almost-commutative spectral triples*, Lett. Math. Phys. 100 (2012), no. 2, 181–202.
10. Branimir Ćaćić, *Moduli spaces of Dirac operators for finite spectral triples*, in "Quantum groups and noncommutative spaces", 9–68, Aspects Math., E41, Vieweg, 2011.
11. Branimir Ćaćić, *Real structures on almost-commutative spectral triples*, arXiv:1209.4832.
12. Branimir Ćaćić, Matilde Marcolli, Kevin Teh, *Coupling of gravity to matter, spectral action and cosmic topology*, arXiv:1106.5473, to appear in Journal of Noncommutative Geometry.
13. Ali Chamseddine, Alain Connes, *The spectral action principle*, Comm. Math. Phys. 186 (1997), no. 3, 731–750.
14. A.Chamseddine, A.Connes, *Why the Standard Model*, J. Geom. Phys. 58 (2008) 38–47
15. Ali Chamseddine, Alain Connes, *The uncanny precision of the spectral action*, Comm. Math. Phys. 293 (2010), no. 3, 867–897.
16. Ali Chamseddine, Alain Connes, *Resilience of the Spectral Standard Model*, arXiv:1208.1030.
17. Ali Chamseddine, Alain Connes, Matilde Marcolli, *Gravity and the standard model with neutrino mixing*, Advances in Theoretical and Mathematical Physics, 11 (2007) 991–1090.
18. A. Codello, R. Percacci, *Fixed points of higher derivative gravity*,

Phys.Rev.Lett. 97 (2006) 221301.
19. Alain Connes, *Noncommutative geometry*, Academic Press, 1994.
20. Alain Connes, *Geometry from the spectral point of view*, Lett. Math. Phys. 34 (1995), no. 3, 203–238.
21. Alain Connes, *Gravity coupled with matter and the foundation of noncommutative geometry*, Comm. Math. Phys. 182 (1996), no. 1, 155–176.
22. Alain Connes, *On the spectral characterization of manifolds*, arXiv:0810.2088.
23. Alain Connes, John Lott, *Particle models and noncommutative geometry*, in "Recent advances in field theory (Annecy-le-Vieux, 1990)". Nuclear Phys. B Proc. Suppl. 18B (1990), 29–47 (1991).
24. Alain Connes, Matilde Marcolli, *Noncommutative Geometry, Quantum Fields and Motives*, Colloquium Publications, Vol.55, American Mathematical Society, 2008.
25. Ludwik Dabrowski, Giacomo Dossena, *Product of real spectral triples*, Int. J. Geom. Methods Mod. Phys. Vol. 8, No. 8 (2011) 1833–1848.
26. [dSHW] A. De Simone, M.P. Hertzberg, F. Wilczek, *Running inflation in the Standard Model*, hep-ph/0812.4946v2
27. J.F. Donoghue, *General relativity as an effective field theory: the leading quantum corrections*, Phys. Rev. D, Vol.50 (1994) N.6, 3874–3888.
28. Koen van den Dungen, Walter D. van Suijlekom, *Electrodynamics from Noncommutative Geometry*, arXiv:1103.2928
29. Christopher Estrada, Matilde Marcolli, *Asymptotic safety, hypergeometric functions, and the Higgs mass in spectral action models*, arXiv:1208.5023, to appear in International Journal of Geometric Methods in Modern Physics.
30. Daniel Kolodrubetz, Matilde Marcolli, *Boundary conditions of the RGE flow in the noncommutative geometry approach to particle physics and cosmology*, Phys. Lett. B, Vol.693 (2010) 166–174.
31. Steven Lord, Adam Rennie, Joseph C. Varilly, *Riemannian manifolds in noncommutative geometry*, J. Geom. Phys. 62 (2012), no. 7, 1611–1638..
32. Matilde Marcolli, *Building cosmological models via noncommutative geometry*, International Journal of Geometric Methods in Modern Physics, Vol.8, N.5 (2011) 1131–1168.
33. Matilde Marcolli, Elena Pierpaoli, *Early universe models from Noncommutative Geometry*, Advances in Theoretical and Mathematical Physics, Vol.14 (2010) 1373–1432.
34. Matilde Marcolli, Elena Pierpaoli, *Noncommutative cosmology*, book in preparation, to be published by World Scientific.
35. Matilde Marcolli, Elena Pierpaoli, Kevin Teh, *The spectral action and cosmic topology*, Commun.Math.Phys.304 (2011) 125–174, arXiv:1005.2256
36. Matilde Marcolli, Elena Pierpaoli, Kevin Teh, *The coupling of topology and inflation in noncommutative cosmology*, Communications in Mathematical Physics, Vol.309 (2012) N.2, 341–369, arXiv:1012.0780.

37. C.P. Martín, J.M. Gracia-Bondía, J. Várilly, *The standard model as a noncommutative geometry: the low-energy regime*, Phys. Rep. 294 (1998), no. 6, 363–406.
38. G. Narain, R. Percacci, *Renormalization group flow in scalar-tensor theories, I*, arXiv:0911.0386v2.
39. I.D. Novikov, A.G. Polnarev, A.A. Starobinsky, Ya.B. Zeldovich, *Primordial black holes*, Astron. Astrophys. 80 (1979) 104–109.
40. R. Percacci, D. Perini, *Asymptotic safety of gravity coupled to matter*, hep-th/0304222.
41. M. Reuter, *Nonperturbative evolution equation for quantum gravity*, Phys Rev D 57 (1998) N.2, 971–985.
42. M. Shaposhnikov, *The νMSM, dark matter and baryon asymmetry of the Universe*, Journal of Physics: Conference Series 39 (2006) 9–11.
43. M. Shaposhnikov, *The νMSM, leptonic asymmetries, and properties of singlet fermions*, JHEP (2008) 0808:008, 56pp.
44. M. Shaposhnikov, I. Tkachev, *The νMSM, inflation, and dark matter*, Phys.Lett. B Vol.639 (2006) 414–417.
45. M. Shaposhnikov, C. Wetterich, *Asymptotic safety of gravity and the Higgs boson mass*, arXiv:0912.0208v2.
46. Kevin Teh, *Nonperturbative Spectral Action of Round Coset Spaces of SU(2)*, arXiv:1010.1827, to appear in Journal of Noncommutative Geometry.
47. J.J. Venselaar, *Morita "equivalences" of equivariant torus spectral triples*, arXiv:1112.6102.
48. S. Weinberg, *Ultraviolet divergences in quantum theories of gravitation*, in "General Relativity: an Einstein centenary survey" (S.W. Hawking, W. Israel, Eds.) Cambridge Univ. Press, 1979, pp.790–831.
49. Dapeng Zhang, *Projective Dirac operators, twisted K-theory and local index formula*, arXiv:1008.0707.

NONCOMMUTATIVE SPACETIMES AND QUANTUM PHYSICS

A. P. BALACHANDRAN

Institute of Mathematical Sciences,
CIT Campus, Taramani, Chennai 600113, India
and
Physics Department,
Syracuse University, Syracuse, NY, 13244-1130, USA
E-mail: balachandran38@gmail.com

In these notes, the motivation for noncommutative spacetimes coming from quantum gravity will be reviewed. The implementation of symmetries coming from theory of Hopf algebras on such spacetimes and the attendant twist of quantum statistics will be discussed in detail. These ideas will be applied to construct covariant quantum fields on such spacetimes. The formalism will then be applied to cosmic microwave background. It will be explained as to how these models predict apparent non-Pauli transitions.

Keywords: Noncommutative geometry; Hopf algebras; quantum physics.

1. Introduction

In these lecture notes I will describe radical ideas which throw doubts on basic concepts. We will learn about challenges to the fundamental principles governing physics, chemistry and indeed life. These challenges question the nature of spacetime, causality as we understand it, principles of relativity and other cherished principles. This is a dangerous endeavor, and may be perceived even as the rantings of reckless revolutionaries. I want to convince you that it is not so. Rather these radical ideas are naturally suggested when we examine nature in the small.

1.1. *On Spacetime in the Small*

There are good reasons to imagine that spacetime at the smallest scales will display some sort of discreteness. They come from *quantum physics* and *the nature of black holes.* Let me outline them.

As we know, the length scales in gravity are far smaller than all nuclear or atomic scales, being 10^{-32} centimetres. This is 10^9 times smaller than the size of an atom. At such small length scales, there is no reason to suppose that the nature of spacetime would be anything like what we are familiar with. In fact, there are good reasons to suppose that it would be radically different. In particular we expect fundamental limitations on probing such small length scales. This comes about from considerations involving gravity and quantum theory. The following arguments were described by Doplicher, Fredenhagen and Roberts.[1]

Probing Planck length-scales In order to probe physics at the Planck scale $L \simeq 10^{-32}$cm, the Compton wavelength λ_C of a probe of mass M must therefore fulfill

$$\lambda_C = \frac{\hbar}{Mc} \leq L \text{ or } M \geq \frac{\hbar}{Lc} \simeq 10^{19}\text{GeV (Planck mass)} \simeq 10^{-5}\text{grams.} \quad (1)$$

Such a huge mass in the small volume $L^3 \simeq (10^{-32}\text{cm})^3$ will strongly affect gravity and can cause *black holes* and their *horizons* to form: gravity can become so large that no signal can escape the volume.

Lessons from quantum mechanics Quantum theory too puts limitations on measurements of position x and momentum p:

$$\Delta x \Delta p \geq \frac{\hbar}{2}, \quad (2)$$

where Δx and Δp represent the uncertainties in measurements of position and momentum, respectively. There is a precise mathematical manner to achieve this, namely, we can "make x and p noncommuting": $x\,p - p\,x = i\hbar$. This suggests we can incorporate limitations on spatial resolution by making position noncommuting as in

$$x\,y - y\,x = i\theta, \quad (3)$$

where θ is a new Planck-like fundamental constant with dimensions of length squared. When spacetime is noncommuting, we say that it obeys "noncommutative" geometry.

Now, from (3) we obtain

$$\Delta x\, \Delta y \geq \frac{\theta}{2}, \quad (4)$$

where Δx and Δy stand for the intrinsic uncertainties in measurements of the x and y coordinates. There are similar equations in any pair of coordinates.

As will be explained in these notes, there is a particular way to implement spacetime noncommutativity, making use of the *Groenewold-Moyal (G-M) plane.* It is defined as the algebra $\mathcal{A}_\theta$ of functions on $\mathbb{R}^{d+1}$ with a twisted product:

$$f * g(x) = f e^{i/2 \overleftarrow{\partial_\mu} \theta^{\mu\nu} \overrightarrow{\partial_\nu}} g. \tag{5}$$

At the level of *coordinates*, this twisted product implies the following commutation relations:

$$\hat{x}_\mu \star \hat{x}_\nu - \hat{x}_\nu \star \hat{x}_\mu = [\hat{x}_\mu, \hat{x}_\nu]_\star = i\theta_{\mu\nu}, \quad \mu, \nu = 0, 1, \ldots, d. \tag{6}$$

A bit of history The idea that spacetime geometry may be noncommutative is old. It goes back to Schrödinger and Heisenberg. Heisenberg raised this possibility in a letter to Rudolph Peierls in the 30's. Heisenberg also complained that he did not know enough mathematics to explore the physical consequences of this possibility. Peierls then mentioned Heisenberg's ideas to Wolfgang Pauli, who in turn explained them to Hartland Snyder. So it was Snyder who published the first paper on the subject in Physical Review in 1947.[2] I should also mention the role of Joe Weinberg in these developments. Joe was a student of Oppenheimer and was a close associate of Pauli and a classmate of Schwinger. He was the person accused of passing nuclear secrets to the Soviets and who lost his job in 1952 at Minnesota for that reason. His wife supported the family for several years. Eventually he got a faculty position at Case in 1958 and from there, he came to Syracuse. Joe was remarkable. He seemed to know everything, from Sanskrit to noncommutative geometry, and published very little. He had done extensive research on this new vision of spacetime. He had showed me his manuscripts. They are now in Syracuse University archives.

The development of quantum field theory on the G-M plane was pioneered by Doplicher, Fredenhagen, Roberts, as mentioned above, and by Julius Wess and his group.[3] Wess was a student of Hans Thirring, father of Walter Thirring in Vienna. Julius passed away in Hamburg in August, 2007 at the age of 72. All of us in this field miss him for his kindness and wisdom.

2. Hopf Algebras in Physics

Although it has been the practice in physics to rely on groups to define symmetries, we now understand that they provide just certain simple possibilities in this direction and that there exist more general possibilities as well. They are based on Hopf algebras and their variants. Such algebras were first encountered in work on integrable models and conformal field theories and later were given a precise mathematical formulation by Drinfeld,[4] Woronowicz,[5] Majid[6] and Mack and Schomerus[7] . We have now encountered them in fundamental theories as well. Thus Hopf algebras and their variants have an emergent importance for physicists.

We therefore present an elementary introduction to Hopf algebras in this section, closely following our book on Group Theory and Hopf Algebras,[8] where more details can be found. Following Mack and Schomerus,[7] we first expose the base features of a group G which let us use it as a symmetry principle. We then formulate them in terms of $\mathbb{C}G$, the group algebra of G. Now $\mathbb{C}G$ is a Hopf algebra of a special kind, having an underlying group G. We can extract its relevant features to serve as a symmetry and formulate the notion of a general Hopf algebra H. The discussion concludes with simple examples of H.

2.1. *On Symmetries and Symmetry Groups*

Symmetries arise in Physics as a way to formulate equivalences between different physical states. For example, if the Hamiltonian for a system is rotationally invariant, then the energy of a state and the energies of all states we get therefrom by rotations are equal. There is another aspect of symmetry which is important in physical considerations. If rotations for example are symmetries of a system and we perceive a state of motion of the system such as the earth going around the sun in an elliptical orbit, then we can rotate this orbit with sun as center and get another possible state of motion of the earth. In this way, we can predict more states of motions, and they can be infinite as in the case of rotations. In quantum theory, this aspect of a symmetry has a more powerful role. Thus even for a finite symmetry group G, if a state vector Ψ changes to a new vector $U(g)\Psi$ under the action of $g \in G$, then we can superpose Ψ and $U(g)\Psi$ with complex coefficients and predict the existence of infinitely many new state vectors. More generally the entire linear manifold spanned by Ψ and its G-transforms are possible state vectors.

Formulation of physical theories is governed by symmetries as well. If observation indicates that the physical world is governed by certain symmetries to a high degree of accuracy, then we can count for these observations by working with Lagrangias and actions exactly or approximately invariant under these symmetries. The Poincaré group is one such symmetry with this role in fundamental theories of nature. Although gauge groups are not symmetries in this sense, they too nowadays provide a metaprinciple controlling the formulation of potential fundamental theories much like the Poincaré group. Thus, symmetries assume many significant roles for a physicist. Although it has been the practice in physics to rely on groups to define symmetries, we now understand that they provide just certain simple possibilities in this direction and that there exist more general possibilities as well. They are based on Hopf algebras and their variants. Such algebras were first encountered in work on integrable models and conformal field theories and later were given a precise mathematical formulation by Drinfel'd, Woronowickz, Majid and Mack and Schomerus. We have now encountered them in fundamental theories as well. Thus Hopf algebras and their variants have an emergent importance for physicists. We therefore give an elementary presentation of Hopf algebras in this part. Following Mack and Schomerus,[7] we first expose the base features of a group G which let us use it as a symmetry principle. We then formulate them in terms of $\mathbb{C}G$, the group algebra of G. Now $\mathbb{C}G$ is a Hopf algebra of a special kind, having an underlying group G. We can extract its relevant features to serve as a symmetry and formulate the notion of a general Hopf algebra H.

2.2. *Symmetries in Many-Particle Systems*

Let us consider a general many-particle quantum system, with $\mathcal{H}^{(1)}$ denoting the Hilbert space of one-particle states and let G be the symmetry group of the theory. N-particle states are created from the vacuum $|0\rangle$ and correspond to state vectors in the N-fold tensor product $\mathcal{H}^{(N)} = \mathcal{H}^{(1)}\otimes\cdots\otimes\mathcal{H}^{(1)}$. Here we are considering the particles to be distinguishable; later on we will consider appropriate symmetrisation postulates. The symmetry group G acts on the Fock space $\mathcal{F} = \bigoplus_N \mathcal{H}^{(N)}$ through a unitary representation of the form

$$U(g) = \bigoplus_N U^{(N)}(g), \tag{1}$$

where each $U^{(N)}$ acts on the corresponding $\mathcal{H}^{(N)}$ and, therefore, does not change the particle number.

Usually, the representation $U^{(N)}$ is given as the N-fold product of $U^{(1)}$, that is,

$$U^{(N)}(g)|\rho_1\rangle\otimes|\rho_2\rangle\otimes\cdots\otimes|\rho_N\rangle = U^{(1)}(g)|\rho_1\rangle\otimes U^{(1)}(g)|\rho_2\rangle\otimes\cdots\otimes U^{(1)}(g)|\rho_N\rangle. \tag{2}$$

It is thus apparent that the representation $U^{(N)}$ is being fixed by the choice of $U^{(1)}$. A crucial fact is the observation that the way this happens is far from unique. By introducing the notion of a *coproduct*, we will discover a very general structure, of which the present example represents the simplest choice.

For the discussion that follows, we will need not only the group G, but the full *group algebra* $\mathbb{C}G$, that we briefly recall. For finite G, the group algebra elements are formal sums of the form

$$\alpha^* = \sum_{g \in G} \alpha(g) g, \tag{3}$$

with α a function on G. The product of two elements α^* and β^* is defined as

$$\alpha^* \beta^* := \sum_{g \in G} (\alpha * \beta)(g)\, g, \tag{4}$$

where $\alpha * \beta$ stands for the *convolution product*

$$(\alpha * \beta)(g) = \sum_{h \in G} \alpha(gh^{-1})\beta(h). \tag{5}$$

For G a Lie group, we make use of the Haar measure in order to replace the sums on the above expressions by integrals over the group. So for instance elements of $\mathbb{C}G$ are now of the form

$$\alpha^* = \int d\mu(g)\alpha(g) g. \tag{6}$$

Definitions (4) and (5) for the product and convolution carry over as well.[7,8] In this way we also obtain a $*$-representation of the group algebra on each $\mathcal{H}^{(N)}$, as well as on $\mathcal{F}$ by means of

$$\begin{aligned} U^{(N)}(\alpha^*) &= \int d\mu(g)\alpha(g) U^{(N)}(g), \\ U(\alpha^*) &= \int d\mu(g)\alpha(g) U(g). \end{aligned} \tag{7}$$

We now examine $U^{(N)}$. We will see that $U^{(1)}$ determines $U^{(N)}$ $(N \geq 2)$ only if we have a homomorphism Δ from $\mathbb{C}G$ to $\mathbb{C}G \otimes \mathbb{C}G$ (and hence from G to $\mathbb{C}G \otimes \mathbb{C}G$ by restriction). The main point is that Δ is not fixed by group theory, but is an independent choice.

(i) The trivial representation $U^{(0)}$:

As the vacuum state is invariant under the group action, we have $U^{(0)}(g)|0\rangle = |0\rangle$.

(ii) The one-particle representation $U^{(1)}$:

We can write it as

$$U^{(1)}(g)|\rho\rangle = |\sigma\rangle D_{\sigma\rho}(g), \tag{8}$$

where $D(g)$ is the matrix for $U^{(1)}(g)$.

(iii) The two-particle representation $U^{(2)}$:

Group theory does not fix the action of G on two-particle states. We need a homomorphism $\Delta : G \to \mathbb{C}G \otimes \mathbb{C}G$ to fix $U^{(2)}$. Given Δ, we can write $U^{(2)}(g) = U^{(1)} \otimes U^{(1)}\Delta(g)$. A standard choice for Δ is

$$\Delta(g) = g \otimes g. \tag{9}$$

in that case,

$$U^{(2)}(g) = U^{(1)}(g) \otimes U^{(1)}(g), \tag{10}$$

but this choice is not mandatory. It is enough to have a homomorphism $\Delta : \mathbb{C}G \to \mathbb{C}G \otimes \mathbb{C}G$ (with further properties discussed below).

(iv) N-particle representation $U^{(N)}$:

Once again we are given a homomorphism Δ. We have seen how $U^{(2)}$ is fixed by $U^{(1)}$. We can then set

$$U^{(3)}(g) = \left(U^{(2)} \otimes U^{(1)}\right) \Delta(g),$$
$$U^{(4)}(g) = \left(U^{(3)} \otimes U^{(1)}\right) \Delta(g),$$

and so on.

For the choice $\Delta(g) = g \otimes g$, we get the familiar answer

$$U^{(N)}(g) = U^{(1)}(g) \otimes U^{(1)}(g) \otimes \cdots \otimes U^{(1)}(g). \tag{11}$$

A very important property of the homomorphism Δ is *coassociativity*:

$$(\Delta \otimes \mathrm{Id})\Delta = (\mathrm{Id} \otimes \Delta)\Delta. \tag{12}$$

The importance of this property is that it provides a way to inductively obtain a unique homomorphism

$$\mathbb{C}G \to \underbrace{\mathbb{C}G \otimes \mathbb{C}G \otimes \cdots \mathbb{C}G}_{N \text{ factors}}. \tag{13}$$

In fact, suppose that we have deduced such a homomorphism (for N factors). Then (12) says that splitting any two adjacent $\mathbb{C}G$'s on the right hand side using Δ with the remaining $\mathbb{C}G$'s invariant will give a homomorphsim

$$\mathbb{C}G \to \underbrace{\mathbb{C}G \otimes \mathbb{C}G \otimes \cdots \otimes \mathbb{C}G}_{(N+1) \text{ factors}}, \tag{14}$$

independently of the splitting we choose.

Example 2.1. Consider the underlying group G to be the Euclidean group $\mathcal{E}_3$. Here the natural choices for state vector labels are momenta and components of spin. Define now, for $g \in \mathcal{E}_3$,

$$\Delta_\theta(g) = F_\theta^{-1}(g \otimes g)F_\theta, \tag{15}$$

where

$$F_\theta = e^{\frac{1}{2}P_\mu \theta^{\mu\nu} \otimes P_\nu},$$

$\theta_{\mu\nu}$ being a real antisymmetric matrix and the P_ν being the translation generators of $\mathcal{E}_3$. Notice that the right hand side of (15) is not an element of $G \otimes G$. Rather it is an element of $\mathbb{C}G \otimes \mathbb{C}G$. This example can be generalized if we replace the Euclidean group by the Poincaré group $\mathcal{P}_+^\uparrow$, thus providing an example of particular interest in quantum field theory, in particular for investigations of quantum physics at the Planck scale, where the appropriate algebra of functions on spacetime is expected to be noncommutative.

Changing the coproduct along the lines of (15) is called the Drinfel'd twist.

Problem 2.1. *Check coassociativity of the coproduct Δ_θ from the previous example.*

2.3. *Further properties of $\mathbb{C}G$*

The group algebra $\mathbb{C}G$ has further important properties besides the coproduct. They are significant for physics. Let us now discuss them.

1) A group has the "trivial" one-dimensional representation ϵ:

$$\epsilon : G \to \mathbb{C}, \quad \epsilon(g) = 1.$$

It extends to $\mathbb{C}G$ to give a one-dimensional representation,

$$\epsilon : \int d\mu(g)\alpha(g)g \to \int d\mu(g)\alpha(g)$$

and is then called a counit.
We need ϵ to define state vectors such as the vacuum and operators invariant by G. Under $\mathbb{C}G$, they transform by the representation ϵ.

2) Every element $g \in G$ has an inverse g^{-1}. We define an antipode S on G as the map $g \to g^{-1}$. It is an antihomomorphism ($S(gh) = S(h)S(g)$) and extends by linearity to an antihomomorphism from $\mathbb{C}G$to $\mathbb{C}G$:

$$S : \hat{\alpha} = \int d\mu(g)\alpha(g)g \to S(\hat{\alpha}) = \int d\mu(g)\alpha(g)S(g).$$

3) There is a compatibility condition fulfilled by multiplication and Δ. It is a generalized version of the requirement that they commute. Thus let m denote the multiplication map:

$$m(g \otimes h) = gh, \quad ; g, h \in G.$$

It extends by linearity to $\mathbb{C}G$:

$$m\left(\int d\mu(g)\alpha(g)g \otimes \int d\mu(h)\beta(h)h\right) = \int d\mu(g)d\mu(h)\alpha(g)\beta(h)gh.$$

Then

$$\Delta \circ m(a \otimes b) = (m_{13} \otimes m_{24})(\Delta \otimes \Delta)(a \otimes b).$$

Here

$$(m_{13} \otimes m_{24})(a_1 \otimes a_2 \otimes a_3 \otimes a_4) = a_1 a_3 \otimes a_2 a_4.$$

4) In quantum theory, we are generally interested in unitary representations U of G, which means that $U(g^{-1}) = U(g)^*$, $*$ denoting the adjoint operation. We can define a $*$-operation on $\mathbb{C}G$ as follows:

$$\left(\int d\mu(g)\alpha(g)g\right)^* = \int d\mu(g)\bar{\alpha}(g)g^{-1}.$$

The unitarity condition on the representation U now translates to the condition $U(\hat{\alpha}^*) = U(\hat{\alpha})^*$. U is thus a "$*$-representation" of $\mathbb{C}G$.

2.4. *Introducing Hopf Algebras*

The definition of a Hopf algebra H can be formulated by extracting the appropriate properties of $\mathbb{C}G$ so that it can be used as a symmetry algebra in quantum theory.

Definition 2.1. A $*$-Hopf algebra H is an associative, $*$-algebra with unit element e, this meaning that

(i) $(\alpha\beta)\gamma = \alpha(\beta\gamma)$ for $\alpha, \beta, \gamma \in H$,

(ii) $* : \alpha \in H \to \alpha^* \in H$,
$(\alpha\beta)^* = \beta^*\alpha^*, \ \ \alpha, \beta \in H$,
$(\lambda\alpha)^* = \bar{\lambda}\alpha^* \ \ \lambda \in \mathbb{C}, \alpha \in H$,

(iii) $e^* = e, \ \ \alpha e = e\alpha = \alpha, \ \forall \alpha \in H$,

together with a coproduct (or comultiplication) Δ, a counit ϵ and an antipode S, with the following properties:

(1) The coproduct

$$\Delta : H \to H \otimes H \tag{16}$$

is a $*$-homomorphism.

(2) The counit $\epsilon : H \to \mathbb{C}$ is a $*$-homomorphism subject to the condition

$$(\mathrm{Id} \otimes \epsilon)\Delta = \mathrm{Id} = (\epsilon \otimes \mathrm{Id})\Delta. \tag{17}$$

(3) The antipode $S : H \to H$ is a $*$-antihomomorphism, $S(\alpha\beta) = S(\beta\alpha)$ subject to the following conditions. Define

$$m_r(\xi \otimes \eta \otimes \rho) = \eta\xi \otimes \rho$$
$$m'_r(\xi \otimes \eta \otimes \rho) = \rho \otimes \eta\xi.$$

Then

$$m_r[(S \otimes \mathrm{Id} \otimes \mathrm{Id})(\mathrm{Id} \otimes \Delta)\Delta(\alpha)] = e \otimes \alpha$$
$$m'_r[(\mathrm{Id} \otimes S \otimes \mathrm{Id})(\mathrm{Id} \otimes \Delta)\Delta(\alpha)] = \alpha \otimes e, \ \ \alpha \in H. \tag{18}$$

(4) It is convenient to write multiplication in terms of the multiplication map m:

$$m(\alpha \otimes \beta) = \alpha\beta.$$

Then, as for $\mathbb{C}G$, we can also define $m_{13} \otimes m_{24}$:

$$(m_{13} \otimes m_{24})(\alpha_1 \otimes \alpha_2 \otimes \alpha_3 \otimes \alpha_4) = \alpha_1\alpha_3 \otimes \alpha_2\alpha_4.$$

This operation extends by linearity to all of $H \otimes H \otimes H \otimes H$. Then we require, as for $\mathbb{C}G$, that

$$\Delta m(\alpha \otimes \beta) = (m_{13} \otimes m_{24})(\Delta \otimes \Delta)(\alpha \otimes \beta). \tag{19}$$

This is a compatibility condition between m and Δ.

Problem 2.2. *Let $\mathcal{A}$ be the algebra generated by an invertible element a and an element b such that $b^n = 0$ and $ab = rba$, where r is a primitive $2n$-th root of unity. Show that it is a Hopf algebra with the coproduct and counit defined by $\Delta(a) = a \otimes a$, $\Delta(b) = a \otimes b + b \otimes a^{-1}$, $\epsilon(a) = 1$, $\epsilon(b) = 0$ and determine the antipode S.*

Problem 2.3. *Let H be a Hopf algebra. A twist F is an element of $H \otimes H$ that is invertible and satisfies*

$$(F \otimes Id)(\Delta \otimes Id)F = (Id \otimes F)(Id \otimes \Delta)F, \quad (\epsilon \otimes Id)F = (Id \otimes \epsilon)F = 1.$$

Show that Δ_F defined by $\Delta_F(h) = F^{-1}\Delta(h)F$ satisfies the coassociativity condition $(\Delta_F \otimes Id)\Delta_F = (Id \otimes \Delta_F)\Delta_F$.

There are several known examples of Hopf algebras with applications in physics. One such well-known example is $U_q(SL(S, \mathbb{C}))$. The Poincar algebra with the twisted coproduct is another such example. It is a non-trivial example as the twisted coproduct of the Poincar group element involves functions of momenta which cannot be expressed in terms of group elements. Other useful references for Hopf algebras and their use in physics, besides our book,[8] are[6,7,9–12] .

We now turn to the notion of statistics of identical particles and its role in the theory of Hopf algebras.

2.5. *Identical Particles and Statistics*

We begin the discussion with a familiar example. Consider a particle in a quantum theory with its associated space V of state vectors. Let a symmetry group G act on V by a representation ρ. The associated two-particle vector space is based on $V \otimes V$. For the conventional choice of the coproduct Δ,

$$\Delta(g) = g \otimes g, \quad g \in G \subset G\mathbb{C}G,$$

the symmetry group G acts on $\xi \otimes \eta \in V \otimes V$ according to

$$\xi \otimes \eta \to (\rho \otimes \rho)\Delta(g)\xi \otimes \eta = \rho(g)\xi \otimes \rho(g)\eta.$$

Let τ be the "flip" operator, defined as the linear extension to all $V \otimes V$ of

$$\tau(\xi \otimes \eta) = \eta \otimes \xi.$$

Problem 2.4. *Show that τ commutes with $(\rho \otimes \rho)\Delta(g)$, for any $g \in G$. Use this to explain how $V \otimes V$ splits as the sum of two eigenspaces $V \otimes_S V$*

and $V \otimes_A V$ for τ (corresponding, respectively, to the eigenvalues 1 and -1) that are invariant under the action of $\mathbb{C}G$. Physically, the space $V \otimes_S V$ describes bosons, whereas $V \otimes_A V$ describes fermions.

We can easily extend the argument of the previous problem to N-particle states. Thus on $V \otimes V \otimes \cdots \otimes V$, we can define transposition operators

$$\tau_{i,i+1} = \underbrace{1 \otimes \cdots \otimes 1}_{(i-1)\text{ factors}} \otimes \tau \underbrace{\otimes 1 \otimes \cdots \otimes 1}_{(N-i-1)\text{ factors}} .$$

They generate the full permutation group S_N and commute with the action of $\mathbb{C}G$. Each subspace of $V \otimes V \otimes \cdots \otimes V$ transforming by an irreducible representation of S_N is invariant under the action of $\mathbb{C}G$. They define particles with definite statistics. Bosons and fermions are obtained by the representations where $\tau_{i,i+1} \to \pm 1$ respectively. The corresponding vector spaces are the symmetrized and anti-symmetrized tensor products $V \otimes_{S,A} V \otimes_{S,A} \cdots \otimes_{S,A} V \equiv V^{\otimes_S}, V^{\otimes_A}$.

In quantum theory, there is the further assumption that *all observables commute with S_N*. Hence the above symmetrized and anti-symmetrized subspaces are invariant under the full observable algebra.

We can proceed as follows to generalize these considerations to any coproduct. The coproduct can written as a series:

$$\Delta(\eta) = \sum_\alpha \eta_\alpha^{(1)} \otimes \eta_\alpha^{(2)} \equiv \eta_\alpha^{(1)} \otimes \eta_\alpha^{(2)}$$
$$\equiv \eta^{(1)} \otimes \eta^{(2)}.$$

Such a notation is called the "Sweedler notation". Then to every coproduct Δ, there is another coproduct Δ^{op} ('op' standing for 'opposite'):

$$\Delta^{\text{op}}(\eta) = \eta^{(2)} \otimes \eta^{(2)}.$$

Suppose that Δ and Δ^{op} are equivalent in the following sense:

There exists an invertible *R-matrix* $R \in H \otimes H$ such that $\Delta^{\text{op}}(\eta) = R\Delta(\eta)R^{-1}$ and fulfills also certain further properties (Yang-Baxter relations) as explained below. Then the Hopf algebra is said to be "quasi-triangular". From these properties it follows that, if *rho* is a representation of H, then $(\rho \otimes \rho)\Delta(\alpha)$ commutes with $\tau(\rho \otimes \rho)R$. This means that, at least for two identical particles, we can use τR instead of τ to define statistics, since we can decompose $V \otimes V$ into irreducible subspaces of τR. These irreducible subspaces describe particles of definite statistics.

We can generalize τR to N-particle sectors. Thus let

$$R_{i,i+1} = \underbrace{1 \otimes 1 \otimes \cdots \otimes 1}_{(i-1)\text{ factors}} \otimes R \otimes \underbrace{1 \otimes 1 \otimes \cdots \otimes 1}_{(N-i-1)\text{ factors}}$$

Then

$$\mathcal{R}_{i,i+1} = \tau_{i,i+1}(\rho \otimes \rho \otimes \cdots \otimes \rho) R_{i,i+1}$$

generalizes $\tau_{i,i+1}$ of $\mathbb{C}G$ on $V \otimes V \otimes \cdots \otimes V$. The group it generates replaces S_N (it is the braid group $\mathcal{B}_N$). Notice that the square of τR is not necessarily the identity. There could be representations where its eigenvalues are phases, leading to "anyons".

Remark 2.1. The Yang-Baxter relations alluded to above read as follows:

$$R_{i+1,i+2} R_{i,i+2} R_{i,i+1} = R_{i,i+1} R_{i,i+2} R_{i+1,i+2}.$$

They can be derived from the requirement that Δ^{op} is $R \Delta R^{-1}$ and both Δ and Δ_{op} satisfy the identity coming from coassociativity. This imposes a condition on R that is fulfilled when R satisfies the Yang-Baxter relation.

Example 2.2. As a simple example, we consider the twisted coproduct

$$\Delta_\theta(g) = F_\theta^{-1}(g \otimes g) F_\theta$$

for the Poincar group (cf. 2.1). Since τ commutes with with $(\rho \otimes \rho)\Delta_0(g)$, one can see from the structure of $\Delta_\theta(g)$ that

$$\tau_\theta = (\rho \otimes \rho) F_\theta^{-1} \tau (\rho \otimes \rho) F_\theta$$

commutes with $(\rho \otimes \rho)\Delta_\theta(g)$. We can also write τ_θ as

$$\tau_\theta = \tau(\rho \otimes \rho) F_\theta^2,$$

from which we obtain, for the R-matrix,

$$R_\theta = F_\theta^2.$$

We can also show this result by examining $\Delta_\theta^{\mathrm{op}}$. It is simple to show that $\Delta_\theta^{\mathrm{op}}(g) = F_\theta(g \otimes g) F_\theta^{-1}$. Clearly, $R_\theta \Delta_\theta(g) = \Delta_\theta^{\mathrm{op}}(g) R_\theta$, as we want. A point to note is that $\tau_\theta^2 = \mathrm{Id}$. Hence the statistics group is still the permutation group S_N.

3. Causality on the Moyal plane

Having introduced Hopf algebras, we now turn our attention to the Moyal plane. As discussed in the introduction, there are strong physical reasons leading us to believe that the structure of spacetime at small scales will be modified. One of the models that have been studied in the context of noncommutative quantum field theory is the Moyal plane. The noncommutativity of spacetime coordinates has very important consequences, affecting, among others, the meanings of causality and of quantum statistics of identical particles. We therefore start this section with a brief discussion on the meaning of causality.

3.1. *What is causality?*

There are different, apparently unrelated, concepts of causality. One of them comes from the Kramers-Kronig relation. Suppose $R(t)$ is a response function for a system that is undisturbed for $t < 0$. The system should not respond before the time at which it is disturbed. Hence we must have:

$$R(t) = 0 \quad t < 0 \tag{1}$$

Its Fourier transform

$$\widetilde{R}(\omega) = \int_{-\infty}^{\infty} dte^{i\omega t} R(t) = \int_{0}^{\infty} e^{i\omega t} R(t) \tag{2}$$

is then holomorphic in

$$\Im\omega > 0. \tag{3}$$

This is a manifestation of causality.

For relativistic systems, causality for events can be described as a partial order: If A is an event that is to the future of an event B, we say that $A > B$. So is O stands for an observer, for an arbitrary event C in the future light cone of O we will have $C > O$ and for any event A in the past light cone we will have $O > A$. An event P that is spacelike relative to O is not causally related to O. Thus, the relation $>$ is a partial order on the set of events. This notion of causality plays a prominent role in Sorkin's approach to quantum gravity.[13]

Yet another manifestation of causality appears in quantum field theory: Local observables $\rho(x)$, $\eta(y)$ commute if x and y are spacelike separated:

$$[\rho(x), \eta(y)] = 0 \quad \text{if} \tag{4}$$

$$(x^0 - y^0)^2 - (\vec{x} - \vec{y})^2 < 0 \tag{5}$$

This is enforced by

$$[\varphi(x), \chi(y)]_- = 0 \quad x \sim y \tag{6}$$

for scalar fields, and by

$$[\psi_\alpha^{(1)}(x), \psi_\alpha^{(2)}(y))]_+ = 0 \quad x \sim y \tag{7}$$

for spinors fields.

Problem 3.1. *Use the real scalar field and a spin 1/2 field as examples in order to illustrate and explain the last claim.*

Now, equations (6) and (7) also express statistics. So causality and statistics are connected.

In the cased of the Moyal Plane, we are confronted with a noncommutative spacetime where commutation relations and hence causality and statistics are deformed.

3.2. *The Moyal plane*

The Groenewold-Moyal (G-M) plane is the algebra of functions $\mathcal{A}_\theta$ on $\mathbb{R}^{d+1}$ with a twisted product:

$$f * g(x) = f e^{\frac{i}{2}\overleftarrow{\partial}_\mu \theta^{\mu N} \overrightarrow{\partial}_\nu} g \tag{8}$$

It implies that spacetime is noncommutative:

$$\widehat{x}_\mu \star \widehat{x}_\nu - \widehat{x}_\nu \star \widehat{x}_\mu = [\widehat{x}_\mu, \widehat{x}_\nu]_\star = i\theta_{\mu\nu} \quad \mu, \nu = 0, 1, \ldots d. \tag{9}$$

Here the coordinate functions x_μ take on an operator meaning: $\widehat{x}_\mu$.

I will now describe a particular approach to the formulation of quantum field theories on the *GM* plane and indicate its physical consequences. The relevance of the *GM*, as explained in the introduction, comes from considerations of physics at the Planck scale,[1] the main conclusion being that at such scales we expect spacetime uncertainties. The *GM* plane *models* such uncertainties.

Recall from the discussion in section 2 that if there is a symmetry group G with elements g acting on single particle Hilbert spaces $\mathcal{H}_i$ by unitary representations $g \to U_i(g)$, then conventionally they act on $\mathcal{H}_1 \otimes \mathcal{H}_2$ by the representation

$$g \to [U_1 \otimes U_2](g \otimes g) \tag{10}$$

As we have seen, the homomorphism

$$\begin{aligned} \Delta : G \to \ & G \otimes G, \\ g \to \ & \Delta(g) := g \otimes g, \end{aligned} \tag{11}$$

underlying these equations is said to be a coproduct on G. So the action of G on multiparticle states involves more than just group theory. It involves the coproduct Δ. Similarly as in example 2.1, let

$$F_\theta = e^{\frac{i}{2}\partial_\mu \otimes \theta^{\mu\nu} \partial_\nu} = \text{“}\mathit{Twist\ element}\text{”} \tag{12}$$

Then

$$f \star g = m_0[F_\theta f \otimes g], \tag{13}$$

where m_0 is the point-wise multiplication map of $\mathcal{A}_0$:

$$m_0(\alpha \otimes \beta)(x) = \alpha(x)\beta(x) \tag{14}$$

Now let Λ be an element of the Poincaré group $\mathcal{P}_+^\uparrow$. For $x \in \mathbb{R}^N$,

$$\Lambda : x \to \Lambda(x) \in \mathbb{R}^N.$$

It acts on functions on $\mathbb{R}^N$ by pull-back:

$$\Lambda : \alpha \to \Lambda^* \alpha, \quad (\Lambda^*\alpha)(x) = \alpha[\Lambda^{-1}(x)].$$

The work of Aschieri *et al.* and Chaichian *et al.* based on Drinfel'd's basic paper shows that $\mathcal{P}_+^\uparrow$ acts on $\mathcal{A}_\theta(\mathbb{R}^N)$ compatibly with m_θ if its coproduct is "twisted" to Δ_θ, where

$$\Delta_\theta(\Lambda) = F_\theta^{-1}(\Lambda \otimes \lambda)F_\theta. \tag{15}$$

But twisting the coproduct implies twisting of statistics in quantum theory. In order to understand how this comes about, let us consider a quantum mechanical two-particle system.

Such a two-particle system, for $\theta^{\mu\nu} = 0$, is a function of two sets variables, “lives” in $\mathcal{A}_0 \otimes \mathcal{A}_0$ and transforms according to the usual coproduct Δ_0. Similarly, in the noncommutative case, the wavefunction lives in $\mathcal{A}_\theta \otimes \mathcal{A}_\theta$ and transforms according to the twisted coproduct Δ_θ.

For $\theta_{\mu\nu} = 0$ we require that the physical wave functions describing identical particles are either symmetric (bosons) or antisymmetric (fermions). That is, we work with either the symmetrized or antisymmetrized tensor product

$$\phi \otimes_{S,A} \chi \equiv \frac{1}{2}(\phi \otimes \chi \pm \chi \otimes \phi) \tag{16}$$

In a Lorentz-invariant theory, these relations have to hold in all frames of reference. But one easily sees that the twisted coproduct action of the Lorentz group is not compatible with the usual symmetrization/antisymmetrization.

Thus let τ_0 be the statistics (flip) operator associated with exchange for $\theta^{\mu\nu} = 0$:

$$\tau_0(\phi \otimes \chi) = \chi \otimes \phi \tag{17}$$

For $\theta^{\mu\nu} = 0$, we have the axiom that τ_0 is superselected. In particular the Lorentz group action, $\Delta_0(\Lambda) = \Lambda \otimes \Lambda$, must and does commute with the statistics operator:

$$\tau_0 \Delta_0(\Lambda) = \Delta_0(\Lambda)\tau_0 \tag{18}$$

Also all the states in a given superselection sector are eigenstates of τ_0 with the same eigenvalue. Given an element $\phi \otimes \chi$ of the tensor product, the physical Hilbert spaces can be constructed from the elements

$$\left(\frac{1 \pm \tau_0}{2}\right)(\phi \otimes \chi). \tag{19}$$

Now

$$\tau_0 F_\theta = F_\theta^{-1} \tau_0, \tag{20}$$

so that

$$\tau_0 \Delta_\theta(\Lambda) \neq \Delta_\theta(\Lambda)\tau_0, \tag{21}$$

showing that the usual statistics is not compatible with the twisted coproduct. But the new statistics operator

$$\tau_\theta \equiv F_\theta^{-1} \tau_0 F_\theta, \qquad \tau_\theta^2 = 1 \otimes 1, \tag{22}$$

does commute with the twisted coproduct Δ_θ:

$$\Delta_\theta(\Lambda) = F_\theta^{-1} \Lambda \otimes \Lambda F_\theta. \tag{23}$$

The states constructed according to

$$\phi \otimes_{S_\theta} \chi \equiv \left(\frac{1 + \tau_0}{2}\right)(\phi \otimes \chi), \tag{24}$$

$$\phi \otimes_{A_\theta} \chi \equiv \left(\frac{1 - \tau_0}{2}\right)(\phi \otimes \chi), \tag{25}$$

form the physical two-particle Hilbert spaces of (generalized) bosons and fermions and obey twisted statistics.

4. Covariant quantum fields on noncommutative spacetimes

4.1. *Poincaré covariance*

As we have seen, the action of the Poincaré group $\mathcal{P}$ on the Minkowski space M^{d+1} is given by

$$(a,\Lambda) \in \mathcal{P} : (a,\Lambda)x = \Lambda x + a. \tag{1}$$

The commutative algebra of functions on M^{d+1} is denoted by $\mathcal{A}_0(M^{d+1})$.

If φ is a quantum relativistic scalar field on M^{d+1}, we require that there exists a unitary representation

$$U : (a,\Lambda) \to U(a,\Lambda) \tag{2}$$

on the Hilbert space $\mathcal{H}$ of state vectors such that

$$U(a,\Lambda)\varphi(x)U(a,\Lambda)^{-1} = \varphi((a,\Delta)x) \tag{3}$$

There are similar requirements on relativistic fields of all spins. They express the requirement that the spacetime transformations can be unitarily implemented in quantum theory. A field φ fulfilling this condition is said to be a (Poincaré) *covariant field.* This is the primitive covariance condition.

Note that for a quantum field φ we have $\varphi \in L(\mathcal{H}) \otimes S(M^{d+1})$ where $L(\mathcal{H})$ are linear operators on $\mathcal{H}$ and $S(M^{d+1})$ are distributions on M^{d+1}. The action of $\mathcal{P}$ on $L(\mathcal{H})$ is given by the adjoint action

$$\mathrm{Ad}U(a,\Lambda)\varphi = U(a,\Lambda)\varphi U(a,\Lambda)^{-1} \tag{4}$$

and its action on $S(M^{d+1})$ is given by

$$(a,\Delta) : \alpha \to (a,\Lambda) \triangleright \alpha, \quad [(a,\Lambda)\alpha](x) \quad \alpha \in S(M^{d+1}) \tag{5}$$

Call this action as V. The coproduct on $\mathcal{P}$ for commutative spacetimes is Δ_0, where

$$\Delta_0((a,\Lambda)) = (a,\Lambda) \otimes (a,\Lambda) \tag{6}$$

Then the covariance property of quantum fields can be expressed as:

$$(\mathrm{Ad}U \otimes V)\Delta_0((a,\Lambda))\varphi = \varphi \tag{7}$$

Products of fields bring in new features which although present for commutative spacetimes, assume prominence on noncommutative spacetimes. Thus, consider a product of quantum fields on M^{d+1}:

$$\varphi(x_1)\varphi(x_2)\ldots\varphi(x_N). \tag{8}$$

This can be understood as the element $\varphi \otimes \varphi \otimes \ldots \otimes \varphi$ belonging to $L(\mathcal{H})\otimes(S(M^{d+1}) \otimes S(M^{d+1}) \otimes \ldots \otimes S(M^{d+1})$ evaluated at $x_1, x_2, \ldots x_N$. Thus for $\varphi^{\otimes N} = \varphi \otimes \varphi \otimes \ldots \otimes \varphi \in L(\mathcal{H})\otimes(S(M^{d+1})^{\otimes N}$, we will have:

$$(\varphi \otimes \ldots \otimes \varphi)(x_1, x_2, \ldots, x_N) = \varphi(x_1)\varphi(x_2)\ldots\varphi(x_N) \tag{9}$$

Note that tensoring refers only to $S(M^{d+1})$, there is no tensoring involving $L(\mathcal{H})$. There is only one Hilbert space $\mathcal{H}$ which for free particles is the Fock space and $U(a, \Lambda)$ acts by conjugation on it for all N. But that is not the case for $S(M^{d+1})^{\otimes N}$. The Poincaré group acts on it by means of the coproduct

$$\underbrace{(\mathbb{I} \otimes \mathbb{I} \otimes \mathbb{I} \ldots \mathbb{I} \otimes \Delta_0}_{N-1}\underbrace{(\mathbb{I} \otimes \mathbb{I} \otimes \mathbb{I} \ldots \mathbb{I} \otimes \Delta_0}_{N-2}) \ldots \Delta_0 \tag{10}$$

of (a, Λ). We will denote it by Δ_0^N in what follows. Thus

$$\Delta_0^N(a, \Lambda) = (a, \Lambda) \otimes (a, \Lambda) \otimes \cdots \otimes (a, \Lambda). \tag{11}$$

Covariance is now the demand

$$U(a, \Lambda)(\varphi^{\otimes N}((a, \Lambda)^{-1}x_1, \ldots, (a, \Lambda)^{-1}x_N))U(a, \Lambda)^{-1} = \varphi^{\otimes N}(x_1, x_2, \ldots x_N) \tag{12}$$

for all $(a, \Lambda) \in \mathcal{P}$.

Free, in and out- fields. For a free real scalar field φ of mass m, we have

$$\varphi = \int d\mu(p)(c_p^\dagger e_p + c_p e_{-p}) = \varphi^{(-)} + \varphi^{(+)} \tag{13}$$

and

$$e_p(x) = \exp -ip \cdot x, \quad |p_0| = (\mathbf{p}^2 + m^2)^{\frac{1}{2}}, \quad d\mu(p) = \frac{d^d p}{2\,|p_0|}, \tag{14}$$

where c_p, $c_p^\dagger$ are the standard annihilation and creation operators, and $\varphi^{(\pm)}$ refer to the annihilation and creation parts of φ. Now $\varphi^{(\pm)}$ must separately fulfill the covariance requirement. Let us consider $\varphi^{(-)}$. We have that

$$\varphi^{(-)}(x_1)\varphi^{(-)}(x_2)\ldots\varphi^{(-)}(x_N)|0\rangle =$$
$$\int \prod_i d\mu(p_i) c_{p_1}^\dagger(x_1) c_{p_2}^\dagger(x_2) \ldots c_{p_N}^\dagger(x_N)|0\rangle e_{p_1}(x_1) e_{p_2}(x_2) \ldots e_{p_N}(x_N)$$

Let us check this for translations. Let P_μ be the translation generators on the Hilbert space

$$[P_\mu, c_p^\dagger] = P_\mu c_p^\dagger, \quad P_\mu|0\rangle = 0 \tag{15}$$

and let $\mathcal{P}_\mu = -i\partial_\mu$ be the representation of the translation generator on $S(M^{d+1})$:

$$\mathcal{P}_\mu e_p = -p_\mu e_p \tag{16}$$

The coproduct Δ_0 gives for the Lie algebra element $\mathcal{P}_\mu$,

$$\Delta_0(\mathcal{P}_\mu) = \mathbb{I} \otimes \mathcal{P}_\mu + \mathcal{P}_\mu \otimes \mathbb{I} \tag{17}$$

If $\underline{\nu}$ is the representation of the Lie algebra of $\mathcal{P}$ on functions, and $\widehat{P}_\mu$ is the Lie algebra generator in the abstract group $\mathcal{P}$ so that $\underline{\nu}(\widehat{P}_\mu) = \mathcal{P}_\mu$, $\mathcal{P}_\mu$ should strictly read $\widehat{P}_\mu$. Acting on functions, $\widehat{P}_\mu$ should be replaced by $\underline{\nu}(\widehat{P}_\mu) = \mathcal{P}_\mu$. It follows that

$$(\mathbb{I}\otimes\mathbb{I}\otimes\mathbb{I}\ldots\mathbb{I}\otimes\Delta_0)\ldots\Delta_0(\mathcal{P}_\mu)e_{p_1}\otimes e_{p_2}\otimes\ldots e_{p_N} = -\sum_i p_{ij}e_{p_1}\otimes e_{p_2}\otimes\ldots e_{p_N}. \tag{18}$$

Covariance for translations is the requirement

$$\begin{aligned} P_\mu c^\dagger_{p_1} c^\dagger_{p_2} \ldots c^\dagger_{p_N}|0\rangle e_{p_1} \otimes e_{p_2} \otimes \ldots e_{p_N} \ + \\ c^\dagger_{p_1} c^\dagger_{p_2} \ldots c^\dagger_{p_N}|0\rangle (-\sum_i P_{i\mu}) e_{p_1} \otimes e_{p_2} \otimes \ldots e_{p_N} = 0, \end{aligned} \tag{19}$$

which is clearly fulfilled.

Next consider Lorentz transformations. A Lorentz transformation Λ acts on e_p according to

$$(\Lambda e_p)(x) = e_p(\Lambda^{-1}x) = e_{\Lambda p(x)}, \tag{20}$$

or $\Lambda e_p = e_{\Lambda p}$. For Lorentz transformations Λ, covariance is thus the identity

$$\begin{aligned} \int \prod_i d\mu(p_i) c^\dagger_{\Delta p_1} c^\dagger_{\Delta p_2} \cdots c^\dagger_{\Delta p_N}|0\rangle e_{\Delta p_1} \otimes e_{\Delta p_1} \otimes \ldots \otimes e_{\Delta p_N} = \\ \int \prod_i d\mu(p_i) c^\dagger_{p_1} c^\dagger_{p_2} \ldots c^\dagger_{p_N}|0\rangle e_{p_1} \otimes e_{p_1} \otimes \ldots \otimes e_{p_N}, \end{aligned} \tag{21}$$

which is true because of the Lorentz invariance of the measure:

$$d\mu(\Lambda^{-1}p_i) = d\mu(p_i) \tag{22}$$

4.2. *Quantum statistics: The Schur-Weyl duality*

The permutation group S_N and its irreducible representations govern statistics of N-particle state vectors on commutative spacetimes for $d \geq 3$. We consider only such d. It is a fundamental axiom of quantum theory that the N-particle observables must commute with the action of S_N so that the action of observation does not affect particle identity. In particular the action of the symmetry group must commute with the action of S_N. If that is the case, we can consistently work with irreducible representations of S_N. (a, Δ) acts on $e_{p_1} \otimes e_{p_2} \otimes \ldots \otimes e_{p_N}$ via the coproduct. This action commutes with the action of S_N if S_N acts by permuting p_i. Thus we can work with irreducible representation of S_N. In particular we can work with bosons and fermions by totally symmetrizing or antisymmetrizing e_{p_i}. The important point here is that the group algebras $\mathbb{C}\mathcal{P}$ and $\mathbb{C}S_N$ are commutants of each other in their action on N-particle states.

We now consider the double commutant theorem and the Schur-Weyl duality. A result of this sort is familiar to particle physicists in case the symmetry group is $U(k)$. Here $U(k)$ can be the k-flavour symmetry group. It acts on $\mathbb{C}^k$. Then to reduce the representation of $U(k)$ on $\mathbb{C}^{k \otimes N}$, we use the fact that $\mathbb{C}S_N$ commutes with $\mathbb{C}U(k)$. That lets us use Young tableaux methods. It is in fact the case that $\mathbb{C}U(k)$ and $\mathbb{C}S_N$ exhaust the commutants of each other. This result and the Young tableaux methods are part of the contents of Schur-Weyl duality. So we are working with aspects of an infinite-dimensional analogue of this duality for a noncompact symmetry group $\mathcal{P}$ when we remark that $\mathbb{C}\mathcal{P}$ and $\mathbb{C}S_N$ mutually commute.

Let us imagine that S_N acts by transforming N objects numbered from 1 to N and let τ_{ij} denote the transposition of objects i and j. Then S_N has the presentation

$$S_N = \langle \tau_{i,i+1} | i \in |1, 2, \ldots, N-1|, \tau_{i,i+1}^2 = \mathbb{I}, \\ \tau_{i,i+1}\tau_{i+1,i+2}\tau_{i,i+1} = \tau_{i+1,i+2}\tau_{i,i+1}\tau_{i+1,i+2} \rangle \tag{23}$$

Multiplication map and self-reproduction The multiplication map involves products of fields at the same point and hence the algebra of the underlying manifold. It is not the same as the tensor product which involves products of fields at different points.

There is a further property of φ, involving now the multiplication map, which is easily understood on commutative spacetimes. It has much importance for both commutative and noncommutative spacetimes. It is the property of self-reproduction. Let us first understand this property for $C^\infty(M)$, the set of smooth functions on a manifold M. If $\alpha : p \to \alpha p$, $p \in M$, is a diffeomorphism of M, it acts on $f \in C^\infty(M)$ by pull-back:

$$(\alpha^* f)(p) = f(\alpha p) \tag{24}$$

But $C^\infty(M)$ has a further property, routinely used in differential geometry:

$\mathbb{C}^\infty(M)$ is closed under point-wise multiplication: If $f_1,\ f_2 \in C^\infty(M)$, then:

$$f_1 f_2 \in C^\infty(M), \tag{25}$$

where

$$(f_1 f_2)(p) = f_1(p) f_2(p). \tag{26}$$

This property is very important for noncommutative geometry: The completion of this algebra under the supremum norm gives the commutative algebra of $C^0(M)$, a commutative $C^\star$-algebra. By the Gelfand-Naimark theorem it encodes the topology of M.

We see that multiplication of functions preserves transformation under diffeomorphisms. This simple property generalizes to covariant quantum fields. The pointwise product of covariant quantum fields is covariant.

We will call this property of quantum fields m-covariance (m standing for multiplication map). That means in particular that

$$U(a,\Lambda)\varphi^2((a,\Lambda)^{-1}x)U(a,\Lambda)^{-1} = \varphi^2(x). \tag{27}$$

This result is obviously true modulo renormalization problems. It is at the basis of writing invariant interactions in quantum field theories on $\mathcal{A}_0(M^{d+1})$.

In quantum field theories on $\mathcal{A}_0(M^{d+1})$, another routine requirement is that covariance and the $\star-$ or the adjoint operation are compatible. Thus if y is a covariant complex field,

$$U(a,\Delta)\Psi((a,\Delta)^{-1}x)U(a,\Delta)^{-1} = \Psi \tag{28}$$

we require that $\Psi^\dagger$ is also a covariant complex field. That is fulfilled if $U(a,\Delta)$ is unitary. Thus $\star$-covariance is linked to unitarity of time-evolution and the S-matrix.

As a brief summary of our covariance requirements on quantum fields for commutative spacetimes (ignoring the possibility of parastatistics of order 2 or more) we can state: A quantum field should be $m-$ and $\star$-covariant with commutation or anticommutation relations (symmetrisation postulates) compatible with m- and $\star$-covariance.

4.3. *Covariance on the Moyal plane*

The Moyal plane $\mathcal{A}_\theta(M^{d+1})$ is the algebra of smooth functions on $M^{(d+1)}$ with the product

$$m_\theta(\alpha\otimes\beta) = m_0\mathcal{F}_\theta(\alpha\otimes\beta) \quad \alpha,\beta\in\mathcal{A}_\theta(M^{d+1}), \quad \mathcal{F}_\theta = e^{\frac{i}{2}\theta_{\mu\nu}\partial_\mu\otimes\partial\nu}, \tag{29}$$

where m_0 is the point-wise product

$$m_\theta(\gamma \otimes \delta) = \gamma(x)\delta(x), \quad \gamma, \delta \in \mathcal{A}_0(M^{(d+1)}). \tag{30}$$

The Poincaré group $\mathcal{P}$ acts on smooth functions α on $M^{(d+1)}$ by pull-back as before:

$$\mathcal{P} \ni (a, \Delta) : \alpha \to (a, \Delta)\alpha, \quad ((a, \Delta)\alpha)(x) = \alpha(((a, \Delta)^{-1}x). \tag{31}$$

This action extends to the algebra $\mathcal{A}_\theta(M^{(d+1)}$ compatibly with the product m_θ only if the coproduct on $\mathcal{P}$ is twisted. The twisted coproduct Δ_θ on $\mathcal{P}$ is

$$\Delta_\theta(g) = F_\theta^{-1}(g \otimes g)F_\theta; \quad F_\theta = \exp^{\frac{i}{2}\hat{P}_\mu \otimes \theta_{\mu\nu}\hat{P}_\nu}, \tag{32}$$

with F_θ the Drinfel'd twist. Here $\hat{P}_\mu$ is as before the translation generator in $\mathcal{P}$ with representatives $\mathcal{P}_\mu = -i\partial_\mu$ and P_μ on functions and $L(\mathcal{H})$ respectively.

Let φ_θ be the twisted analogue of the field φ and let U_θ be the unitary operator implementing $\mathcal{P}$ in $L(\mathcal{H})$. Covariance then is the requirement

$$U_\theta(a, \Lambda)\varphi_\theta((a, \Lambda)^{-1}x)U_\theta(a, \Lambda)^{-1} = \varphi_\theta(x) \tag{33}$$

and its multifield generalization, while compatibility with $\star$ or unitarity requires that $\varphi_\theta^\dagger$ is also covariant. There is also one further requirement, namely compatibility with symmetrization postulate. The analysis of these requirements becomes transparent on working with the mode expansion of φ_θ which is assumed to exist:

$$\varphi = \int d\mu(p)[a_p^\dagger e_p + a_p e_{-p}] = \varphi_\theta^{(-)} + \varphi_\theta^{(+)}, \quad d\mu(p) = \frac{d^d p}{2\,|p_0|}. \tag{34}$$

The expansion can refer to in-, out- or free fields. We also assume the existence of vacuum $|0\rangle$:

$$a_p|0\rangle = 0, \forall p. \tag{35}$$

We require that

$$U_\theta(a, \Lambda)a_p^\dagger U_\theta(a, \Lambda)^{-1} = a_{\Lambda p}^\dagger, \quad U_\theta(a, \Lambda)a_p U_\theta(a, \Lambda)^{-1} = a_{\Delta p} \tag{36}$$

single particle states transform for all θ in the same manner or assuming that $U_\theta(a, \Lambda)|0\rangle = |0\rangle$:

$$U_\theta(a, \Lambda)|0\rangle = a_{\Lambda p}^\dagger|0\rangle. \tag{37}$$

New physics can be expected only in multi-particle sectors.

Covariance in multi-particle sectors. On the Moyal plane, multi-particle wave functions $e_{p_1} \otimes e_{p_2} \otimes \ldots \otimes e_{p_N}$ transform under $\mathcal{P}$ with the twisted coproduct. This affects the properties of a_p, $a_p^\dagger$ in a $\theta_{\mu\nu}$-dependent manner. Let us focus on the two-particle sector:

$$\int \prod_i d\mu(p_i) a_{p_1}^\dagger a_{p_2}^\dagger |0\rangle e_{p_1} \otimes e_{p_2} \tag{38}$$

Since translations act in the usual way on $e_{p_1} \otimes e_{p_2}$,

$$\Delta_\theta(P_\mu) e_{p_1} \otimes e_{p_2} = (\mathbb{I} \otimes P_\mu + P_\mu \otimes \mathbb{I}) e_{p_1} \otimes e_{p_2} = -(\sum_i p_{i\mu} e_{p_1} \otimes e_{p_2}). \tag{39}$$

Translational covariance requires the standard transformation of $a_{p_1}^\dagger$:

$$[P_\mu^\theta, a_p^\dagger] = P_\mu a_p^\dagger, \tag{40}$$

P_μ^θ is the possibly θ dependent translation generator. Lorentz transformations are more interesting. We have that

$$\Delta_\theta(\Lambda) \triangleright e_{p_1} \otimes e_{p_2} = \mathcal{F}_\theta^{-1} (\Lambda \otimes \Lambda) \mathcal{F} e_{p_1} \otimes e_{p_2}. \tag{41}$$

Covariance thus requires that

$$\int \prod_i d\mu(p_i) U_\theta(\Lambda) a_{p_1}^\dagger a_{p_2}^\dagger |0\rangle \exp^{\frac{i}{2}(\Lambda p_1)} \wedge (\Lambda p_2) \exp^{\frac{i}{2}(p_1)\wedge(p_2)} e_{\Lambda p_1} \otimes e_{\Lambda p_2} =$$

$$\int \prod_i d\mu(p_i) U_\theta(\Lambda) a_{p_1}^\dagger a_{p_2}^\dagger |0\rangle e_{\Lambda p_1} \otimes e_{\Lambda p_2}$$

The covariance requirement is solved by writing $a_p^\dagger$ in terms of the $c_p^\dagger$ and P_μ:

$$a_p^\dagger = c_p^\dagger e^{\frac{i}{2} p \wedge P} \tag{42}$$

and setting

$$U_\theta(a, \Lambda) = U_0(a, \Lambda) = U(a, \Lambda). \tag{43}$$

Its adjoint is

$$a_p = e^{-\frac{i}{2} p \wedge P} c_p = c_p e^{-\frac{i}{2} p \wedge P}, \tag{44}$$

where the equality in the last step uses the anti-symmetry of $\theta_{\mu\nu}$. As we can twist c_p on left *or* on right, we can write φ_θ as a twist applied to $\varphi_0 \equiv \varphi$:

$$\varphi_\theta = \varphi_0 e^{-\frac{i}{2} \overleftarrow{\partial} \wedge P}. \tag{45}$$

The transformation $\varphi_0 \to \varphi_\theta$ is an example of a *dressing transformation.* This is well-defined for a fully interacting Heisenberg field φ_0 if P_μ stands for the total four momentum of the interacting theory. In that case φ_θ is the twisted Heisenberg field.

We can now check that

$$U(a,\Lambda)\varphi_\theta(x_1)\varphi_\theta(x_2)\ldots\varphi_\theta(x_N)U(a,\Lambda)^{-1}|0\rangle = \varphi_\theta((a,\Lambda)(x_1))\varphi_\theta((a,\Lambda)(x_2))\ldots\varphi_\theta((a,\Lambda)(x_N))|0\rangle \tag{46}$$

with a similar equation for the vacuum $\langle 0|$ put on the left. Since the vacuum is a cyclic vector, we can then be convinced that the dressing transformation fully solves the problem of constructing a covariant quantum field on the Moyal plane at the multi-field level as well.

We will now show that the dressing transformations are exactly what we need to be compatible with appropriate symmetrisation postulates. At the level of the particle dynamics (functions on $M^{(d+1)}$ and their tensor products), it is known that for the coproduct Δ_θ, symmetrisation and antisymmetrisation should be based on the twisted flip operator

$$\tau_\theta = \mathcal{F}_\theta^{-1}\tau_0\mathcal{F}, \tag{47}$$

$$\tau_0\alpha\otimes\beta := \beta\otimes\alpha, \tag{48}$$

where α, β are single particle wave functions. As defined, τ_0 and τ_θ act on two-particle wave functions and generate S_2 since

$$\tau_0^2 = \mathbb{I} \quad \tau_\theta^2 = \mathbb{I}. \tag{49}$$

But soon we will generalise them to N-particles to get S_N. Thus twisted bosons (fermions) have the two-particle plane wave states

$$e_{p_1}\otimes S_\theta e_{p_2} = \frac{\mathbb{I}\pm\tau_\theta}{2}e_{p_1}\otimes e_{p_2} \tag{50}$$

Let us focus on S_θ:

$$\begin{aligned} e_{p_1}\otimes_{S_\theta} e_{p_2} &= \frac{1}{2}[e_{p_1}\otimes e_{p_2} + \mathcal{F}_\theta^{-2}e_{p_2}\otimes e_{p_1}] \\ &= \frac{1}{2}[e_{p_1}\otimes e_{p_2} + e^{ip_2\wedge p_1}e_{p_2}\otimes e_{p_1}] \\ &= e^{ip_2\wedge p_1}e_{p_2}\otimes e_{p_1}. \end{aligned} \tag{51}$$

This gives

$$\int\prod_i^2 d\mu(p_i)a_{p_1}^\dagger a_{p_2}^\dagger|0\rangle e_{p_1}\otimes S_\theta e_{\Lambda p_2}$$

$$= \int\prod_i^2 d\mu(p_i)a_{p_1}^\dagger a_{p_2}^\dagger|0\rangle e^{ip_2\wedge p_1}e_{\Lambda p_2}\otimes S_\theta e_{\Lambda p_1}$$

$$= \int \prod_{i}^{2} d\mu(p_i)(\exp^{ip_1\wedge p_2} a^\dagger_{p_2} a^\dagger_{p_1})|0\rangle e^{ip_1\wedge p_2} e_{\Lambda p_1} \otimes S_\theta e_{\Lambda p_2}. \tag{52}$$

Thus we require that

$$a^\dagger_{p_1} a^\dagger_{p_2} = e^{ip_1\wedge p_2} a^\dagger_{p_2} a^\dagger_{p_1}. \tag{53}$$

We can extend this demonstration regarding the consistency of the twist to multinomials in $a^\dagger$'s and as. In the N-particle sector, call $\mathcal{F}^{ij}_\theta$ the Drinfeld twist where in $\partial_\mu \otimes \partial_\nu$, ∂_μ acts on the ith and ∂_ν on the jth factor in the tensor product.

Define

$$\tau^{ij}_\theta = \mathcal{F}^{-1}_\theta \tau^{ij}_0 \mathcal{F} = \mathcal{F}^{-2}_\theta \tau^{ij}_0, \tag{54}$$

where τ^{ij}_0 flips the entries of an N-fold tensor product by flipping the ith and j^{th} entries. Then

$$(\tau^{ij}_0)^2 = \mathbb{I} \tag{55}$$

which is obvious and

$$\tau^{i,i+1}_\theta \tau^{i+1,i+2}_\theta \tau^{i,i+1}_\theta = \tau^{i+1,i+2}_\theta \tau^{i,i+1}_\theta \tau^{i+1,i+2}_\theta \tag{56}$$

which is not obvious. It follows from (23) that the $\tau^{i,i+1}_\theta$s generate S_N in this sector. One can check that the Poincaré group action with the twisted coproduct commutes with this action of S_N.

Covariance requirements on the Moyal plane has led us to the dressed field. We now require it to be compatible with the $\star$-operation. That is if $\varphi^*_0 = \varphi_0$, we want that $\varphi^*_\theta = \varphi_\theta$ Now

$$\varphi^*_\theta = \exp \frac{i}{2} \partial \wedge P \varphi_0, \tag{57}$$

where ∂_μ acts just on φ_0, P_ν acts on φ_0 and all that may follow. But since P_ν acting on φ_0 $-\partial_\nu \varphi_0$ and $\partial \wedge \partial = 0$, we see that

$$\varphi^*_\theta = \varphi^*_0 e - \frac{i}{2} \overleftarrow{\partial} \wedge P. \tag{58}$$

So the dressing transformations preserves $\star$-covariance. The antisymmetry of θ plays a role in this process.

We finish this section with some remarks. We have seen that covariance is the conceptual basis behind the dressing transformation. This means that symmetry transformations on spacetime and associated structures like suitable symmetrization postulates of particle wave functions are implementable in the quantum Hilbert space. It remains a challenge to locate potential signals of Planck scale spacetime effects at presently accessible energy scales.

5. Phenomenological implications

In this last section we (very briefly) mention some of the most striking phenomenological implications of spacetime noncommutativity. They follow from the general principles developed in these lecture notes. The interested reader may consult our papers[14–17] where all details can be found.

5.1. *The Pauli principle*

In an interesting paper[18] , the statistical potential V_{STAT} between fermions has been computed:

$$\exp(-\beta V_{STAT}(x_1,x_2)) = \langle x_1,x_2|e^{-\beta H}|x_1,x_2\rangle, \tag{1}$$

$$H = \frac{1}{2m}(p_1^2+p_2^2). \tag{2}$$

Here $|x_1,x_2\rangle$ has twisted antisymmetry:

$$\tau_\theta|x_1,x_2\rangle = -|x_1,x_2\rangle. \tag{3}$$

It is explicitly shown not to have an infinitely repulsive core, establishing the violation of Pauli principle, as we had earlier suggested. The result has phenomenological consequences. It affects the Chandrasekhar limit on stars. It will also predict Pauli forbidden transitions on which there are stringent limits.

For example, in the Borexino and Super Kamiokande experiments, the forbidden transitions from $O^{16}(C^{12})$ to $\tilde{O}^{16}(\tilde{C}^{12})$ where the tilde nuclei have an extra nucleon in the filled $1S_{1/2}$ level are found to have lifetimes greater than 10^{27} years.

There are also experiments on forbidden transitions to filled K-shells of crystals done in Maryland which give branching ratios less than 10^{-25} for such transitions. The consequences of these results to noncommutative models are yet to be studied. A more detailed discussion of this issues can be found in[19] .

5.2. *Causality on the Moyal plane*

On the Moyal plane the fields ϕ_θ are not local, as can be seen from

$$\phi_\theta(x)\phi_\theta(y) = e^{-\frac{1}{2}\frac{\partial}{\partial x^\mu}\theta^{\mu\nu}\frac{\partial}{\partial y^\nu}}\phi_0(x)\phi_0(y)e^{\frac{1}{2}\left(\frac{\overleftarrow{\partial}}{\partial x^\mu}+\frac{\overleftarrow{\partial}}{\partial y^\mu}\right)\theta^{\mu\nu}P_\nu} \neq \phi_0(x)\phi_0(y). \tag{4}$$

In general observables are not local, but for translationally invariant states locality and causality are restored.

Let us consider some examples:

a) For ρ_θ, η_θ observables,

$$\begin{aligned}\langle 0|\rho_\theta(x)\eta_\theta(y)|0\rangle &= e^{-\frac{1}{2}\frac{\partial}{\partial x^\mu}\theta^{\mu\nu}\frac{\partial}{\partial y^\nu}}\langle 0|\rho_0(x)\eta_0(y)|0\rangle \\ &= e^{-\frac{1}{2}\frac{\partial}{\partial x^\mu}\theta^{\mu\nu}\frac{\partial}{\partial y^\nu}}F(x-y) = F(x-y) \\ &= \langle 0|\rho_0(x)\eta_0(y)|0\rangle \end{aligned} \tag{5}$$

Thus,

$$\langle [0|\rho_\theta(x), \eta_\theta(y)]|0\rangle = 0. \tag{6}$$

b) The same holds for the Gibbs state ω_β given by:

$$\omega_\beta(\rho_\theta(x)\eta_\theta(y)) = \frac{Tr\left[e^{-\beta H}\rho_\theta(x)\eta_\theta(y)\right]}{Tre^{-\beta H}}, \tag{7}$$

if H is space-time translationally invariant.

Now the FRW metric

$$ds^2 = dt^2 - a(t)^2\left[dr^2 + r^2 d\Omega^2\right] \tag{8}$$

has explicit time-dependence, leading to a causality violation.

Let us now consider causality in relation to Lorentz invariance. For the noncommutative theories we are considering, the S-matrix is not Lorentz invariant. The reason is loss of causality. We can see this as follows. Let H_I be the interaction Hamiltonian density in the interaction representation. The interaction representation S-matrix is

$$S = T\exp\left(-i\int d^4x H_I(x)\right). \tag{9}$$

Bogoliubov and Shirkov and then Weinberg long ago deduced from causality (locality) and relativistic invariance that H_I must a local field:

$$[H_I(x), H_I(y)] = 0, \tag{10}$$

where $x \sim y$ means x and y are space-like separated. That is because the step functions in time-ordering are Lorentz invariant only for time-like separations. Thus in second order

$$S^{(2)} = \frac{-i^2}{2!}\int d_x^4 d^4y T(H_I(x), H_I(y)) \tag{11}$$

$$= \frac{-i^2}{2!}\int d_x^4 d^4y\left(\theta(x_0 - y_0)[H_I(x), H_I(y)] + H_I(y)H_I(x)\right), \tag{12}$$

this leading to

$$U(\Lambda)S^{(2)}U^{-1}(\Lambda) = \frac{-i^2}{2!}\int d_x^4 d^4y\,(\theta(x_0 - y_0)[H_I(x), H_I(y)] + H_I(y)H_I(x)) \tag{13}$$

This is $S^{(2)}$ only if

$$\theta(x_0 - y_0)[H_I(\Lambda x), H_I(\Lambda y)] = \theta((\Lambda x)_0 - (\Lambda y)_0)[H_I(\Lambda x), H_I(\Lambda y)] \tag{14}$$

Now, for time-like separations we have

$$\theta(x_0 - y_0) = \theta((\Lambda x)_0 - (\Lambda y)_0), \tag{15}$$

but not for space-like separations, hence we need causality (locality),

$$[H_I(\Lambda x), H_I(\Lambda y)] = 0, \quad x \sim y, \tag{16}$$

for Lorentz invariance.

But noncommutative theories are nonlocal and violate this condition: This is the essential reason for Lorentz noninvariance. The effect on scattering amplitudes is striking. They depend on total incident momentum $\vec{P}_{inc}$ through

$$\theta_{0i}\left(\vec{P}_{inc}\right)_i \tag{17}$$

So effects of $\theta_{\mu\nu}$ disappear in the center-of-mass system, or more generally if

$$\theta_{0i}\left(\vec{P}_{inc}\right)_i = 0 \tag{18}$$

but otherwise there is dependence on θ_{0i}. The violation is of order $theta_{0i}\left(\vec{P}_{inc}\right)_i$ in cross sections.

Let us remark that even with noncommutativity the decay $Z^0 \to 2\gamma$ is forbidden in the approach of Aschieri *et al.* More generally, a massive particle of spin j does not decay into two massless particles of same helicity if j is odd.

5.3. *Cosmic microwave background*

In 1992, the COBE satellite detected anisotropies in the CMB radiation, which led to the conclusion that the early universe was not smooth; there were small perturbations in the photon-baryon fluid. The perturbations could be due to the quantum fluctuations in the inflaton (the scalar field driving inflation). The quantum fluctuations were transmitted into the metric as primordial perturbations.

The temperature field in the sky can be expanded in spherical harmonics:

$$\frac{\Delta T(\widehat{n})}{T} = \sum_{lm} a_{lm} Y_{lm}(\widehat{n}). \tag{19}$$

The a_{lm} can be written in terms of perturbations to Newtonian potential Φ

$$a_{lm} = 4\pi(-i)^l \int \frac{d^3k}{(2\pi)^3} \Phi(k) \Delta_l^T(k) Y_{lm}^*(\widehat{k}) \tag{20}$$

where $\Delta_l^T(k)$ are called the transfer functions. We find

$$\begin{aligned}
&\langle a_{lm} a_{l'm'}^* \rangle_\theta \\
&= *\pi^2 \int dk \sum_{l''=0, l'':even}^{\infty} i^{l+l'}(-1)^{l+m}(2l''+1)k^2 \Delta_l(k)\Delta_l'(k) P_\Phi(k) i_{l''}(\theta k H) \\
&\quad \times \sqrt{(2l+1)(2l'+1)} \begin{bmatrix} l & l' & l'' \\ 0 & 0 & 0 \end{bmatrix} \begin{bmatrix} l & l' & l'' \\ -m & m' & 0 \end{bmatrix},
\end{aligned}$$

where $\vec{\theta^0} = \theta(0,0,1)$. P_Φ is the power spectrum

$$P_\Phi = \frac{8\pi G H^2}{9\varepsilon k^3}|_{aH=k} \tag{21}$$

with H the Hubble parameter and ε the slow-roll parameter. The angular correlator $\langle a_{lm} a_{l'm'}^* \rangle_\theta$ is θ dependent indicating a preferred direction. Correlation functions are not invariant under rotations. They are not Gaussian either.

It clearly breaks isotropy in the CMB radiation. On fitting data, one finds, $\sqrt{\theta} \leq 10^{-32}$ meters or energy scale $\geq 10^{16}$ GeV.

If φ_θ is a (twisted) inflaton field and

$$\varphi(\alpha, x) = \left(\frac{\alpha}{3}\right)^{\frac{3}{2}} \int d^3x' \varphi_\theta(x) \exp^{-\alpha(x'-x)^2}, \tag{22}$$

then for mean square fluctuations

$$\Delta\varphi_\theta(\alpha, x_1)\Delta\varphi_\theta(\alpha, x_2) \geq \langle 0|\varphi_\theta(\alpha, x_1), \varphi_\theta(\alpha, x_2)|0\rangle \tag{23}$$

by uncertainty principle, and it is non-zero by causality violation, for $x_1 \sim x_2$. This implies a new uncertainty principle.

References

1. S. Doplicher, K. Fredenhagen and J. Roberts, *Communications in Mathematical Physics* **172**, 187 (1995).
2. H. S. Snyder, *Phys. Rev.* **71**, 38 (1947).
3. P. Aschieri, M. Dimitrijevic, P. Kulish, F. Lizzi and J. Wess, *Noncommutative Spacetimes*, Lecture Notes in Physics, Vol. 774 (Springer Verlag, 2009).
4. V. Drinfel'd, *Journal of Soviet Mathematics* **41**, 898 (1988).
5. S. L. Woronowicz, *Symétries quantiques (Les Houches, 1995)* **845**, p. 884 (1998).
6. S. Majid, *Foundations of Quantum Group Theory* (Cambridge University Press, 1995).
7. G. Mack and V. Schomerus, Quantum symmetry in quantum theory, in *New Symmetry Principles in Quantum Field Theory*, eds. J. Frhlich, G. Hooft, A. Jaffe, G. Mack, P. Mitter and R. Stora, NATO ASI Series, Vol. 295 (Springer US, 1992) pp. 329–353.
8. A. P. Balachandran, S. G. Jo and G. Marmo, *Group Theory and Hopf Algebras: Lectures for Physicists* (World Scientific, 2010).
9. J. Milnor and J. Moore, *Annals of Math.* **81**, 211 (1965).
10. M. Sweedler, *Hopf Algebra* (W. A. Benjamin, New York, 1969).
11. V. Chari and A. Pressley, *A Guide to Quantum Groups* (Cambridge University Press, 1994).
12. S. Dascalescu, C. Nastasescu and S. Raianu, *Hopf Algebra, an Introduction* (M. Dekker, New York, 2001).
13. L. Bombelli, J. Lee, D. Meyer and R. D. Sorkin, *Phys. Rev. Lett.* **59**, 521(Aug 1987).
14. A. Balachandran, A. Pinzul and A. Queiroz, *Physics Letters B* **668**, 241 (2008).
15. E. Akofor, A. Balachandran, S. Jo, A. Joseph and B. Qureshi, *Journal of High Energy Physics* **2008**, p. 092 (2008).
16. E. Akofor, A. P. Balachandran, A. Joseph, L. Pekowsky and B. A. Qureshi, *Phys. Rev. D* **79**, p. 063004(Mar 2009).
17. A. P. Balachandran, A. Joseph and P. Padmanabhan, *Phys. Rev. Lett.* **105**, p. 051601(Jul 2010).
18. B. Chakraborty, S. Gangopadhyay, A. G. Hazra and F. G. Scholtz, *Journal of Physics A: Mathematical and General* **39**, p. 9557 (2006).
19. S. Vaidya, A. Pinzul, A. P. Balachandran and G. Mangano, *International Journal of Modern Physics A* **21**, 3111 (2006).

INTEGRABILITY AND THE AdS/CFT CORRESPONDENCE

Matthias Staudacher

Humboldt-Universität zu Berlin
Institut für Mathematik und Institut für Physik, IRIS-Haus
Zum Großen Windkanal 6, 12489 Berlin, Germany
Email: matthias@aei.mpg.de

We present a brief, pedagogical introduction to the field of integrability in gauge and string theory, which has been an intensely researched topic in the last few years. We point the reader intending to learn the subject to more extensive reviews of this vast and technically demanding subject.

Keywords: Integrability, AdS/CFT Correspondence.

1. Introduction

Integrability is frequently confused with exact solubility. In classical mechanics integrability occurs when the number of conserved charges matches the number of degrees of freedom. This allows to express the physical variables of the system as functions of the conserved parameters. These might be combinations of elementary or higher transcendental functions, which are solutions of differential or integral equations. In turn, the quantum case is much harder, because generically the dimension of the Hilbert space of a quantum mechanical system is either infinite dimensional, or, as in the case of spin chains, at least very large. It is in any case much larger than the naive number of degrees of freedom. For now nearly fifty years the Yang-Baxter equation has been studied as the master equation for integrable models in quantum field theory and statistical mechanics. It can be considered to be a consistency condition for reducing the n-body scattering problem to a sequence of two-body processes. Hence, in the quantum case it is more useful to define integrability through Yang-Baxter symmetry. For an extensive in-depth discussion, *cf.*[1]

In this lecture we will show how integrability appears in the framework of the AdS/CFT correspondence. CFT stands for Conformal Field Theory, which is a theory invariant under the conformal group, i.e. invariant under Poincaré symmetry and inversion symmetry. This then leads to special conformal transformations as a sequence of inversion, translation, and inversion, and finally scaling symmetry appearing in the commutators of translations and special transformations. One rather trivial example is free massless scalar field theory. Here we will consider the most famous instance of AdS/CFT, where the conformal field theory is $\mathcal{N} = 4$ supersymmetric Yang-Mills theory. AdS stands for a type IIB a theory, where the embedding space is the maximally supersymmetric completion of the $5 + 5 = 10$ dimensional product spaces $AdS_5 \times S^5$, the 5-dimensional Anti-de-Sitter space and the 5-sphere. The fermionic coordinates can be intuitively considered to be "brackets" stapling the AdS_5 and S^5 spaces together. The reason is that all supercharges carry one index referring to the AdS space and a second index referring to the sphere S^5.

In 1997 Juan Maldacena[2] conjectured an exact correspondence between this Yang-Mills and this string theory. See[3] and[4] for particularly pedagogical introductions to this subject, and[5] for the historic roots of this approach in mathematics and physics. The conjecture predicts that the string theory in this background at the boundary of AdS_5 spacetime is a 4-dimensional Conformal Field Theory. The AdS/CFT duality then relates the string partition function at the boundary of AdS_5 to the CFT_4 partition function. Here the sources ϕ for string vertex operators at the boundary fix the sources J for local operators

$$Z_{str}[\phi\ |_{\partial\, AdS}= J] = Z_{CFT}[J]$$

More colloquially: For every string observable at the boundary of AdS_5 there is a corresponding observable in CFT_4 and vice versa. Finding a proof of AdS/CFT integrability will likely lead to a proof of Maldacena's conjecture in the so-called free/planar limit (the string theory is free, and the planar limit of the gauge theory is taken).

Integrability is a phenomenon which is typically confined to two-dimensional models, but oddly here it helps in solving a four-dimensional QFT. An old dream of QFT is to express the masses of the particles as functions of the parameters of the theory. $\mathcal{N} = 4$ SYM has a richer set of symmetries: Supersymmetry and Conformal Symmetry. This leads to simplifications and even allows for exact statements about both models. Being massless, in the gauge theory there nevertheless exists a characteristic set of numbers replacing the masses, the scaling dimensions. In the planar model one is able to express the scaling dimensions as functions of the coupling constant λ. These functions are given as solutions of a set of equations following from a Bethe Ansatz. Thus it appears that planar $\mathcal{N} = 4$ SYM can be solved exactly. This miracle which leads to the solution of this planar model is quantum integrability. By means of the AdS/CFT duality it translates to integrability of the string worldsheet model, a two-dimensional non-linear sigma model on a symmetric coset space, for which integrability is a common phenomenon.

In the first part of this lecture we discuss the discovery of quantum integrability by Bethe in 1931.[6] We begin with the one dimensional spin chain followed by the Heisenberg spin chain, XXX model. After that we present the coordinate Bethe Ansatz and the Algebraic Bethe Ansatz. We then exhibit in a simple case at one-loop integrability of planar $\mathcal{N} = 4$ SYM by sketching the emergence of spin chains in this model. We end with remarks on the symmetry structure of IIB string theory on AdS and sketch the excitation spectrum. These lectures are meant to merely excite the interest of the reader. Many crucial topics are not discussed here, such as the structure of integrability at higher loops and for the full set of operators in the gauge theory, and the perturbative quantization of the string sigma model. A much more detailed review series of this field is presented in the collection of articles.[7]

We thank Alejandro Gonzales, Diego Rengifo, Jan Fokken, Rouven Frassek, Nils Kanning, and Hernan Ocampo who either took notes and/or significantly contributed to the writing of these lectures.

2. The one-dimensional Heisenberg spin chain

Let us recall spin-$\frac{1}{2}$ particles from quantum mechanics like for example the electron. The spin part of the Hilbert space of the system is $\mathbb{C}^2$. The operator $\vec{S}$ associated to the spin can be expressed via the Pauli matrices

$$\vec{S} = \frac{1}{2}\vec{\sigma}\,, \tag{1}$$

with

$$\sigma^1 = \begin{pmatrix} 0 & +1 \\ +1 & 0 \end{pmatrix}, \quad \sigma^2 = \begin{pmatrix} 0 & -i \\ +i & 0 \end{pmatrix}, \quad \sigma^3 = \begin{pmatrix} +1 & 0 \\ 0 & -1 \end{pmatrix}. \tag{2}$$

These matrices generate the simplest non-trivial (fundamental) representation of the Lie algebra $\mathfrak{su}(2)$ defined by the commutation relations

$$[\sigma^i, \sigma^j] = 2i\epsilon^{ijk}\sigma^k\,,$$

or equivalently

$$[S^i, S^j] = i\epsilon^{ijk} S^k \,. \tag{3}$$

The unitary irreducible representations are characterized by the spin

$$s = 0, \frac{1}{2}, 1, \frac{3}{2}, \dots \,, \tag{4}$$

where the trivial case with $s = 0$ is called the singlet representation. In this case the state is annihilated by the spin operator

$$\vec{S} \cdot |\text{state}\rangle = 0 \,, \tag{5}$$

and we say that a state is invariant under the Lie group SU(2). The Lie algebra consists of the elements of the tangent space at the identity of the group, equipped with a product defined by the above commutator. The exponential map extends the tangent space to the component of the group manifold connected to the identity. Schematically, for SU(2)

$$\text{“}Group\text{”} \approx e^{i\text{“}Algebra\text{”}} \,,$$

$$SU(2) \approx e^{i\mathfrak{su}(2)} \,.$$

Similar to the case of the quantum harmonic oscillator it is possible to define $\mathfrak{su}(2)$ in terms of ladder operators

$$S^{\pm} = S^1 \pm iS^2 \,. \tag{6}$$

From (6) immediately follows that

$$[S^3, S^{\pm}] = \pm S^{\pm}, \qquad [S^+, S^-] = 2S^3 \,.$$

In the case $s = \frac{1}{2}$ the ladder operators take the form

$$S^+ = \frac{1}{2}\sigma^+ = \begin{pmatrix} 0 & 1 \\ 0 & 0 \end{pmatrix}, \qquad S^- = \frac{1}{2}\sigma^- = \begin{pmatrix} 0 & 0 \\ 1 & 0 \end{pmatrix},$$

A convenient basis for $\mathbb{C}^2$ that diagonalizes σ^3 is

$$|\uparrow\rangle = \begin{pmatrix} 1 \\ 0 \end{pmatrix}, \qquad |\downarrow\rangle = \begin{pmatrix} 0 \\ 1 \end{pmatrix}.$$

For this basis one has

$$S^3|\uparrow\rangle = +\frac{1}{2}|\uparrow\rangle, \qquad S^3|\downarrow\rangle = -\frac{1}{2}|\downarrow\rangle,$$

and the non-diagonal operators act as

$$S^+|\uparrow\rangle = 0, \qquad S^-|\uparrow\rangle = |\downarrow\rangle,$$
$$S^+|\downarrow\rangle = |\uparrow\rangle, \qquad S^-|\downarrow\rangle = 0 \,.$$

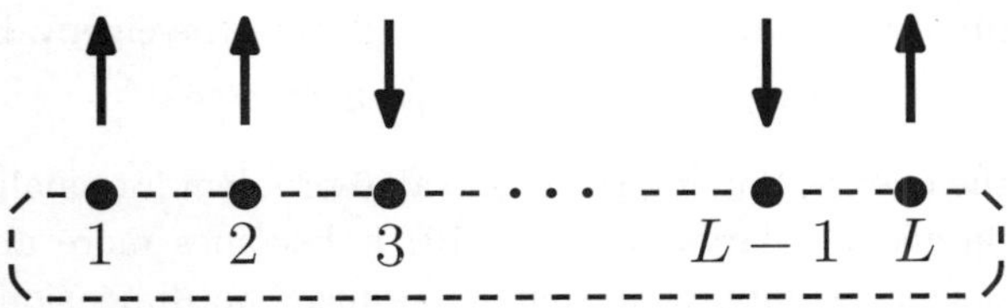

Fig. 1. Spin Chain

From these relations the name ladder operator becomes obvious, S^+ raises the spin and S^- lowers it. Usually $|\uparrow\rangle$ is called highest weight state and $|\downarrow\rangle$ lowest weight state.

Let us consider a one-dimensional spin chain with L sites (see figure 1). The Hilbert space of this system is

$$\mathcal{H} = \mathbb{C}^2 \otimes \cdots \otimes \mathbb{C}^2 = (\mathbb{C}^2)^{\otimes L} \tag{7}$$

with total dimension 2^L. An element of this space in the introduced basis is denoted as e.g.

$$|\uparrow\uparrow \cdots \downarrow\uparrow\rangle = \begin{pmatrix} 1 \\ 0 \end{pmatrix} \otimes \begin{pmatrix} 1 \\ 0 \end{pmatrix} \otimes \cdots \otimes \begin{pmatrix} 0 \\ 1 \end{pmatrix} \otimes \begin{pmatrix} 1 \\ 0 \end{pmatrix} .$$

In the case $L = 2$ the Hilbert space is just $\mathbb{C}^2 \otimes \mathbb{C}^2$. The tensor product decomposition into irreducible representations of the Lie algebra reads

$$\frac{1}{2} \otimes \frac{1}{2} = 1 \oplus 0 .$$

Or in our basis

$$\{|\uparrow\uparrow\rangle, \frac{1}{\sqrt{2}}(|\uparrow\downarrow\rangle + |\downarrow\uparrow\rangle, |\downarrow\downarrow\rangle, \frac{1}{\sqrt{2}}(|\uparrow\downarrow\rangle - |\downarrow\uparrow\rangle\} .$$

Here, the last element is the singlet state and the others build the triplet in the decomposition above. The interaction between the spins is governed by the Heisenberg Hamiltonian which only contains nearest-neighbour interactions. It is given by

$$\hat{H} = 4 \sum_{l=1}^{L} \left(\frac{1}{4} - \vec{S}_l \cdot \vec{S}_{l+1} \right) , \tag{8}$$

with boundary conditions corresponding to a compact topology (closed spin chain)

$$\vec{S}_{L+1} = \vec{S}_1 .$$

The spectrum of the Hamiltonian is obtained from the eigenvalue equation

$$\hat{H}|\psi\rangle = E|\psi\rangle, \qquad |\psi\rangle \in (\mathbb{C}^2)^{\otimes L}, \tag{9}$$

where E is the energy. Solving the eigenvalue problem for small L is rather trivial. For larger and larger L ($L \sim 10$) it becomes more and more involved (even using computer systems) because the dimension of the the Hamiltonian is $2^L \times 2^L$.

To proceed we rewrite (8) using (1) as

$$\begin{aligned}\hat{H} &= \sum_{l=1}^{L} (1 - \vec{\sigma}_l \cdot \vec{\sigma}_{l+1}) \\ &= \sum_{l=1}^{L} \left(1 - \sigma_l^3 \sigma_{l+1}^3 - \frac{1}{2}\sigma_l^+ \sigma_{l+1}^- - \frac{1}{2}\sigma_l^- \sigma_{l+1}^+\right) \\ &= 2\sum_{l=1}^{L} (\mathbb{I}_{l,l+1} - \mathbb{P}_{l,l+1}) \, . \end{aligned} \tag{10}$$

Here

$$\mathbb{P}_{l,l+1} = \begin{pmatrix} 1\,0\,0\,0 \\ 0\,0\,1\,0 \\ 0\,1\,0\,0 \\ 0\,0\,0\,1 \end{pmatrix},$$

is the permutation operator acting on the lth and $l+1^{th}$ space. The Hamiltonian density has the matrix representation

$$\mathcal{H}_{l,l+1} = 2\,(\mathbb{I}_{l,l+1} - \mathbb{P}_{l,l+1}) = \begin{pmatrix} 0 & 0 & 0 & 0 \\ 0 & +2 & -2 & 0 \\ 0 & -2 & +2 & 0 \\ 0 & 0 & 0 & 0 \end{pmatrix},$$

and the full Hamiltonian can be written, compare (10), as

$$\hat{H} = \sum_{l=1}^{L} \mathcal{H}_{l,l+1} \, .$$

An important symmetry for such systems is given by

$$[\vec{S}_{\rm tot}, \hat{H}] = 0\,,$$

where $\vec{S}_{\rm tot}$ is the total spin operator $\vec{S}_{\rm tot} = \sum_{l=1}^{L} \vec{S}_l$ and $\vec{S}_l$ is the local spin operator that only acts non-trivially on the lth site. It means that the Hamiltonian operator preserves the number of up and down spins and therefore is block-diagonal. In total it contains $L+1$ blocks of size $\binom{L}{M} \times \binom{L}{M}$, where M is the total number of down spins (magnon number). Another important operator is the shift operator defined as

$$\hat{U} = \mathbb{P}_{1,2} \cdots \mathbb{P}_{L-1,L} \,.$$

When it acts on any spin chain state, its name becomes obvious[a]. In the case of the closed chain, the shift operator satisfies $[\hat{H}, \hat{U}] = 0$. If one acts L-times with $\hat{U}$ on some state one obtains back the same state

$$\hat{U}^L = \mathbb{I} \,. \tag{11}$$

It follows that the spectrum of the shift operator is given by

$$U = e^{2\pi i n/L} \,, \tag{12}$$

where $n = 0, 1, \ldots, L-1$.
The problem of how one can diagonalize the Hamiltonian and also the shift operator was beautifully answered by Hans Bethe.[6] One just has to solve the equations that carry his name (Bethe equations)

$$\left(\frac{u_k + \frac{i}{2}}{u_k - \frac{i}{2}}\right)^L = \prod_{j \neq k}^{M} \frac{u_k - u_j + i}{u_k - u_j - i}, \qquad k = 1, \ldots, M \,. \tag{13}$$

Here the complex variables u_k are called Bethe roots and one of them is introduced for each spin down (magnon) in the spin chain. From the solutions of these equations we find the spectrum of (8) and (12) from the relations

$$E = 2\sum_{k=1}^{M} \frac{1}{u_k^2 + \frac{1}{4}}, \qquad U = \prod_{k=1}^{M} \frac{u_k + \frac{i}{2}}{u_k - \frac{i}{2}} \,.$$

3. Coordinate Bethe ansatz

To diagonalize the Hamiltonian Bethe proposed an ansatz for the eigenvector in (9). In this way the energy and the eigenfunctions are obtained as functions of the Bethe roots. First we note that any state can be written as

$$|\,\psi\rangle = \sum_{1 \leq l_1 < \cdots < l_M \leq L} \psi(l_1, \ldots, l_M) S_{l_1}^- \cdots S_{l_M}^- \,|\, 0\rangle \,, \tag{1}$$

[a]Strictly speaking, we could define a left-shift and a right-shift operator, but as we are interest in closed chains this distinction is not important here.

where l_i labels the positions of the ith down spins and $\mid 0\rangle = \mid\uparrow \cdots \uparrow\rangle$ is the vacuum state. Now, Bethe's ansatz for the wave function is

$$\psi(l_1, \ldots, l_M) = \sum_{\tau \in S_n} A(\tau) e^{ip_{\tau_1} l_1 + \cdots + ip_{\tau_M} l_M}\,, \tag{2}$$

where S_n is the permutation group of a set containing n elements. The amplitude $A(\tau)$ only depends on the permutation and not on the positions l_i. Acting with the Hamiltonian (8) and neglecting boundary terms one finds that the wave function (1) is a solution of (9) with the amplitudes given by

$$A(\tau) = \text{sgn}(\tau) \prod_{j<k} (e^{ip_k + ip_j} - 2e^{ip_k} + 1)\,,$$

where $\text{sgn}(\tau)$ gives the signature of the permutation τ. The corresponding energy eigenvalue is determined to be

$$E = \sum_{k=1}^{M} 8 \sin^2(\frac{p_k}{2})\,.$$

This is the solution for the infinite system. To obtain the solution of the finite case, periodicity must be imposed on the wave function

$$e^{ip_k L} = \prod_{j \neq k}^{M} S(p_k, p_j) \qquad \text{for} \qquad k = 1, \ldots, M\,,$$

with

$$S(p_k, p_j) = -\frac{e^{ip_k + ip_j} - 2e^{ip_k} + 1}{e^{ip_k + ip_j} - 2e^{ip_j} + 1}\,.$$

After a change of variables

$$e^{ip_k} = \frac{u_k + \frac{i}{2}}{u_k - \frac{i}{2}} \qquad \Longleftrightarrow \qquad \frac{1}{2} \cot(\frac{p_k}{2}) = u_k\,,$$

we arrive at the famous Bethe equations (13). A good reference on this part is e.g.[8]

4. Algebraic Bethe ansatz

The coordinate Bethe ansatz, as discussed in the previous section, solves the problem of diagonalizing the Heisenberg Hamiltonian. However, the reason why this solution works remains mostly unclear in this approach. A beautiful algebraic structure underlying integrable quantum systems originates in work of R. Baxter in the 1970's and was systematized and generalized in the late 1970's and early 1980's leading to the *quantum inverse scattering program* initiated by the "Leningrad School" around L. D. Faddeev. Within this framework there is an alternative method for the diagonalization of the Hamiltonian: the *algebraic Bethe ansatz.*

In this approach the starting point is not the Hamiltonian itself, but a so-called *R-matrix*

$$R_{a,b}(u) = u1_{a,b} + iP_{a,b} = \begin{pmatrix} u+i & 0 & 0 & 0 \\ 0 & u & i & 0 \\ 0 & i & u & 0 \\ 0 & 0 & 0 & u+i \end{pmatrix},$$

which is an operator acting in the tensor product $\mathbb{C}^2 \otimes \mathbb{C}^2$ of two spaces a and b. The complex parameter u is called *spectral parameter.* The characteristic property of this R-matrix is that it solves the *Yang-Baxter equation*[b]

$$R_{a,b}(u-u')R_{a,c}(u)R_{b,c}(u') = R_{b,c}(u')R_{a,c}(u)R_{a,b}(u-u')$$

acting on three spaces $\mathbb{C}^2 \otimes \mathbb{C}^2 \otimes \mathbb{C}^2$.

The R-matrix may be graphically represented as

$R_{a,b}(u-u') =$ 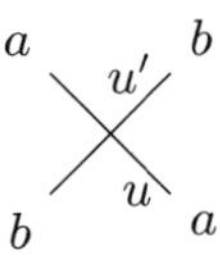,

where each space is corresponds to a line and has an associated spectral parameter. In this language the Yang Baxter equation becomes

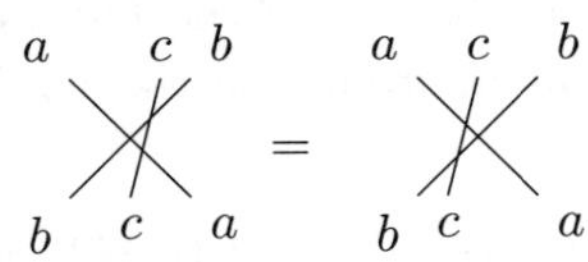

where we left out the spectral parameters for notational convenience. We read the diagrams from the right to the left, which can in some sense be thought of as "time" direction. Then these diagrams are reminiscent of Feynman diagrams, where the intersection of lines describes a scattering event.

To make contact with the Heisenberg spin chain we also introduce the *quantum Lax operator* $L_{a,l}(u) = R_{a,l}(u - \frac{i}{2})$, which is just the R-matrix

[b]This is an equation for 8×8 matrices and can be easily verified by hand (see also[15])!

with a shifted spectral parameter. Later the space l will be identified with the Hilbert space $\mathbb{C}^2$ of the l-th site of the spin chain ("quantum space"). a is an auxiliary space needed for the construction.

Using this terminology the Yang-Baxter equation reads

$$R_{a,b}(u-u')L_{a,l}(u)L_{b,l}(u') = L_{b,l}(u')L_{a,l}(u)R_{a,b}(u-u').$$

Next we build the *monodromy matrix* as a product of Lax operators

$$M_a(u) = L_{a,L}(u)L_{a,L-1}(u)\cdots L_{a,2}(u)L_{a,1}(u)$$

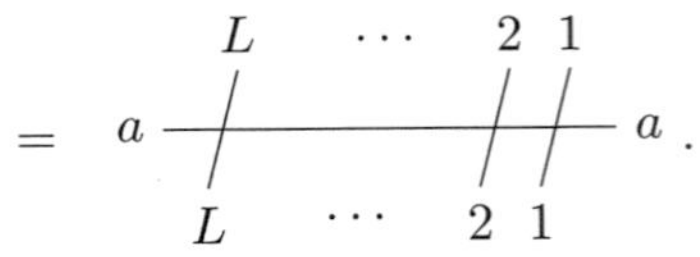

From this the *transfer matrix* is obtained by taking the trace over the two-dimensional auxiliary space a:

$$\hat{T} = \mathrm{tr}_a M_a(u)$$

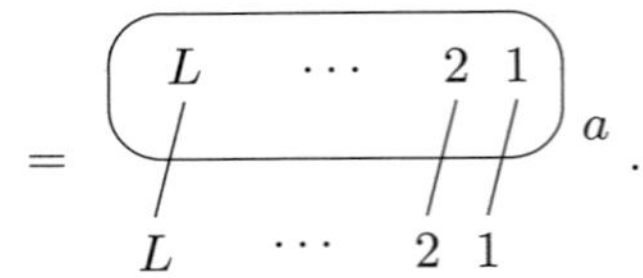

This transfer matrix is an operator acting on the tensor product $(\mathbb{C}^2)^{\otimes L}$ of the quantum spaces $1, \ldots, L$, which is the Hilbert space of the whole spin chain.

Using the Yang-Baxter relation for the Lax operators at the different sites, we see that also the monodromy matrix satisfies a Yang-Baxter equation:

$$R_{a,b}(u-u')M_a(u)M_b(u') = M_b(u')M_a(u)R_{a,b}(u-u').$$

Graphically this equation reads

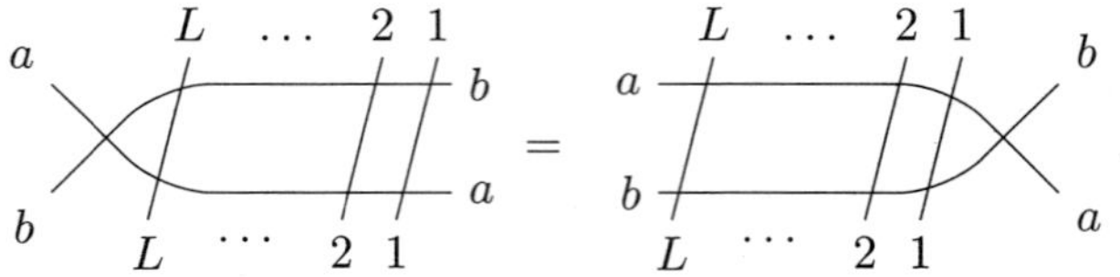

From this pictorial representation the proof of the equation is immediate: one just has to move the lines $1, 2, \ldots, L$ through the intersection of lines a

and b. After multiplying this equation with the inverse of $R_{a,b}(u-u')$ from the left and taking the trace over both auxiliary spaces a and b we obtain

$$\hat{T}(u)\hat{T}(u') = \hat{T}(u')\hat{T}(u)$$

or equivalently

$$[\hat{T}(u), \hat{T}(u')] = 0.$$

This means that transfer matrices with different values of the spectral parameter u can be diagonalized simultaneously.

In order to finally make contact with the Heisenberg spin chain we note that at the special point $u = \frac{i}{2}$ of the spectral parameter the Lax operator reduces essentially to a permutation: $L_{a,l}(\frac{i}{2}) = iP_{a,l}$. Using this observation and properties of the permutation operator we obtain

$$\tfrac{1}{i^L}\hat{T}(\tfrac{i}{2}) = \mathrm{tr}_a(P_{a,L}\cdots P_{a,L-1}\cdots P_{a,1}) = P_{12}P_{23}\cdots P_{L-1,L} = \hat{U} = e^{i\hat{P}},$$

which is just the *shift operator* that can also be expressed in terms of the total momentum $\hat{P}$. Next we compute

$$i\hat{T}(u)^{-1}\frac{\mathrm{d}}{\mathrm{d}u}\hat{T}(u)\bigg|_{u=\frac{i}{2}} = \sum_{l=1}^{L} P_{l-1,l}.$$

This identity can be used to obtain the Hamiltonian of the Heisenberg spin chain as a logarithmic derivative of the transfer matrix at the special point $u = \frac{i}{2}$:

$$\hat{H} = 2\sum_{l=1}^{L}(1_{l,l+1} - P_{l,l+1}) = 2L - 2i\,\frac{\mathrm{d}}{\mathrm{d}u}\log\hat{T}(u)\bigg|_{u=\frac{i}{2}}.$$

More generally we can also look at higher logarithmic derivatives of the transfer matrix at this point. This motivates the expansion

$$\log\hat{T}(u) = \hat{Q}_1 + (u-\tfrac{i}{2})\hat{Q}_2 + (u-\tfrac{i}{2})^2\hat{Q}_3 + (u-\tfrac{i}{2})^3\hat{Q}_4 + \ldots,$$

where we already know the physical interpretation of the first two of these *charges*

$$\hat{Q}_1 \propto \hat{P}, \quad \hat{Q}_2 \propto \hat{H} - 2L.$$

From $[\hat{T}(u), \hat{T}(u')] = 0$, we conclude that all charges have to commute

$$[\hat{Q}_n, \hat{Q}_m] = 0.$$

Consequently the transfer matrix is a *generating function* for this "tower" of commuting charges. Because the Hamiltonian is among the $\{\hat{Q}_n\}$, we also refer to them as "conserved" charges and because the charges commute, one might think of them as being "in involution". A sufficient amount of conserved charges in involution is the root of integrability in classical mechanics. In the sense of this formal analogy the construction presented here leads from the Yang-Baxter equation to a *quantum integrable system.*

However, a substitute for the coordinate Bethe ansatz is still missing, i.e. a method to actually diagonalize the Hamiltonian: $\hat{H}|\psi\rangle = E|\psi\rangle$. This is achieved by the *algebraic Bethe ansatz*, which diagonalizes the whole "tower" of commuting charges at once:

$$\hat{T}(u)|\psi\rangle = T(u)|\psi\rangle.$$

To construct the eigenvectors $|\psi\rangle$ we look at the structure of the monodromy matrix. We write the monodromy as a matrix in the auxiliary space a,

$$M_a(u) = \begin{pmatrix} \hat{A}(u) & \hat{B}(u) \\ \hat{C}(u) & \hat{D}(u) \end{pmatrix},$$

where the matrix elements $\hat{A}(u)$, $\hat{B}(u)$, $\hat{C}(u)$ and $\hat{D}(u)$ are operators acting on the quantum space $(\mathbb{C}^2)^{\otimes L}$ of the whole spin chain. In addition we choose the vacuum state $|0\rangle = |\uparrow \cdots \uparrow\rangle$, which satisfies $\hat{C}(u)|0\rangle = 0$. Then one can show that the ansatz

$$|\psi\rangle = \hat{B}(u_1) \cdots \hat{B}(u_M)|0\rangle$$

yields eigenvectors of the transfer matrix, if the *Bethe roots* $\{u_k\}$ satisfy the *Bethe equations*

$$\left(\frac{u_k + \frac{i}{2}}{u_k - \frac{i}{2}} \right)^L = \prod_{\substack{j=1 \\ j \neq k}}^{M} \frac{u_k - u_j + i}{u_k - u_j - i}.$$

The corresponding eigenvalue of the transfer matrix $\hat{T}(u)$ is

$$T(u) = (u + \tfrac{i}{2})^L \prod_{j=1}^{M} \frac{u - u_j - i}{u - u_j} + (u - \tfrac{i}{2})^L \prod_{j=1}^{M} \frac{u - u_j + i}{u - u_j}.$$

Having achieved the goal of diagonalizing the transfer matrix in this algebraic framework, we make some further remarks.

By construction $\hat{T}(u)$ (and thus also its eigenvalues $T(u)$) is a polynomial in u. This means that the residues of $T(u)$ at the apparent poles $u = u_k$ have to vanish. Demanding $\text{res}_{u=u_k} T(u) = 0$ yields again the Bethe equations.

Using the Bethe roots $\{u_k\}$, one defines the *Baxter function*

$$Q(u) = \prod_{j=1}^{M} (u - u_j).$$

This function satisfies the so-called *TQ-relation*

$$T(u)Q(u) = (u + \tfrac{i}{2})^L Q(u - i) + (u - \tfrac{i}{2})^L Q(u + i).$$

The construction of an operatorial version of this equation, i.e. the *Baxter Q-operator*, is an interesting story!

Finally there are some excellent pedagogical articles on both "flavors" of the Bethe ansatz. An introduction to the coordinate Bethe ansatz can be found in.[8] [9] provides an authoritative review of the algebraic Bethe ansatz and somewhat complementary material can be found in.[10] Further aspects of quantum integrability are covered in.[11] A construction of the Baxter Q-operator is given in.[14] The role of the Bethe ansatz in the AdS/CFT correspondence is highlighted in[13] and an application of the Bethe ansatz in this context can be found e.g. in.[12]

5. $\mathcal{N} = 4$ Super Yang-Mills

In this part we introduce the $\mathcal{N} = 4$ super Yangs-Mills theory (SYM), a local SU(N) gauge theory with the maximal amount of supersymmetry in four space-time dimensions. One of the remarkable properties of this theory is its conformal symmetry, meaning that it does not contain any inherent mass scale. While many theories exibit a conformal invariance on the classical level, $\mathcal{N} = 4$ SYM stays conformal even at quantum level. In the language of quantum field theory, this means that its β-function vanishes to all orders in perturbation theory. For a more detailed introduction to this fascinating model and further references see the article[16] in the collection.[7]

5.1. *General properties and conventions*

In SYM all fields (scalars, fermions, gauge- and ghost-fields) come in the adjoint representation of the gauge group. Thus an arbitrary field, Y, can be expanded as $Y_\mu = \sum_{a=1}^{(N^2-1)} Y_\mu^a T_a$, where $\mu \in \{0, 1, 2, 3\}$ (if present) is a generic space-time index and T_a are the generators of the symmetry algebra in the adjoint representation. They are $N \times N$ matrices that fulfil the usual Lie algebra relation $[T_a, T_b] = i f_{abc} T_c$. Under gauge transformations U, fermions and scalars transform as $Y \to UYU^{-1}$, where $U \equiv U(x) \in \mathrm{SU}(N)$

with $x \in \mathbb{R}^{1,3}$. The gauge field, A_μ, transforms as $A_\mu \to U A_\mu U^{-1} - i(\partial_\mu U) \cdot U^{-1}$. This transformation ensures the gauge invariance of the theory. The field content of SYM consists of:

- 6 massless, real scalars, φ_m with $m \in \{1, ..., 6\}$ with classical mass-dimension $\Delta_0 = 1$. Equivalently, these may be represented by 3 complex scalars: $X = \varphi_1 + i\varphi_2$, $Y = \varphi_3 + i\varphi_4$, $Z = \varphi_5 + i\varphi_6$ and their complex conjugates, denoted by $\bar{X}$, $\bar{Y}$, $\bar{Z}$. Yet another, important, notation is to express the scalars as antisymmetric bi-spinors φ_{ab}, where $a, b \in \{1, ..., \mathcal{N} = 4\}$.
- 4 massless Weyl-fermions, $\psi_{a\alpha}$, and their 4 conjugates $\bar{\psi}^a_{\dot{\alpha}}$. Here $a \in \{1, ...\mathcal{N} = 4\}$ labels the flavour of the (anti-) fermion and α or $\dot{\alpha}$ are connected to their helicities. Concretely $\alpha \in \{1, 2\}$ labels the $\mathfrak{su}_L(2)$ algebra and $\dot{\alpha} \in \{\dot{1}, \dot{2}\}$ the $\mathfrak{su}_R(2)$ in the Lie algebra given by the isomorphism with the complexification of the Lorentz algebra, $\mathfrak{so}(1,3) \simeq \mathfrak{su}_L(2) \oplus \mathfrak{su}_R(2)$. The fermions have classical mass-dimension $\Delta_0 = 3/2$.
- 1 field strength tensor, $F_{\mu\nu} = i[D_\mu, D_\nu] = \partial_\mu A_\nu - \partial_\nu A_\mu - i[A_\mu, A_\nu]$, with $D_\mu = \partial_\mu - iA_\mu$. The field strength has classical mass-dimensions $\Delta_0 = 2$.
- In addition to these fundamental fields infinitely many local fields can be generated by acting with covariant derivatives on all the fields just discussed. For this note that a covariant derivative acts on a field in the adjoint representation as $D_\mu \star = \partial_\mu \star - i[A_\mu, \star]$. The classical mass-dimension of the covariant derivative is $\Delta_0 = 1$, as can be seen already from the field strength tensor.

With these ingredients the action of SYM is given by

$$\begin{aligned} S = \frac{N}{\lambda} \int \frac{\mathrm{d}^4 x}{(2\pi)^4} \mathrm{Tr} \Big(& \frac{1}{4} F^{\mu\nu} F_{\mu\nu} + \frac{1}{2} (D^\mu)^* \varphi_m D_\mu \varphi_m + \bar{\psi}^a_{\dot{\alpha}} (\sigma_\mu)^{\dot{\alpha}\beta} D^\mu \psi_{a\beta} \\ & - \frac{1}{4} [\varphi^m, \varphi^n][\varphi_m, \varphi_n] \\ & - \frac{i}{2} \psi_{a\alpha} (\Sigma_m)^{ab} \varepsilon^{\alpha\beta} [\varphi^m \psi_{a\beta}] - \frac{i}{2} \bar{\psi}^a_{\dot{\alpha}} (\Sigma_m)^{ab} \varepsilon^{\dot{\alpha}\dot{\beta}} [\varphi^m \bar{\psi}_{a\dot{\beta}}] \Big) \\ & + \text{ghosts} + \text{gauge fixing} + \text{topological term}\,. \end{aligned} \tag{1}$$

Here $\lambda = N g^2_{YM}$ is the 't Hooft coupling and $(\Sigma_m)^{ab}$ is the 6-dimensional analogon to the 4-dimensional $(\sigma_\mu)^{\alpha\dot{\beta}}$-symbol. The free parameters of this action are λ and N^{-1}.

In the 't Hooft limit (also known as planar limit) the perturbative evaluation of this theory can be assessed in an efficient way. It is associated with taking

$N \to \infty$, $g_{YM} \to 0$ while keeping λ fixed. In contrast to the total number of Feynman diagrams contributing to a particular process, which exhibits a factorial growth, the advantage of this limit is that the number of diagrams only grows exponentially with the loop order. Thus perturbation theory has a nonzero radius of convergence. The only contributing diagrams in the planar limit have a topology that allows to draw them in a planar fashion (on a piece of paper). This statement can easily be verified by counting the powers of N in each loop calculation that arise from traces in colour space. To do so, one should use 't Hooft's double line notation, in which the Feynman diagrams can be represented via so called fat graphs. Each line then corresponds to one matrix index of the generator matrices, e.g. $Y_\mu^{ij} = Y_\mu^a (T_a)^{ij}$, with $i, j \in \{1, 2, ..., N\}$.

5.2. *Symmetries*

By construction the action Eq. (1) has a SU(N) gauge symmetry, but in addition it is invariant under the action of several other symmetry operations. For a more detailed exposition on symmetries and further references see the article[17] in the collection.[7]

- Lorentz group: 3 rotations + 3 boosts, $\mathfrak{sl}(2, \mathbb{C}) \simeq \mathfrak{so}(1, 3)$, generators: $M_{\mu\nu}$.
- 4 (covariant) translations, $\partial_\mu \to D_\mu$, generators: $P_\mu = iD_\mu$.
- The two above algebras form the Poincaré algebra, $(M_{\mu\nu}, P_\rho)$. With metric $\eta_{\mu\nu} = \text{diag}(1, -1, -1, -1)$ it is given by the following symmetry algebra

$$[M_{\mu\nu}, M_{\rho\sigma}] = i(\eta_{\mu\rho} M_{\nu\sigma} + \eta_{\nu\sigma} M_{\mu\rho} - \eta_{\mu\sigma} M_{\nu\rho} - \eta_{\nu\rho} M_{\mu\sigma}) , \quad (2)$$

$$[M_{\mu\nu}, P_\rho] = i(\eta_{\rho\nu} P_\mu - \eta_{\mu\rho} P_\nu) , \quad (3)$$

$$[P_\mu, P_\nu] = 0 . \quad (4)$$

- Another 4 symmetry operations are the special conformal transformations, K_μ. To find these first define the inversion, $I : x_\mu \to x_\mu / x^2$. With this one can define $K_\mu = I P_\mu I$ and finds the remaining two non-trivial algebra relations

$$[M_{\mu\nu}, K_\rho] = i(\eta_{\rho\nu} K_\mu - \eta_{\mu\rho} K_\nu) , \quad (5)$$

$$[P_\mu, K_\nu] = 2i(\eta_{\mu\nu} \mathcal{D} - M_{\mu\nu}) . \quad (6)$$

- In the last equation the symmetry generator $\mathcal{D}$ has been introduced. It is called the dilatation operator and generates scale transformations. The remaining two commutation relations are

$$[\mathcal{D}, P_\mu] = iP_\mu \qquad , \qquad [\mathcal{D}, K_\mu] = -iK_\mu \tag{7}$$

The combination of the $6+4+4+1 = 15$ symmetry generators listed above forms the *conformal algebra*, $\mathfrak{so}(2,4) \simeq \mathfrak{su}(2,2)$. Under the given set of commutation relations (Eq.(2) - Eq. (7)) the conformal algebra is closed. For a historical discussion with deep insights into the meaning of conformal symmetry see.[5] Note that the complete conformal symmetry is not manifest in the action Eq. (1). Manifest subgroups of it are only:

- the Lorentzgroup,
- the $\mathfrak{so}(6)$ symmetry transforming the 6 scalars into each other,
- the $\mathfrak{su}(4)$ symmetry acting on the 4 fermionic flavours.

In fact the $\mathfrak{so}(6)$ and the $\mathfrak{su}(4)$ symmetry are the same in the sense that there exists an isomorphism transforming one into the other (this has already been used in form of the $(\Sigma_m)^{ab}$ symbol).

Apart from the conformal symmetry the action Eq. (1) exibits even more symmetry, namely a *supersymmetry*, which exchanges the role of scalars and fermions. To get familiar with this type of symmetry lets start with the relatively simple $\mathfrak{su}(2)$ sector. Grouping its generators into a 2×2 table gives

$$L = \begin{array}{|c|c|} \hline S^3 & S^+ \\ \hline S^- & -S^3 \\ \hline \end{array} \,.$$

Now by gluing together a $\mathfrak{su}(2)$ and a $\bar{\mathfrak{su}}(2)$ one obtains the 4×4 table

$$\begin{array}{|c|c|} \hline L & \begin{array}{cc} p & p \\ p & p \end{array} \\ \hline \begin{array}{cc} k & k \\ k & k \end{array} & \bar{L} \\ \hline \end{array} \,,$$

where L corresponds to the table for $\mathfrak{su}(2)$ and $\bar{L}$ to the $\bar{\mathfrak{su}}(2)$. This table incorporates the generators of a $\mathfrak{su}(2,2)$.
In the next and final step one glues the $\mathfrak{su}(2,2)$ together with an $\mathfrak{su}(4)$ and arrives at the $\mathfrak{psu}(2,2|4)$ algebra, expressed in the the 8×8 table

$$\begin{array}{|c|c|c|} \hline L & P & Q \\ \hline K & \bar{L} & \bar{S} \\ \hline S & \bar{Q} & R \\ \hline \end{array} \,.$$

This algebra is called a *super algebra* and it is a symmetry of the $\mathcal{N} = 4$ SYM! Although this symmetry is not manifest, it can be shown from the action Eq. (1) is invariant under it. In the above table $Q_{a\alpha}$ and $\bar{Q}^a_{\dot{\alpha}}$ (with $a \in \{1, ..., \mathcal{N} = 4\}$) are the Poincaré supercharges, fulfilling the (anti-) commutation relations:

$$\{Q_{a\alpha}, \bar{Q}^b_{\dot{\alpha}}\} = (\sigma^\mu)_{\alpha\dot{\alpha}} P_\mu \delta^b_a \tag{8}$$

$$[\mathcal{D}, Q_{a\alpha}] = \frac{i}{2} Q_{a\alpha} \tag{9}$$

and their action on the fundamental fields is proportional to

$$Q \cdot \varphi \sim \psi \quad , \quad \bar{Q} \cdot \varphi \sim \bar{\psi}$$
$$Q \cdot \psi \sim F + g_{YM}[\varphi, \varphi] \quad , \quad \bar{Q} \cdot \psi \sim \mathcal{D}\varphi$$

This concludes the brief introduction on the symmetries of $\mathcal{N} = 4$ SYM. For further reading consider for example.[7,18,19]

5.3. *Local composite operators and the spectral problem of $\mathcal{N} = 4$ SYM*

In $\mathcal{N} = 4$ SYM it is very natural to study the behaviour of composite operators. They can be constructed from elementary fields by merging them in a gauge invariant fashion into local composite objects, for instance

$$\mathcal{O} \equiv \mathcal{O}_{(x)} = \mathrm{Tr}\Big(ZX\bar{Z}(DY)(D^5Y)\psi(D^3\psi)\bar{\psi}F(D^7\bar{F})...\Big)_{(x)} .$$

At the origin the action of the dilatation operator on $\mathcal{O}$ takes the form of an eigenvalue equation

$$[\mathcal{D}, \mathcal{O}] = i\Delta\mathcal{O} \,,$$

where $\Delta = \Delta_0 + \gamma$ now consists of the classical scaling-dimension, Δ_0 and an anomalous contribution, γ. While Δ_0 is simply given by the sum of classical scaling-dimensions of the constituent field in $\mathcal{O}(x)$, the anomalous dimension has to be calculated perturbatively:

$$\gamma = g^2_{YM} \underbrace{\gamma_1}_{1-loop} + g^4_{YM} \underbrace{\gamma_2}_{2-loop} + ...$$

With the notion of composite operators the spectral problem of $\mathcal{N} = 4$ SYM amounts to finding *all* sets of eigenstates, $\mathcal{O}$, and their scaling dimensions, Δ. In other words one would like to diagonalise the dilatation operator.

The spectral problem of planar $\mathcal{N} = 4$ SYM has a striking similarity to the diagonalisation problem of a Heisenberg magnet, where the analogous set consists of the eigenfunction $|\psi\rangle$ and its energy, E. The single trace operators, $\mathcal{O}$, of $\mathcal{N} = 4$ SYM can be viewed as spin chains and the fields within $\mathcal{O}$ correspond to the spins. In the $N = \infty$ limit the fusion and fission of composite operators is suppressed and thus multi trace operators don't need to be considered in the 't Hooft limit.
For the lowest weight states the following generators of $\mathfrak{psu}(2,2|4)$ commute with $\mathcal{O}$:

$$[K, \mathcal{O}] = 0 \quad , \quad [S, \mathcal{O}] = 0 \quad , \quad [\bar{S}, \mathcal{O}] = 0.$$

To find the anomalous dimension of the operator $\mathcal{O}$ consider its two-point correlation function. As the $\mathcal{N} = 4$ SYM is conformal, it is given by

$$\langle \mathcal{O}(x), \overline{\mathcal{O}}(y) \rangle = \frac{1}{(x-y)^{2\Delta}} \sim \frac{1}{(x-y)^{2\Delta_0}} (1 - g_{\mathrm{YM}}^2 \gamma_1 \log\left[(x-y)^2 \Lambda^2\right]) ,$$

where Λ is a suitable ultraviolet cutoff.
The simplest example of such a correlator is given by the operator $\mathcal{O} = \mathrm{Tr}(Z^j)$. This operator corresponds to a Heisenberg spin chain with all spins pointing in the same direction, namely $|\uparrow^j\rangle$. In this case $\mathcal{O}$ is an BPS state and thus one has $\Delta_0 = j$ and $\gamma = 0$.

$$\langle (\bar{Z}^j)(x)(Z^j)(y) \rangle = \frac{1}{(x-y)^{2j}}$$

For a non-trivial example take the operator $\mathcal{O} = \mathrm{Tr}(Z^{j-1}X)$. this situation corresponds to the same Heisenberg spin chain as before, only that this time one spin is flipped down, $|\uparrow^{j-1}\downarrow\rangle$. The solution to the spectral problem of this operator in $\mathcal{N} = 4$ SYM is then obtained in the same way as one can arrive at the energy of the state $|\uparrow^{j-1}\downarrow\rangle$ in the XXX spin chain model. So, perturbatively the action of the dilatation operator on such a state is given by

$$\mathcal{D} = \mathcal{D}_0 + \frac{g_{\mathrm{YM}}^2}{16\pi^2} \hat{H}_{XXX} + O(g_{\mathrm{YM}}^4) .$$

Again, more detailed lectures on AdS/CFT integrability are found in.[7]

6. IIB Superstring Theory on $AdS_5 \times S^5$

This section gives a short overview over the crucial aspects of type IIB string theory on the maximally supersymmetric background, $AdS_5 \times S^5$. For more details, see chapter II of,[7] as well as[17] and.[20] Morally such a theory can be constructed by embedding the strings on the 2D worldsheet into a space which is constructed by the direct product of $AdS_5 \times S^5$. Where Ads_5 is the Lorentzian hyperboloid Anti de Sitter and S_5 is a compact space. The inner products in these spaces are characterized by the metrics $(-,+,+,+,+,-)$ and $(+,+,+,+,+,+)$, or more explicitly:

$$\begin{aligned} AdS_5: \qquad & -R^2 = -Z_0^2 + Z_1^2 + Z_2^2 + Z_3^2 + Z_4^2 - Z_5^2\,, \\ S^5: \qquad & R^2 = Y_1^2 + Y_2^2 + Y_3^2 + Y_4^2 + Y_5^2 + Y_6^2\,. \end{aligned}$$

More generally type IIB string theory can be represented by a non-linear sigma model on a coset space. As for many two-dimensional sigma models on coset spaces this model is classically integrable. For the supercoset construction the background space can be viewed as $AdS_5 = \mathrm{SO}(4,2)/\mathrm{SO}(1,4)$ and $S^5 = \mathrm{SO}(6)/\mathrm{SO}(5)$. With this one can construct

$$\mathrm{Super}\Big[AdS_5 \times S^5\Big] := \frac{\widetilde{\mathrm{PSU}}(2,2|4)}{\mathrm{SO}(1,4)\times \mathrm{SO}(5)}\,,$$

where the tilde signifies an infinite cover. $\mathrm{PSU}(2,2|4)$ is the super-isometry group of this space. Its super Lie algebra has the following structure

$\mathfrak{psu}(2,2|4) =$

$\mathfrak{u}(2,2)$	supercharges
supercharges	$\mathfrak{u}(4)$

where the supercharges are physicist's connotation for the fermionic generators. The number of fermionic and bosonic dimensions of this theory can directly be obtained from the dimension of the involved groups:

$$\frac{30|32}{10|0 \oplus 10|0} = 10|32\,,$$

where the left number of the vertical line labels the bosonic and the right one the fermionic dimension.
The action in the present context is given by

$$S = \frac{R^2}{\alpha'}\int d\tau \frac{d\sigma}{2\pi}\left(\frac{1}{2}\partial_a Z^M \partial^a Z_M + \frac{1}{2}\partial_a Y^N \partial^a Y_N\right) + \text{fermions}\,, \qquad (1)$$

with $M \in \{0,1,...,5\}$, $N \in \{1,...,6\}$, the radius R of AdS_5 and S^5 and with the string tension $1/\alpha'$. The partial derivatives act on the worldsheet with metric $(+,+)$ and are identified as $\partial_0 = \partial_\tau$, $\partial_1 = \partial_\sigma$. The string tension effectively influences the size of appearing strings. For small α' the tension is large and thus strings are short and vice versa. In the limit $\alpha' \to 0$ strings become pointlike objects. This limit corresponds to the super gravity limes of string theory (as always in string theory there are massless spin$=2$ particles, which can be identified with gravitons). Amazingly, in this limit the action Eq. (1) simply is a two-dimensional QFT! In addition it preserves all super symmetries, as the ten-dimensional super string preserves SUSY by construction. So the analysis of IIB string theory can be done via perturbative methods (Feynman diagrams) on curved manifolds with non-trivial topology.

7. The AdS/CFT correspondence

In 1997 Maldacena made the remarkable observation that a type IIB string theory on an $AdS_5 \times S^5$ background is strictly equivalent to $\mathcal{N} = 4$ SYM,[2] see e.g.[3,4] for explanations accessible to students. Note that this means a theory with gravity can be translated into another one without gravity! In order to get a grasp on the behaviour of this correspondence, lets investigate how the parameters in the two theories are linked. On the SYM side there is the 't Hooft coupling constant $\lambda = g_{\rm YM} N^2$, where $N-1$ is the rank of the gauge group $SU(N)$, and $g_{\rm YM}$ the coupling constant. On the string theory side there is the radius of AdS_5 (or S^5 respectively), R and the string tension α', which can also be connected to the string coupling constant $g_{\rm s}$. The AdS/CFT correspondence now states that theses parameters are connected as

$$\sqrt{\lambda} = \frac{R^2}{\alpha'} \quad , \quad 4\pi g_{\rm s} = \frac{\lambda}{N} \, .$$

Comparing the SYM action with the one of the string theory σ-model one sees

$$S_{SYM} = \int \frac{N}{\lambda} \frac{d^4x}{4\pi^2} Tr\{\dots\} \longleftrightarrow S_{\sigma-model} = \sqrt{\lambda} \int d\tau \frac{d\sigma}{2\pi} \{\dots\} \, ,$$

that the correspondence is a *weak-strong duality*. The weakly coupled field theory is connected to a strongly coupled string theory and vice versa. While this particular relation provides a very powerful tool for calculations, it also means that the proposed correspondence is very hard to prove. In the planar limit the degree of the gauge group goes to infinity, $N \to \infty$, which means that the string coupling goes to zero, $g_{\rm s} \to 0$. So the planar SYM is connected to a non-interacting string theory.

While this is very convenient, it is still necessary to give a more direct way of comparing results of gauge and string theory. For this one should come back to the spectral problem and make the connection to the spectral problem of $AdS_5 \times S^5$ superstrings. In Poincaré coordinates one has:

$$Z_4 + Z_5 = \frac{R}{z}, \quad (Z_0, Z_1, Z_2, Z_3) = \frac{R}{z} x_\mu, \quad ds^2_{AdS_5} = \frac{R^2}{z^2}(dx^\mu dx_\mu + dz^2) \, .$$

Focussing on the 5-sphere, S^5, use the parametrisation

$$Y_1 + iY_2 = r_1 e^{i\phi_1}, \qquad Y_3 + iY_4 = r_2 e^{i\phi_2}, \qquad Y_5 + iY_6 = r_3 e^{i\phi_3},$$

which is strikingly similar to the parametrisation of 3 complex scalars in the SYM case. The three angles ϕ_1, ϕ_2, ϕ_3, are associated with the string angular momenta J_1, J_2, J_3 and correspond to the 3 *Cartan generators* of $\mathfrak{so}(6) \simeq \mathfrak{su}(4)$ (generate rotations around three orthogonal equators). For the AdS_5 space a similar parametrisation gives

$$Z_0 - iZ_5 = \rho_3 e^{it}, \qquad Z_1 + iZ_2 = \rho_1 e^{i\alpha_1}, \qquad ZY_3 + iZ_4 = \rho_2 e^{i\alpha_2}.$$

This time the three angles t, α_1, α_2 give the three conserved quantities E, S_1 and S_2 and they come from the three *Cartan generators* of $\mathfrak{so}(2,4) \simeq \mathfrak{su}(2,2)$. In the table from section 5.2 they belong to

L	$p\,p$ $p\,p$
$k\,k$ $k\,k$	$\bar{L}$

.

Since E is a *non-compact* generator one has to go to the *universal cover* (which can be view as the fact that there are no closed time-like loops). E gives the *string energy*. The *key prediction of AdS/CFT* is now

$$\text{string energy} \longleftrightarrow \text{scaling dimension}$$
$$E(\lambda) \quad = \quad \Delta(\lambda).$$

While this prediction is not proven yet there are strong indications that it is correct:

- It is certain that the $AdS_5 \times S^5$ string σ-model is classically integrable, as was first shown by Bena, Polchinski, Roiban in 2003. For an in-depth discussion of the classical integrability of this model along with a discussion of the difficulties of its quantization see.[20]
- On the gauge theory side it is also verified that the full one-loop dilatation operator of $\mathcal{N} = 4$ SYM can be mapped to a quantum integrable spin chain. It has been completely diagonalized by means of the Bethe ansatz as was first shown by Beisert and the author in 2003, following the initial insight of Minahan and Zarembo in 2002.

The conjecture is that the free planar AdS/CFT correspondence is quantum integrable at non-zero λ. The AdS/CFT scattering picture has 30|32 generators which can be organised in a 8×8 matrix to make a connection to spin chains.

$\mathfrak{su}(2\|2)$	create $8_B + 8_F$ magnons
destroy $8+8$ magnons	$\overline{\mathfrak{su}(2\|2)}$

So one finds

$$\uparrow = Z = \text{nothing}\,,$$
$$\downarrow = M_{A\dot{A}} = 8+8 \text{ magnons transforming under } \mathfrak{su}(2|2) \otimes \overline{\mathfrak{su}}(2|2)\,,$$

where A and $\dot{A}$ are fundamental super indices. Everything else are composite states, e.g.: $\bar{Z} \sim X\bar{X} + Y\bar{Y}$ or $F \sim DD$.

References

1. B. Sutherland, "*Beautiful Models*", World Scientific 2004.
2. J. Maldacena, "*The Large N limit of superconformal field theories and supergravity,*" Adv. Theor. Math. Phys. 2 **231** (1998) arXiv:hep-th/9711200.
3. J. Maldacena, *"The Gauge/gravity duality,"* arXiv:1106.6073 [hep-th].
4. H. Nastase, *"Introduction to AdS-CFT,"* arXiv:0712.0689 [hep-th].
5. H. A. Kastrup, *"On the Advancements of Conformal Transformations and their Associated Symmetries in Geometry and Theoretical Physics,"* Annalen Phys. **17** (2008) 631 [arXiv:0808.2730 [physics.hist-ph]].
6. H. Bethe, "*Zur Theorie der Metalle. I. Eigenwerte und Eigenfunktionen der linearen Atomkette,*" Z. Phys. **71** (1931) 205–226.
7. N. Beisert et. al. "*Review of AdS/CFT Integrability: An Overview,*", Lett. Math. Phys. **99**, 3 (2012) arXiv:hep-th/10123982
8. M. Karbach and G. Müller, "*Introduction to the Bethe ansatz I*", Comput. Phys. **11** (1997) 36, arXiv:cond-mat/9809162.
9. L. D. Faddeev, "*How Algebraic Bethe Ansatz works for integrable model*", arXiv:hep-th/9605187.
10. L. D. Faddeev, "*Algebraic aspects of Bethe Ansatz*", Int. J. Mod. Phys. A **10** (1995) 1845, arXiv:hep-th/9404013.
11. A. Doikou, S. Evangelisti, G. Feverati, and N. Karaiskos, "*Introduction to Quantum Integrability*", Int. J. Mod. Phys. A **25** (2010) 3307, arXiv:0912.3350 [math-ph].

12. J. Escobedo, N. Gromov, A. Sever, and P. Vieira, "*Tailoring Three-Point Functions and Integrability*", J. High Energ. Phys. (2011) 028, arXiv:1012.2475 [hep-th].
13. M. Staudacher, "*Review of AdS/CFT Integrability, Chapter III.1: Bethe Anstze and the R-Matrix Formalism*", Lett. Math. Phys. **99** (2012) 191, arXiv:1012.3990 [hep-th].
14. V. V. Bazhanov, T. Lukowski, C. Meneghelli and M. Staudacher, "*A short-cut to the Q-operator*", J. Stat. Mech. (2010) P11002, arXiv:1005.3261 [hep-th].
15. R. I. Nepomechie, "*A Spin Chain Primer*", Int. J. Mod. Phys. B **13** (1999) 2973, arXiv:hep-th/9810032.
16. J. A. Minahan, *"Review of AdS/CFT Integrability, Chapter I.1: Spin Chains in N=4 Super Yang-Mills,"* Lett. Math. Phys. **99** (2012) 33 [arXiv:1012.3983 [hep-th]].
17. N. Beisert, *"Review of AdS/CFT Integrability, Chapter VI.1: Superconformal Symmetry,"* Lett. Math. Phys. **99** (2012) 529 [arXiv:1012.4004 [hep-th]].
18. E. D'Hoker and D. Phong, "*Lectures on supersymmetric Yang-Mills theory and integrable systems*," arXiv:hep-th/9912271 [hep-th].
19. S.J. Gates, Grisaru, T. Marcus, M. Rocek, and W. Siegel, "*Superspace Or One Thousand and One Lessons in Supersymmetry*," arXiv:0108200 [hep-th].
20. G. Arutyunov and S. Frolov, "Foundations of the $AdS_5 \times S^5$ Superstring. Part I," J. Phys. A **42** (2009) 254003 [arXiv:0901.4937 [hep-th]].

COMPACTIFICATIONS OF STRING THEORY AND GENERALIZED GEOMETRY

Mariana Graña and Hagen Triendl

Institut de Physique Théorique, CEA/ Saclay
91191 Gif-sur-Yvette Cedex, France
E-mail: mariana.grana@cea.fr, hagen.triendl@cea.fr

In these lecture notes we give a brief introduction to generalized geometry and the role it plays in flux compactifications of string theory. We review the quantization of the bosonic and superstring and derive their massless spectrum and the low-energy effective supergravity. We then introduce the mathematical framework of generalized complex geometry and pure spinors, and finally discuss flux compactifications of string theory and the reformulation of its supersymmetry equations in terms of pure spinors.

Keywords: Compactifications, string theory, generalized geometry.

1. Introduction

The search for a unified theory of elementary particles and their interactions has culminated in the last two decades to the spectacular development of string theory. String theory reconciles general relativity with quantum mechanics, and contains the main ingredients of the Standard Model, *i.e.* gauge interactions and parity violation. It furthermore incorporates naturally most of the theoretical ideas for physics beyond the Standard Model – gauge unification, supersymmetry and extra dimensions – while also inspiring new ones such as the possibility of confining our observed world on a brane. Some of these ideas could be tested in the coming years, either in accelerator experiments or by future cosmological and astrophysical observations.

String theory is consistently defined in ten dimensions (see for example[1]-[4]). In order to extract any information about four-dimensional physics, we need to understand the way it is compactified to these four dimensions, or in other words, we need to know the precise way in which the Standard

Model and Einstein's gravity are embedded as low-energy limits in string theory. Early attempts involved mostly compactification of the heterotic string on Calabi-Yau manifolds or on exact (2,0) backgrounds (some of the early constructions are[5]-[7]), and exceptionally type I theory. With the introduction of D-branes, compactifications of the type-II string theory involving orientifolds and intersecting D-branes became the center of attention (some references are[8]-,[13] for a review and more references, see for example[14]).

The current state of the art is that one can find semi-realistic models in both frameworks, but several key issues remain open. Among them, one is the problem of moduli stabilization: in any of these compactifications, the four-dimensional low-energy action has a number of massless fields with no potential. These would lead to long-range scalar forces unobserved in nature. Furthermore, the couplings of other fields (like Yukawa couplings) depend on their vacuum expectation values (VEV's). As a consequence, no predictions can be made in these scenarios since the VEV of the moduli can take any value. Therefore, there should be a mechanism that generates a potential for the moduli, fixing (or "stabilizing") their VEV's. The only known mechanism within perturbative string theory that we know of today is via fluxes: turning on fluxes for some of the field strengths available in the theory (these are generalizations of the electromagnetic field to ten dimensions) generates a non trivial potential for the moduli, which stabilize at their minima. The new "problem" that arises is that fluxes back-react on the geometry, and whatever manifold was allowed in the absence of fluxes, will generically be forbidden in their presence.

Much of what we know about stabilisation of moduli[15] is done in two different contexts: in Calabi-Yau compactifications under a certain combination of three-form fluxes whose back-reaction on the geometry just makes them conformal Calabi-Yau manifolds, and in the context of parallelizable manifolds, otherwise known as "twisted tori". In the former, fluxes stabilize the moduli corresponding to the complex structure of the manifold, as well as the dilaton.[17] To stabilize the other moduli, stringy corrections are invoked. The result is that one can stabilize all moduli in a regime of parameters where the approximations can be somehow trusted,[18] but it is very hard to rigorously prove that the corrections not taken into account do not destabilize the full system. In the later, one takes advantage of the fact that, similarly to tori, there is a trivial structure group. However, the vectors that are globally defined obey some non-trivial algebra, and therefore compactness is ensured by having twisted identifications. What makes these manifolds amenable to the study of moduli stabilisation is the possibility of analyzing them as if they were a torus subject to twists, or in other words a torus in the presence of "geometric fluxes", which combined to the electromagnetic fluxes can give rise to solutions to the equations of motion. One can then use the bases of cycles of the tori, and see which ones of them get a fixed size (or gets "stabilized"), due to a balance of forces between the gravitational and the electromagnetic ones. It turns out that even in this simple situations, it is impossible to stabilize all moduli in a Minkowski vacuum without the addition of other "non-geometric" fluxes,[16] which arise in some cases as duals to the known fluxes, but whose generic string theory interpretation is still under discussion.

In the last years the framework of generalized complex geometry has turned out to be an excellent tool to study flux backgrounds in more detail, in particular going away from the simplest cases of parallelizable manifolds. In these lectures, we review these tools and discuss the allowed manifolds in the presence of fluxes.

Generalized complex geometry is interesting from a mathematical viewpoint on its own, as it incorporates complex and symplectic geometry into a larger framework[19]-,[21] and thereby finds a common language for the two sectors of geometric scalars that typically arise in compactifications. It does so by introducing a new bundle structure that covariantizes symmetries of string compactifications, T-duality among others. In particular, mirror symmetry between the type II string theories appears naturally in this setup.

These lectures shall give a brief introduction into generalized geometry and its appearance in string theory. For more comprehensive reviews and lecture notes, see for instance,[24] while introductions to string theory and D-branes can for example be found in[1]-.[4]

These lectures are organized as follows: In Section 1 we give a short introduction to string theory. We focus on its massless spectrum (of the purely bosonic string in Section 2.1 and of the superstring in Section 2.2), which is the sector that plays a role in the low-energy effective description. In Section 3 we briefly introduce generalized complex geometry and and its pure spinor techniques. In Section 4 we discuss the solutions to the equations of motion whose metric splits into a four-dimensional Minkowski metric and that of a six dimensional space. We show first in Section 4.1 the flux-less solutions, and then go on to describe flux compactification in Section 4.2. Finally in Section 4.3 we interpret the manifolds arising in flux compactification from the the point of view of generalized complex geometry. We conclude and spell out some open problems in Section 5.

2. Introduction to String Theory

In this section we give a basic introduction to string theory, focusing on its low energy effective description. We start with the bosonic string, work out the massless spectrum, and introduce the low energy effective action governing its dynamics. We then continue with the superstring, showing the corresponding massless spectrum and action. More details can be found in[1]-.[4]

2.1. *Bosonic strings*

String theory is a quantum theory of one-dimensional objects (*strings*) moving in a D-dimensional spacetime. Strings sweep a two dimensional surface, the "worldsheet", labeled by the coordinates σ along the string ($0 \leq \sigma \leq \pi$ for an open string and $0 \leq \sigma \leq 2\pi$ for a closed string with 0 and 2π identified), and τ.

The evolution of the worldsheet is given by the "Polyakov" action

$$S = -\frac{1}{4\pi\alpha'} \int d\sigma d\tau \sqrt{\gamma}\, \gamma^{\alpha\beta} \eta_{MN} \partial_\alpha X^M \partial_\beta X^N , \tag{1}$$

$M = 0, ..., D-1$; $\alpha = \tau, \sigma$, which just measures the area of the string worldsheet inside D-dimensional spacetime. In this equation, X^M are the

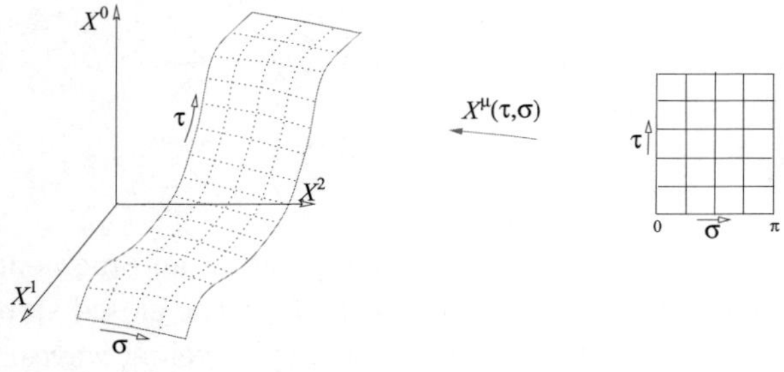

Fig. 1. Embedding of an open string in spacetime by the function $X^\mu(\tau, \sigma)$.

functions defining the embedding of the worldsheet in spacetime (see Figure 1); $\gamma^{\alpha\beta}$ is the worldsheet metric, η_{MN} is the spacetime Minkowski metric and α' is related to the string tension T by

$$T = \frac{1}{2\pi\alpha'} = \frac{1}{2\pi l_s^2} \, . \tag{2}$$

l_s, which has units of length, is called the "string scale". Its inverse gives the "string mass" M_s (we are using units in which $c = \hbar = 1$, to obtain a mass from the inverse of the string scale we should multiply by $\hbar/c$), which defines the typical energy scale of strings.

Varying the action (1) with respect to X^M we get the following equations of motion

$$\left(\frac{\partial^2}{\partial\sigma^2} - \frac{\partial^2}{\partial\tau^2}\right) X^M(\tau,\sigma) = 0 \,. \tag{3}$$

By introducing left- and right-moving worldsheet coordinates $\sigma^\pm = \tau \pm \sigma$, the above equation can be rewritten as

$$\frac{\partial}{\partial\sigma^+}\frac{\partial}{\partial\sigma^-} X^M(\tau,\sigma) = 0 \,, \tag{4}$$

which means that the embedding vector X^M is the sum of left- and right-moving degrees of freedom, i.e.

$$X^M(\tau,\sigma) = X_R^M(\sigma^-) + X_L^M(\sigma^+) \,. \tag{5}$$

We will concentrate on closed strings from now on, and discuss very briefly the open string at the end of this section. Imposing the closed string boundary conditions $X^M(\tau,0) = X^M(\tau,2\pi)$, $X'^M(\tau,0) = X'^M(\tau,2\pi)$ (with prime indicating a derivative along σ) we get the following mode decomposition

$$\begin{aligned} X^M(\tau,\sigma) &= X_R^M(\sigma^-) + X_L^M(\sigma^+) \\ X_R^M(\sigma^-) &= \frac{1}{2}x^M + \alpha' p^M \sigma^- + \mathrm{i}\left(\frac{\alpha'}{2}\right)^{1/2} \sum_{n\neq 0} \frac{1}{n}\alpha_n^M e^{-2\mathrm{i}n\sigma^-} \\ X_L^M(\sigma^+) &= \frac{1}{2}x^M + \alpha' p^M \sigma^+ + \mathrm{i}\left(\frac{\alpha'}{2}\right)^{1/2} \sum_{n\neq 0} \frac{1}{n}\tilde{\alpha}_n^M e^{-2\mathrm{i}n\sigma^+} \,, \end{aligned} \tag{6}$$

where x^M and p^M are the centre of mass position and momentum, respectively. We can see that the mode expansion for the closed string is that of a pair of independent left- and right- moving traveling waves. To ensure a real solution we impose $\alpha_{-n}^M = (\alpha_n^M)^*$ and $\tilde{\alpha}_{-n}^M = (\tilde{\alpha}_n^M)^*$.

Varying (1) with respect to the worldsheet metric $\gamma_{\alpha\beta}$ gives the extra constrains that the energy-momentum tensor is vanishing

$$T^{\alpha\beta} \equiv -\frac{2\pi}{\sqrt{\gamma}}\frac{\delta S}{\delta\gamma_{\alpha\beta}} = -\frac{1}{\alpha'}\left(\partial^\alpha X^M \partial^\beta X_M - \frac{1}{2}\gamma^{\alpha\beta}\gamma_{\lambda\rho}\partial^\lambda X^M \partial^\rho X_M\right) = 0 \,. \tag{7}$$

This enforces the additional conditions

$$\eta_{MN}\partial_{\sigma^+}X_L^M \partial_{\sigma^+}X_L^N = \eta_{MN}\partial_{\sigma^-}X_R^M \partial_{\sigma^-}X_R^N = 0 \,. \tag{8}$$

The system governed by the action (1) can be quantized in a canonical way in terms of left- and right-moving oscillators, resulting in the following commutators

$$[\alpha_n^P, \alpha_m^Q] = [\tilde{\alpha}_n^P, \tilde{\alpha}_m^Q] = n\delta_{n+m}\eta^{PQ} \ , \quad [x^P, p^Q] = i\eta^{PQ} \ . \tag{9}$$

We can therefore interpret $\alpha_n^M, \tilde{\alpha}_n^M$ as creation operators and $\alpha_{-n}^M, \tilde{\alpha}_{-n}^M$ with $n > 0$ as annihilation operators, which create or annihilate a left or right moving excitation at level n.[a] Each mode carries an energy proportional to the level. The mass (energy) of a state is obtained using the operator

$$M^2 = \frac{2}{\alpha'}\left(\sum_{n=1}^{\infty} \alpha_n \cdot \alpha_{-n} + \tilde{\alpha}_n \cdot \tilde{\alpha}_{-n} - 2\right) \tag{10}$$

where the -2 comes from normal ordering the operators (corresponding to the zero point energy of all the oscillators). The classical conditions (8) in the quantum theory become the vanishing of the so-called Virasoro operators on the physical spectrum. We will not go into more details here (these can be found, for example, in[1-3]) but only mention the most important of these contraints, the level-matching condition that says that the operator

$$\hat{L}_0 = \tfrac{1}{2}\left(\sum_{n=1}^{\infty} \alpha_n \cdot \alpha_{-n} - \tilde{\alpha}_n \cdot \tilde{\alpha}_{-n}\right) \tag{11}$$

vanishes on the physical states. Thus, we have to impose $N = \tilde{N}$, where N, $\tilde{N}$ are the total sums of oscillator levels excited on the left and on the right, respectively. The massless states have therefore one left-moving and one right-moving excitation, namely

$$|\xi_{MN}\rangle \equiv \xi_M \tilde{\xi}_N \alpha_1^M \tilde{\alpha}_1^N |0\rangle \tag{12}$$

and a center of mass momentum k^M, $k \cdot k = 0$. It is not hard to check that the norm of the state is positive only if $\xi, \tilde{\xi}$ are space-like vectors. The classical conditions (8) impose $\xi \cdot k = \tilde{\xi} \cdot k = 0$, i.e., the polarization vectors have to be orthogonal to the center of mass momentum. Choosing a frame where $\vec{k} = (k, k, 0, ..., 0)$, we get that $\xi_M, \tilde{\xi}_M$ belong to the $D-2$-dimensional space parametrized by the coordinates $2, ..., D-1$. The states are therefore classified by their SO$(D-2)$ representations. The tensor $\xi_{MN} \equiv \xi_M \tilde{\xi}_N$ decomposes into

$$\xi_{MN} = \xi^s_{MN} + \xi_t\, \eta_{MN} + \xi^a_{MN} \ , \tag{13}$$

[a]Note that the convention used here for creation (positive modes) and annihilation operators (negative modes) is opposite to the one used most often (for example in[2]).

where we have defined

$$\begin{aligned}\xi_t &\equiv \frac{1}{D}\eta^{MN}\xi_{MN} \ , \\ \xi^s_{MN} &= \frac{1}{2}(\xi_{MN} + \xi_{NM} - 2\xi_t\,\eta_{MN}) \ , \\ \xi^a_{MN} &= \frac{1}{2}(\xi_{MN} - \xi_{NM}).\end{aligned} \tag{14}$$

The state corresponding to the polarization ξ_{MN} is a massless state of spin 2: the graviton. The state corresponding to the scalar ξ_t is the dilaton, while the one given by an antisymmetric tensor is called the B-field.

The ground state of Hilbert space $|0\rangle$ has a negative mass square (it is $-\frac{4}{\alpha'}$). The appearance of this tachyon means that the bosonic string is unstable and will condensate to the true vacuum of the theory. We will see in the discussion of the superstring below how to remove the tachyon from the spectrum to obtain a well-behaved theory (this cannot be done for the bosonic string).

Strings moving in a curved background can be studied by modifying the action (1) by the following "σ-model" action[25]

$$S = -\frac{1}{4\pi\alpha'}\int \mathrm{d}\sigma\mathrm{d}\tau\sqrt{\gamma}\,\gamma^{\alpha\beta}G_{MN}(X)\partial_\alpha X^M\partial_\beta X^N \ , \tag{15}$$

$M = 0, ..., D-1$; $\alpha = \tau, \sigma$. This action looks like (1) but it now has field dependent couplings given by the spacetime metric $G_{MN}(X)$. The curved background can be seen as a coherent state of gravitons in the following sense: If we consider a small deviation from flat space, $G_{MN} = \eta_{MN} + h_{MN}(X)$, with h small, and expand the path integral[26]

$$\begin{aligned}Z &= \int \mathcal{D}X\mathcal{D}\gamma e^{-S} \\ &= \int \mathcal{D}X\mathcal{D}\gamma e^{-S_0}\left(1 + \frac{1}{4\pi\alpha'}\int \mathrm{d}\sigma\mathrm{d}\tau\sqrt{\gamma}\,\gamma^{\alpha\beta}h_{MN}\partial_\alpha X^M\partial_\beta X^N + \cdots\right),\end{aligned} \tag{16}$$

we find that the additional terms correspond to inserting a graviton emission vertex operator with $h_{MN} \propto \xi^s_{MN}$. Thus, a background metric is generated by a condensate (or coherent state) of strings with certain excitations. It is natural to try and further generalize the σ-model action (15) to include coherent states of the B-field and the dilaton. The natural reparametrization-invariant σ-model action is

$$\begin{aligned}S = -\frac{1}{4\pi\alpha'}\int \mathrm{d}\sigma\mathrm{d}\tau\sqrt{\gamma}\ &\left[\left(\gamma^{\alpha\beta}g_{MN}(X) + \mathrm{i}\epsilon^{\alpha\beta}B_{MN}(X)\right)\partial_\alpha X^M\partial_\beta X^N\right. \\ &\left. +\alpha'\Phi R\right] \ .\end{aligned} \tag{17}$$

The power of α' in the last term should be there in order to get the right dimensions. Note that α' in this action behaves like $\hbar$: the action is large in the limit $\alpha' \to 0$, which makes it a good limit to expand around.

Something remarkable happens with the dilaton Φ. Since it appears multiplied by the Euler density, it couples to the Euler number of the worldsheet

$$\chi = \frac{1}{4\pi} \int d^2\sigma (-\gamma)^{1/2} R \,. \tag{18}$$

In the path integral (16) the resulting amplitudes are weighted by a factor $e^{-\Phi\chi}$. On a 2-dimensional surface with h handles, b boundaries and c cross-caps, the Euler number is $\chi = 2-2h-b-c$. An emission and reabsorption of a closed string amounts to adding an extra handle on the worldsheet (see Figure 2), and therefore results in $\delta\chi = -2$. Therefore, relative to the tree level closed string diagram, the amplitudes are weighted by $e^{2\Phi}$. This means that the closed string coupling is not a free parameter, but it is given by the VEV of one of the background fields: $g_s = e^{<\Phi>}$. Therefore, the only free parameter in the theory is the string tension (2), or the string energy scale $(1/\sqrt{\alpha'})$. However, without the introduction of fluxes the VEV of the dilaton is undetermined, as the action does not contain a potential term for it. This is actually the main motivation for considering backgrounds with fluxes, as the latter generate a potential whose minimum determines the dilaton VEV.

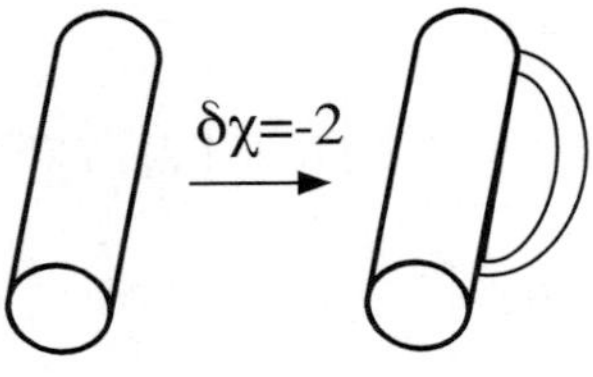

Fig. 2. Worldsheet topology change due to the emission and absorption of a closed string.

Although the σ-model action (17) is invariant under Weyl transformations at the classical level, it is not automatically Weyl invariant in the quantum theory. Weyl invariance (also called conformal invariance) can be translated as a tracelessness condition on the energy momentum tensor (7). The one loop expectation value of the trace of the energy momentum is given by[27]

$$T^{\alpha}_{\ \alpha} = -\frac{1}{2\alpha'}\beta^G_{MN} g^{\alpha\beta}\partial_\alpha X^M \partial_\beta X^N - \frac{i}{2\alpha'}\beta^B_{MN}\epsilon^{\alpha\beta}\partial_\alpha X^M \partial_\beta X^N - \frac{1}{2}\beta^\Phi R\,. \tag{19}$$

where we have defined

$$\begin{aligned}
\beta^G_{MN} &= \alpha'\left(R_{MN} + 2\nabla_M\nabla_N\Phi - \frac{1}{4}H_{MPQ}H_N^{\ \ PQ}\right) + O(\alpha'^2),\\
\beta^B_{MN} &= \alpha'\left(-\frac{1}{2}\nabla^P H_{PMN} + \nabla^P\Phi H_{PMN}\right) + O(\alpha'^2), \qquad (20)\\
\beta^\Phi &= \alpha'\left(\frac{D-26}{6\alpha'} - \frac{1}{2}\nabla^2\Phi + \nabla_M\Phi\nabla^M\Phi - \frac{1}{24}H_{MNP}H^{MNP}\right) + O(\alpha'^2),
\end{aligned}$$

and H_{MPQ} is the field strength of B_{PQ}, i.e. $H_{MPQ} = \partial_{[M}B_{PQ]}$. By setting these beta-functions to zero, we obtain *spacetime* equations for the fields. Therefore, Weyl invariance at one loop (in α') gives the spacetime equation of motion for the massless closed fields. The first equation in (20) resembles Einstein's equation, and we can see that a gradient of the dilaton as well as H-flux carry energy-momentum. The second equation is the equation of motion for the B-field, and it is the antisymmetric tensor generalization of Maxwell's equation. The third equation is the dilaton equation of motion. The first term in this last equation is striking: if the spacetime dimension D is not 26, then the dilaton or the B-field must have large gradients, of the order $1/\sqrt{\alpha'}$. If we do not allow for this, then the spacetime dimension must be 26.[b] The equations of motion (20) can actually be derived from the following spacetime action

$$\begin{aligned}
\mathrm{S} = \frac{1}{2\kappa_0^2}\int \mathrm{d}^D X(-G)^{1/2}e^{-2\Phi}\bigg[&R + 4\nabla_M\Phi\nabla^M\Phi - \frac{1}{12}H_{MNP}H^{MNP}\\
&-\frac{2(D-26)}{3\alpha'} + O(\alpha')\bigg]\,.
\end{aligned} \tag{21}$$

Summarizing, we have found that the massless closed string spectrum contains the graviton, a scalar and an antisymmetric two-form. Demanding Weyl invariance of the the worldsheet action, we have obtained their equations of motion. The equation for the metric resembles Einstein's equation, with the gradient of the other fields acting as sources. The B-field obeys a Maxwell-type equation, while from the equation of motion for the dilaton we have fixed the spacetime dimension to be 26.

[b]There are solutions with $D \neq 26$, like the "linear dilaton"[28] (an exact solution with $\Phi = V_M X^M$, $|V|^2 = (D-26)/\alpha'$), but we will not discuss them here.

One first remark is that the massless closed string spectrum does not contain a regular (one-form) gauge field. There is however a gauge field in the spectrum of massless open strings. The solution to the (worldsheet) equations of motion (3) imposing Neumann boundary conditions at $\sigma = 0, \pi$ (i.e. $\partial_\sigma X^M(\tau, 0) = \partial_\sigma X^M(\tau, \pi) = 0$) gives the following mode decomposition

$$X^M(\tau, \sigma) = x^M + 2\alpha' p^M \tau + \mathrm{i}(2\alpha')^{1/2} \sum_{n \neq 0} \frac{1}{n} \alpha_n^M e^{-\mathrm{i}n\tau} \cos(n\sigma) \, . \tag{22}$$

The mass operator is given by (cf. (10)

$$M^2 = \frac{1}{\alpha'} \left(\sum_{n=1}^{\infty} \alpha_n \cdot \alpha_{-n} - 1 \right) \, . \tag{23}$$

The massless states are therefore given by

$$|\xi_M\rangle \equiv \xi_M \alpha_1^M |0\rangle \tag{24}$$

which is a gauge field A_M. The σ-model action for the gauge field is boundary action

$$\int_{\partial M} d\tau A_M \partial_\tau X^M \tag{25}$$

It's equation of motion can be derived from the spacetime action

$$\mathrm{S} = -\frac{1}{4} \int \mathrm{d}^D X \, e^{-\Phi} F_{MN} F^{MN} + O(\alpha') \, , \tag{26}$$

where $F = dA$. This is the Yang-Mills action, with a field dependent coupling constant $g_o = e^{\Phi/2}$ which is the square root of the closed string coupling constant g_s.

Similarly, one can replace some of the Neumann boundary conditions by Dirichlet boundary conditions, restraining the endpoint of the open string to $p+1$-dimensional surface, a so-called Dp-brane (D for Dirichlet).[29] The sigma-model action then depends on a $p+1$ dim. vector plus $D-p-1$ scalars on the boundary of the open string, similar to (25). It turns out that similar to the closed string, the VEV of these scalars determines the position of the Dp-brane. In other words, the dynamics of the open string corresponds to the dynamics of the Dp-brane, which therefore is a dynamical object.

Open strings can be dressed with extra quantum numbers at their ends, called "Chan-Paton" factors. If one does so, also the vertex operators have Chan-Paton factors and the quantum numbers turn out to form the adjoint of a non-Abelian gauge group. In the simplest case, the Chan-Paton factors give rise to a $U(n)$ gauge group. Such non-Abelian gauge groups can actually be better understood in the dual picture of D-branes. When there are many D-branes, there can be open strings beginning and ending on the same D-brane, or they can also begin on one brane and end on another one. With N D-branes on top of each other, the end points are labelled by the Chan-Paton factors and the gauge field represented by open strings is a U(N) gauge field. In that case, the action (26) becomes non-Abelian. Furthermore, it is the $\alpha' \to 0$ limit of the Dirac-Born-Infeld (DBI) action (given below in (39)), which describes the open string dynamics to all orders in α'.

2.2. *Superstrings*

So far we have found a graviton, a gauge field, a scalar and a two-form in the massless spectrum of open and closed strings. None of these is a fermion like the electron, or the quarks. In order to have fermions in the spectrum, we should introduce fermions in the worldsheet. We therefore introduce superpartners of X^M (i.e., the field related to X^M by a supersymmetry transformation): Ψ^M and $\tilde{\Psi}^M$. Each of them is a Majorana-Weyl spinor on the worldsheet, which has only one component (times $D-2$ for the index M). The combined action (1) for X^M and $\Psi^M, \tilde{\Psi}^M$ is

$$S = \frac{1}{4\pi}\int_M \mathrm{d}^2\sigma\, \eta_{MN}\left[\frac{1}{\alpha'}\partial X^M \bar{\partial} X^N + \Psi^M \bar{\partial}\Psi^N + \tilde{\Psi}^M \bar{\partial}\tilde{\Psi}^N\right] \tag{27}$$

where we have used the Euclidean worldsheet by sending $\tau \to \mathrm{i}\tau$, and introduced the complex coordinates $z = e^{\tau - \mathrm{i}\sigma}$. The equations of motion for Ψ and $\tilde{\Psi}$ admit two possible boundary conditions, denoted "Ramond" (R) and "Neveu-Schwarz" (NS)

$$\begin{aligned} \mathrm{R}: \quad & \Psi^M(\tau,0) = \Psi^M(\tau,2\pi) \\ \mathrm{NS}: \quad & \Psi^M(\tau,0) = -\Psi^M(\tau,2\pi) \end{aligned} \tag{28}$$

and similarly for $\tilde{\psi}$. When expanding in modes as in (6), the first type of boundary condition (R) will give rise to an expansion in integer modes, while the second type (NS) gives an expansion in half-integer modes. The (anti-)commutation relation for the modes are

$$\{\psi_r^M, \psi_s^N\} = \{\tilde{\psi}_r^M, \tilde{\psi}_s^N\} = \eta^{MN}\delta_{r+s}\,. \tag{29}$$

In a similar fashion as for the bosonic string, in the superstring conformal anomaly cancellation imposes a fixed dimension for spacetime: $D = 10$.

The quantized states have masses

$$M^2 = \frac{1}{\alpha'}\left(\sum_{n=1}^{\infty}\alpha_n\cdot\alpha_{-n} + \sum_r r\psi_r\cdot\psi_{-r} - a + \text{same with tilde}\right) , \quad (30)$$

where the normal ordering constant a is zero for Ramond modes, and 1/2 for NS modes. Furthermore, the level-matching condition says that physical states vanish under the action of the operator

$$\tilde{L}_0 = \left(\sum_{n=1}^{\infty}\alpha_n\cdot\alpha_{-n} + \sum_r r\psi_r\cdot\psi_{-r} - a - \text{same with tilde}\right) . \quad (31)$$

The massless states are therefore

$$\begin{aligned} \mathrm{R-R}:\ \xi_{MN}\psi_0^M\tilde{\psi}_0^N|0\rangle\ , \qquad & \mathrm{NS-R}:\xi_{MN}\psi_{1/2}^M\tilde{\psi}_0^N|0\rangle\ , \\ \mathrm{NS-NS}:\xi_{MN}\psi_{1/2}^M\tilde{\psi}_{1/2}^N|0\rangle\ , \qquad & \mathrm{R-NS}:\xi_{MN}\psi_0^M\tilde{\psi}_{1/2}^N|0\rangle\ . \end{aligned} \quad (32)$$

In the same way as for the bosonic string, the polarizations ξ_M have to be orthogonal to the center of mass momentum $\vec{k}$, and are therefore classified by SO(8) representations.

Let us now first discuss the left-moving states. We combine them with the right-moving states afterwards. Let us start with the Ramond states. Since Ψ_0 obeys a Clifford algebra, we can form raising and lowering operators $d^i_\pm = \frac{1}{\sqrt{2}}(\psi_0^{2i} \pm \psi_0^{2i+1})$, with $i = 1,..,4$. The Ramond ground states form a representation of this algebra labeled by s:

$$R:\psi_0^M|0\rangle \to |s\rangle = |\pm\frac{1}{2},\pm\frac{1}{2},\pm\frac{1}{2},\pm\frac{1}{2}\rangle \quad (33)$$

where $\pm\frac{1}{2}$ are the chiralities in the four 2-dimensional planes i.[c] The state $\psi_0^M|0\rangle$ is therefore a *spacetime* fermion, in the representation **16**. The **16** reduces into $\mathbf{8}_s\oplus\mathbf{8}_c$. In order to get spacetime supersymmetry, we need only 8 physical fermions (to be the superpartners of the 8 X^M). We therefore would like to project out half of the fermions. This is consistently done by an operation called "GSO" projection (named after Gliozzi, Scherk, and Olive[30]), which requires $\sum_{i=1}^4 s_i = 0 \pmod 2$. One can view the GSO projection as the projection to the (Majorana-)Weyl spinor subspace of definite chirality.

Let us now turn to the NS states. From the above definition one can see that the GSO projector anti-commutes with single creation and annihilation

[c]The RR massless state $\xi_{MN}\psi_0^M\tilde{\psi}_0^N|0\rangle$ should therefore be understood as $|s\rangle\otimes|\tilde{s}\rangle$.

operators. This means that the GSO projection either projects out the (tachyonic) vacuum state $|0\rangle$ or the massless state

$$NS : \psi^M_{1/2}|0\rangle \ , \tag{34}$$

which forms a vector in ten-dimensional spacetime. This means that we can choose the GSO projection such that there is no tachyon in the superstring spectrum while we keep the massless NS states. The latter is actually identical to the massless left-moving sector of the bosonic string. We therefore expect the same field content in the NS-NS sector as for the bosonic string.

Note that in ten-dimensional spacetime the irreducible spin 1/2 and spin 1 representations have the same number of degrees of freedom: 8. We label the spinors of positive and negative chirality by $\mathbf{8}_s$ and $\mathbf{8}_c$, respectively, while the vector representation is called $\mathbf{8}_v$. Actually, there is a discrete symmetry interchanging the three of them. This is called "triality" (for the trio $\mathbf{8}_s$, $\mathbf{8}_c$, $\mathbf{8}_v$), and plays a very important role in string theory.

We are ready to build the whole closed superstring massless spectrum. In the R-R sector, we have the product of two spinorial **8** representations, one for the left and one for the right movers. We can take the same GSO projection for the left movers as for the right movers, but we could as well take opposite projections. The first choice leads to the so-called type IIB superstring (which is chiral, since we chose the same chiralities on the left and on the right), while the other choice leads to the type IIA superstring. Decomposing the products into irreducible representations we get

$$\begin{aligned} \text{type IIB R} - \text{R}: \ \ & \mathbf{8}_s \otimes \mathbf{8}_s = \mathbf{1} \oplus \mathbf{28} \oplus \mathbf{35}_+ = C_0 \oplus C_2 \oplus C_{4+} \\ \text{type IIA R} - \text{R}: \ \ & \mathbf{8}_s \otimes \mathbf{8}_c = \mathbf{8}_v \oplus \mathbf{56}_t = C_1 \oplus C_3 \ . \end{aligned} \tag{35}$$

In this equation, we have defined C_n to be an n-form, called a R-R potential. In type IIB there are even-degree R-R potentials (with C_4 satisfying a duality condition $dC_4 = *dC_4$). In type IIA, the R-R potentials are odd.

From the NS-R and R-NS sectors we find the fermionic states

$$\begin{aligned} \text{type IIB NS} - \text{R}: \ \ & \mathbf{8}_v \otimes \mathbf{8}_s = \mathbf{8}_s \oplus \mathbf{56}_s = \lambda^1 \oplus \Psi^{1\,M} \ , \\ \text{R} - \text{NS}: \ \ & \mathbf{8}_s \otimes \mathbf{8}_v = \mathbf{8}_s \oplus \mathbf{56}_s = \lambda^2 \oplus \Psi^{2\,M} \ , \\ \text{type IIA NS} - \text{R}: \ \ & \mathbf{8}_v \otimes \mathbf{8}_c = \mathbf{8}_c \oplus \mathbf{56}_c = \lambda^1 \oplus \Psi^{1\,M} \ , \\ \text{R} - \text{NS}: \ \ & \mathbf{8}_s \otimes \mathbf{8}_v = \mathbf{8}_s \oplus \mathbf{56}_s = \lambda^2 \oplus \Psi^{2\,M} \ . \end{aligned} \tag{36}$$

The fermions in the **8** representation λ^A (A=1,2) are called the dilatini, and the ones in the **56**, $\Psi^{A,M}$ are the gravitini. The name "type II" refers to the fact that there are two gravitini (and consequently, two spacetime supersymmetry parameters).

Finally, in the NS-NS sector we get

$$NS-NS: \text{ type IIA and IIB } \mathbf{8}_\mathrm{v} \otimes \mathbf{8}_\mathrm{v} = \mathbf{1} \oplus \mathbf{28} \oplus \mathbf{35} = \Phi \oplus \mathrm{B}_2 \oplus \mathrm{G}_{\mathrm{MN}} \quad (37)$$

This sector is common to both theories and it has the same matter content the as the massless bosonic string: a scalar, the dilaton; a symmetric traceless tensor, the graviton; and a two-form, the B-field.

In the open strings, from the R and NS sectors we get respectively a fermion and a gauge field A_M, whose spacetime action is the supersymmetrized version of (26), i.e. the action for an Abelian (U(1)) gauge field and fermionic matter. The discussion of Chan-Paton factors, D-branes etc. carries over from the discussion for the bosonic superstring.

Note the magic that has happened: We started with worldsheet fermions and worldsheet supersymmetry and we ended up with massless spacetime fermions, spacetime bosons and (two) spacetime supersymmetries. This is known as the NSR formulation of superstrings. The spectrum of massive states has masses of order $1/\sqrt{\alpha'}$. At energies much lower than the string scale, these states can be neglected and one can just consider the massless spectrum, which forms the field content of the type II supergravities in ten dimensions.[d] When we discuss compactifications in Section 4, we will work in this limit.

Type IIA and IIB are not two independent theories, but they are related by a symmetry called T-duality (for a review see[31]). This symmetry is inherent to 1-dimensional objects: if we compactify one direction on an S^1 of radius R, the center of mass momentum for the string along that direction will be quantized, in units of $1/R$. On the other hand, there are "winding states", namely states with boundary conditions $X^M(\tau, 2\pi) = X^M(\tau, 0) + 2\pi m R$. A string with n units of momentum, m units of winding and N, $\tilde{N}$ total number of oscillators on the left and on the right has a total mass given by (cf. Eq(10)

$$M^2 = \frac{n^2}{R^2} + \frac{m^2 R^2}{\alpha'^2} + \frac{2}{\alpha'}(N + \tilde{N} - 2) \quad (38)$$

This formula is symmetric under the exchange of n and m (or in other words winding and momentum states) if we also exchange R with α'/R! This means that large and small radius of compactification are dual, on one side the momentum modes are light, and winding modes are heavy,

[d] Note that the discussion of Weyl anomalies of the superstring is analogous to the one in (20) for the bosonic string. The vanishing of the beta functions correspond to the equations of motion of type IIA and IIB supergravity in ten dimensions.

while on the T-dual picture winding modes will be light, and momentum would be the heavy modes. The exchange between winding and momentum amounts to exchanging $X_L^M + X_R^M$ with $X_L^M - X_R^M$, and in the superstring ψ^M with $\bar{\psi}^M$. This exchanges representations $\mathbf{8}_s$ with $\mathbf{8}_c$ and therefore the GSO projections, or in other words type IIA with IIB!

T-duality has also another important consequence for type II string theories: Since T-duality maps type IIA and IIB theory into each other, the composition of two T-dualities on different circles maps the theory non-trivially into itself and therefore provides a symmetry of type II string theory on an n-dimensional torus. These symmetries combine with the translational and rotational symmetries of such a torus into the discrete symmetry group $O(n, n, \mathbb{Z})$, the so-called T-duality group.

At the classical level, the massless superstring spectrum is symmetric even under *continuous* transformations in the group $O(8, 8)$. Apart from Lorentz-transformations in $Gl(8)$ that act equally on left- and right-moving excitations, this group incorporates rotations between left- and right-movers. More precisely, both left- and right-movers admit a Clifford algebra for $O(8)$ on their own. Together, they combine into the Clifford algebra of $O(8, 8)$. Both the R-R and NS-NS sector form irreducible representations of this group: The R-R fields combine into an $O(8, 8)$ spinor, while the metric and the B-field form a symmetric tensor, the $O(8, 8)$ metric. The dilaton remains a singlet under $O(8, 8)$.

The group $O(8, 8)$ is not only a classical symmetry of the massless fields, but is also locally a symmetry of their field equations in ten dimensions.[e] This means that we can make the corresponding type II supergravity and its field equations covariant under this group. The program to understand geometry in this covariant language is called generalized complex geometry and will be discussed in more detail in the next section. Generalized complex geometry turns out to be extremely helpful for understanding more complicated backgrounds of string theory, as we will see in Section 4.

Under T-duality, the Neumann boundary conditions along the T-dualized directions for the open string are switched to Dirichlet boundary conditions, or the other way around. This means that under T-duality, a Dp-brane changes its dimension and becomes a D$(p \pm 1)$-brane.

[e] The symmetry group of the supergravity theory is in fact $O(10, 10)$ and thus larger, but the orthogonality of the oscillators reduces the symmetry on the spectrum to $O(8, 8)$, cf. the paragraph below (12).

The Dp-brane action is the Born-Infeld action (39) (see[32] for a review)

$$S_{DBI} = -T_p \int \mathrm{d}^{p+1}\xi e^{-\Phi} \sqrt{\det(G_{ab} + B_{ab} + 2\pi\alpha' F_{ab})} \tag{39}$$

(where $T_p = 8\pi^7 \alpha'^{7/2}(4\pi^2\alpha')^{-p/2}$ supplemented by the Chern-Simons term depending on RR potentials

$$S_{CS} = \mathrm{i}T_p \int_{p+1} e^{2\pi\alpha' F+B} \wedge C \tag{40}$$

where C is the sum of the RR potentials, and the integral picks only the $p+1$-rank forms. In these expressions, the background fields should be pulled-back to the D-brane world-volume. For example, the pulled-back metric is

$$G_{ab}(\xi) = \frac{\partial X^M}{\partial \xi^a}\frac{\partial X^N}{\partial \xi^b} G_{MN}(X) \tag{41}$$

With N D-branes on top of each other, the open strings transform in adjoint representations of U(N). The excitations along the brane represent a U(N) gauge field and gaugino, while excitations orthogonal to the brane are bosons and fermions in the adjoint representation of U(N). The DBI action can actually be generalized to the case of a non-Abelian gauge group.[32] With stacks of D-branes intersecting at angles, or D-branes placed at special singularities, the U(N) symmetry can for instance be broken to the gauge group of the Standard Model of particle physics, namely $SU(3) \times SU(2) \times U(1)$ (see for example[33] for a review).

In summary, we have seen that string theory relies on supersymmetry (the spectrum of the purely bosonic string has a tachyon, which signals an instability). We showed that in the massless spectrum of string theory there is a graviton, and there are also gauge fields and fermions in representations that can be that of the Standard Model of particle physics. As well as type IIA and type IIB, there are other string "theories", consisting, for example, of a mixture of the bosonic string on the left movers and the superstring on the right movers. These other "theories" are connected as well to type IIA and type IIB by dualities. There might therefore be a single string theory and various corners of it, or various low energy versions. One can move from one corner to another by varying the VEVs of scalar fields (in a similar fashion as the string coupling constant, or the radius of a circle). The space of VEVs of massless scalar fields is called moduli space, which is shown in Figure 3.

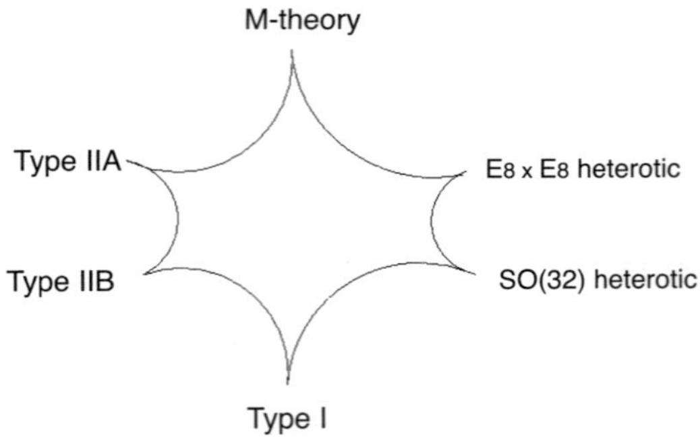

Fig. 3. String theories as limits of one theory.

3. Generalized Complex Geometry

In the last section we explained that the massless type II fields and their field equations are locally symmetric under the group $O(10, 10)$. We will see in Section 4 that it is often useful to consider the group $O(n, n)$, where n is smaller than ten. The formalism that geometrically covariantizes such symmetry groups is called generalized complex geometry. Generalized complex geometry was first proposed by Hitchin and his students in[19,20] in order to describe complex and symplectic geometry in a unifying formalism, before being utilized to describe flux compactifications of string theory.[22,23] In this section we introduce the mathematical formalism and discuss its relevance in string theory in the subsequent section.

Usual complex geometry deals with the tangent bundle of a manifold T, whose sections are vector fields X, and separately, with the cotangent bundle T^*, whose sections are 1-forms ζ. In generalized complex geometry the tangent and cotangent bundle are joined as a single bundle, $T \oplus T^*$. Its sections are the sum of a vector field plus a one-form $X + \zeta$. The bundle $T \oplus T^*$ transforms under a more general group of transformations. While diffeomorphisms act on the tangent and cotangent bundle independently, there are more general transformations that mix both components. However, all these tranformations should preserve the canonical pairing of tangent and cotangent space, represented by the split-signature metric

$$\mathcal{G} = \begin{pmatrix} 0 & 1 \\ 1 & 0 \end{pmatrix} . \tag{1}$$

The transformations that preserve this metric are $O(n,n)$, where n is the dimension of the manifold. In the following we will concentrate on manifolds of dimension $n=6$.

Of particular interest is the nilpotent subgroup of $O(6,6)$ defined by the generator

$$\mathcal{B} = \begin{pmatrix} 0 & 0 \\ B & 0 \end{pmatrix} , \tag{2}$$

with B an antisymmetric 6×6 matrix, or equivalently a two-form. This transformation is used to patch the generalized tangent bundle. To be more precise the generalized tangent bundle E is a particular extension of T by T^*

$$0 \longrightarrow T^*M \longrightarrow E \overset{\pi}{\longrightarrow} TM \longrightarrow 0 \, . \tag{3}$$

In going from one coordinate patch U_α to another U_β, we do not only patch vectors and one-forms by a diffeomorphism-induced Gl(6)-matrix $a_{(\alpha\beta)}$, but also use such two-form shifts $b_{(\alpha\beta)}$ to fiber T^* over T. In total, the patching reads

$$x_{(\alpha)} + \xi_{(\alpha)} = a_{(\alpha\beta)} \cdot x_{(\beta)} + \left[a^{-T}_{(\alpha\beta)} \xi_{(\beta)} - \iota_{a_{(\alpha\beta)} x_{(\beta)}} b_{(\alpha\beta)} \right] , \tag{4}$$

where $a^{-T} = (a^{-1})^T$. In fact, for consistency the two-form shift must satisfy $b_{(\alpha\beta)} = -\mathrm{d}\Lambda_{(\alpha\beta)}$, where $\Lambda_{(\alpha\beta)}$ are required to satisfy

$$\Lambda_{(\alpha\beta)} + \Lambda_{(\beta\gamma)} + \Lambda_{(\gamma\alpha)} = g_{(\alpha\beta\gamma)} \mathrm{d} g_{(\alpha\beta\gamma)} \tag{5}$$

on any triple intersection $U_\alpha \cap U_\beta \cap U_\gamma$ and $g_{\alpha\beta\gamma} := \mathrm{i}\alpha$ is a $U(1)$ element. This is analogous to the patching of a $U(1)$ bundle, except that the transition "functions" are one-forms, $\Lambda_{(\alpha\beta)}$. Formally it is called the "connective structure" of a gerbe.

The standard machinery of complex geometry can be generalized to this even-dimensional bundle. One can construct a generalized almost complex structure $\mathcal{J}$, which is a map of E to itself that squares to $-\mathbb{I}_{2d}$ (d is real the dimension of the manifold). This is analogous to an almost complex structure $I_m{}^n$, which is a bundle map from T to itself that squares to $-\mathbb{I}_d$. As for an almost complex structure, $\mathcal{J}$ must also satisfy the hermiticity condition $\mathcal{J}^t \mathcal{G} \mathcal{J} = \mathcal{G}$, with respect to the natural metric $\mathcal{G}$ on $T \oplus T^*$ defined in (1).

Usual complex structures I are naturally embedded into generalized ones $\mathcal{J}$: take $\mathcal{J}$ to be

$$\mathcal{J}_1 \equiv \begin{pmatrix} I & 0 \\ 0 & -I^t \end{pmatrix} , \tag{6}$$

with $I_m{}^n$ a regular almost complex structure (i.e. $I^2 = -\mathbb{I}_d$). This $\mathcal{J}$ satisfies the desired properties, namely $\mathcal{J}^2 = -\mathbb{I}_{2d}$, $\mathcal{J}^t \mathcal{G} \mathcal{J} = \mathcal{G}$. Another example of generalized almost complex structure can be built using a non degenerate two–form J_{mn},

$$\mathcal{J}_2 \equiv \begin{pmatrix} 0 & -J^{-1} \\ J & 0 \end{pmatrix} . \tag{7}$$

Given an almost complex structure $I_m{}^n$, one can build holomorphic and antiholomorphic projectors $\pi_\pm = \frac{1}{2}(\mathbb{I}_d \pm \mathrm{i}I)$. Correspondingly, projectors can be built out of a generalized almost complex structure, $\Pi_\pm = \frac{1}{2}(\mathbb{I}_{2d} \pm \mathrm{i}\mathcal{J})$. There is an integrability condition for generalized almost complex structures, analogous to the integrability condition for usual almost complex structures. For the usual complex structures, integrability, namely the vanishing of the Nijenhuis tensor, can be written as the condition $\pi_\mp[\pi_\pm X, \pi_\pm Y] = 0$, i.e. the Lie bracket of two holomorphic vectors should again be holomorphic. For generalized almost complex structures, integrability condition reads exactly the same, with π and X replaced respectively by Π and $X + \zeta$, and the Lie bracket replaced by the Courant bracket on $TM \oplus T^*M$, which is defined as follows

$$[X + \zeta, Y + \eta]_C = [X, Y] + \mathcal{L}_X \eta - \mathcal{L}_Y \zeta - \frac{1}{2}\mathrm{d}(\iota_X \eta - \iota_Y \zeta) . \tag{8}$$

The Courant bracket does not satisfy Jacobi identity in general, but it does on the i–eigenspaces of $\mathcal{J}$. In case these conditions are fulfilled, we can drop the "almost" and speak of generalized complex structures.

For the two examples of generalized almost complex structure given above, $\mathcal{J}_1$ and $\mathcal{J}_2$, integrability condition turns into a condition on their building blocks, $I_m{}^n$ and J_{mn}. Integrability of $\mathcal{J}_1$ enforces I to be an integrable almost complex structure on T, and hence I is a complex structure, or equivalently the manifold is complex. For $\mathcal{J}_2$, which was built from a two-form J_{mn}, integrability imposes $dJ = 0$, thus making J into a symplectic form, and the manifold a symplectic one.

These two examples are not exhaustive, and the most general generalized complex structure is partially complex, partially symplectic. Explicitly, a generalized complex manifold is locally equivalent to the product $\mathbb{C}^k \times (\mathbb{R}^{d-2k}, J)$, where $J = \mathrm{d}x^{2k+1} \wedge \mathrm{d}x^{2k+2} + ... + \mathrm{d}x^{d-1} \wedge \mathrm{d}x^d$ is the standard symplectic structure and $k \leq d/2$ is called rank, which can be constant or even vary over the manifold (jump by two at certain special points or planes).

On this generalized tangent bundle E we can also define a positive definite metric $\mathcal{H}$. On $T \oplus T^*$ a standard choice would be to combine the ordinary metric g and its inverse in a block-diagonal metric. However, this metric can be rotated by some two-form shift B as given in (2). Indeed it turns out that locally a generalized metric on the gerbe E can always be written in the form

$$\mathcal{H} = \begin{pmatrix} g - Bg^{-1}B & Bg^{-1} \\ -g^{-1}B & g^{-1} \end{pmatrix} . \tag{9}$$

If we have now a pair of *commuting* generalized almost complex structures $\mathcal{J}_i$, $i = 1, 2$, we can define from this a generalized metric via

$$\mathcal{H} = \mathcal{G}\mathcal{J}_1\mathcal{J}_2 . \tag{10}$$

Commutation and Hermiticity of the $\mathcal{J}_i$ ensures that $\mathcal{H}$ is indeed a generalized metric.

There is an algebraic one-to-one correspondence between generalized almost complex $\mathcal{J}$ structures and (lines of) pure spinors of Clifford(6,6) Φ.[f] It maps the $+i$ eigenspace of $\mathcal{J}$ to the annihilator space of the spinor Φ. In string theory, the picture of generalized almost complex structures emerges naturally from the worldsheet point of view,[34] while that of pure spinors arises on the spacetime side, as we discuss in Section 4.2.

One can build Clifford$(6,6)$Weyl pure spinors by tensoring Clifford(6) Weyl spinors (which are automatically pure), namely

$$\Phi_\pm = \eta^1_+ \otimes \eta^{2\dagger}_\pm . \tag{11}$$

Using Fierz identities, this tensor product can be written in terms of bilinears of the spinors by

$$\eta^1_+ \otimes \eta^{2\dagger}_\pm = \frac{1}{4}\sum_{k=0}^{6} \frac{1}{k!} \eta^{2\dagger}_\pm \gamma_{i_1\ldots i_k} \eta^1_+ \gamma^{i_k \ldots i_1} \tag{12}$$

A Clifford$(6,6)$ spinor can also be mapped to a formal sum of forms via the Clifford map

$$\sum_k \frac{1}{k!} C^{(k)}_{i_1\ldots i_k} \gamma^{i_i \ldots i_k}_{\alpha\beta} \qquad \longleftrightarrow \qquad \sum_k \frac{1}{k!} C^{(k)}_{i_1\ldots i_k} \mathrm{d}x^{i_i} \wedge \ldots \wedge \mathrm{d}x^{i_k} . \tag{13}$$

The tensor products in (11) are then identified with sums of regular forms. From now on, we will use $\Phi_\pm$ to denote just the forms. The subindices plus and minus in $\Phi_\pm$ denote the Spin(6,6) chirality: positive corresponds to an

[f] A spinor is said to be pure if its anihilator space is maximal (i.e. 6-dimensional in this case)

even form, and negative to an odd form. Irreducible Spin(6,6) representations are actually "Majorana-Weyl", namely they are of definite parity ("Weyl") and real ("Majorana"). The B-transform (2) on spinors amounts to the exponential action

$$\Phi_\pm \to e^{-B}\Phi_\pm \equiv \Phi_\pm^D \tag{14}$$

where on the polyform associated to the spinor, the action is $e^{-B}\Phi = (1 - B\wedge + \frac{1}{2}B\wedge B\wedge + ...)\Phi$. We will refer to Φ as naked pure spinor, while Φ^D will be called dressed pure spinor. The former lives in the spinor bundle over $T \oplus T^*$, the latter lives in the spinor bundle over E.

For the particular case $\eta^1 = \eta^2 \equiv \eta$, the pure spinors in (11) reduce to

$$\Phi_+ = e^{-\mathrm{i}J} \ , \qquad \Phi_- = -\mathrm{i}\Omega \ , \tag{15}$$

where the coefficients of J and Ω are the bilinears with two and three gamma matrices

$$J_{mn} = \mp 2\mathrm{i}\, \eta_\pm^\dagger \gamma_{mn} \eta_\pm \qquad \Omega_{mnp} = -2\mathrm{i}\, \eta_-^\dagger \gamma_{mnp} \eta_+ \ . \tag{16}$$

In particular, J is real while Ω is complex. From the Fierz identities that one can derive from the Clifford algebra and the above definitions one finds

$$J \wedge \Omega = 0 \ , \qquad J \wedge J \wedge J = \tfrac{3\mathrm{i}}{4}\Omega \wedge \bar{\Omega} \ , \tag{17}$$

where each side of the latter equation is nowhere vanishing. Equivalently, the existence of a nowhere-vanishing spinor η or of the forms just described each defines an $SU(3)$ structure on the manifold, as we explain now. On a manifold with $SU(3)$ structure, the spinor representation in six dimensions, in the **4** of $SO(6)$, can be further decomposed in representations of $SU(3)$ as $\mathbf{4} \to \mathbf{3} + \mathbf{1}$. There is therefore an $SU(3)$ singlet in the decomposition, which means that there exists a connection (with torsion, if the structure is not integrable), such that there is a spinor that is invariant under parallel transport by this connection. Such a spinor is therefore well defined and non-vanishing. The converse is also true: A six-dimensional manifold that has a globally well-defined non-vanishing spinor has structure group $SU(3)$.

We can decompose other $SO(6)$ representations, such as the vector **6**, two-form **15** and three-form (self-dual and anti-self-dual) $\mathbf{10} + \mathbf{10}^*$ in representations of $SU(3)$. This yields

$$\begin{aligned} \mathbf{6} &\to \mathbf{3} + \bar{\mathbf{3}} \ , \\ \mathbf{15} &\to \mathbf{8} + \mathbf{3} + \bar{\mathbf{3}} + \mathbf{1} \ , \\ \mathbf{10} + \mathbf{10}^* &\to \mathbf{6} + \bar{\mathbf{6}} + \mathbf{3} + \bar{\mathbf{3}} + \mathbf{1} + \mathbf{1} \ . \end{aligned} \tag{18}$$

We can see that there are also singlets in the decomposition of two- and three-forms. This means that there is also a non-vanishing globally well-defined real 2-form and complex 3-form. These are called respectively J and Ω. We can also see that there are no invariant vectors (or equivalently five-forms), which implies the first equation in (17).

One can actually define the SU(3) structure in terms of a pair of real forms (J, ρ) where $\rho = Re\Omega$. The forms cannot be arbitrary but must be *stable*.[35] This means that they live in an open orbit under the action of general transformations $\mathrm{GL}(6,\mathbb{R})$ in the tangent space at each point. A stable two-form J then defines a $Sp(6,\mathbb{R})$ structure while a stable real form ρ defines a $\mathrm{SL}(3,\mathbb{C})$ structure. Together they define an SU(3) structure provided the embeddings of $Sp(6,\mathbb{R})$ and $\mathrm{SL}(3,\mathbb{C})$ in $\mathrm{GL}(6,\mathbb{R})$ are compatible. This requires $J \wedge \rho = 0$. Note that since $\mathrm{SU}(3) \subset \mathrm{SO}(6)$ the pair (J, ρ) satisfying (17) also defines an SO(6) structure and hence a metric.

Under the one-to-one correspondence between spinors and generalized almost complex structures we have

$$\begin{aligned} \Phi_- &= -\frac{\mathrm{i}}{8}\Omega \leftrightarrow \mathcal{J}_1 \\ \Phi_+ &= \frac{1}{8}e^{-\mathrm{i}J} \leftrightarrow \mathcal{J}_2 \end{aligned} \tag{19}$$

where $\mathcal{J}_1$ and $\mathcal{J}_2$ are defined in (6, 7). These psinors are of special type: they correspond to a purely complex or purely symplectic generalized almost complex structure. In a more general case, such as the one obatined by doing tensor products of two different Clifford(6) spinors, one obtains pure spinors which are a mixture of a complex and a symplectic structure. For example, if the two Clifford(6) spinors are orthogonal, they define a complex vector (i.e., one can write $\eta^2 = (v + \mathrm{i}v')_m \gamma^m \eta^1$, and the corresponding pure spinors would read

$$\begin{aligned} \Phi_+ &= \eta^1_+ \otimes \eta^{2\,\dagger}_+ = -\frac{\mathrm{i}}{8}\omega \wedge e^{-\mathrm{i}v \wedge v'} , \\ \Phi_- &= \eta^1_+ \otimes \eta^{2\,\dagger}_- = -\frac{1}{8}e^{-\mathrm{i}j} \wedge (v + \mathrm{i}v') . \end{aligned} \tag{20}$$

These are given in terms of the local SU(2) structure defined by (η^1, η^2): j and ω are the (1,1) and (2,0)-forms on the local four dimensional space orthogonal to v and v'. Φ_+ describes therefore a generalized almost complex structure "of rank 2" (i.e., of complex type in two dimensions), while the rank of Φ_- is one. In the most generic case, the rank of the pure spinors need not be constant over the manifold, but can be point-dependent, and jump across the manifold.

Integrability condition for the generalized complex structure corresponds on the pure spinor side to the condition

$$\mathcal{J} \text{ is integrable} \Leftrightarrow \exists \text{ vector } v \text{ and 1-form } \zeta \text{ such that } \mathrm{d}\Phi = (v\llcorner + \zeta\wedge)\Phi$$

A generalized Calabi-Yau[19] is a manifold on which a closed pure spinor exists:

$$\text{Generalized Calabi-Yau} \Leftrightarrow \exists\ \Phi \text{ pure such that } \mathrm{d}\Phi = 0$$

From the previous property, a generalized Calabi-Yau has obviously an integrable generalized complex structure. Examples of Generalized Calabi-Yau manifolds are symplectic manifolds and complex manifolds with trivial torsion class W_5 (i.e. if $\exists\ f$ such that $\Phi = e^{-f}\Omega$ is closed). More generally, if the integrable pure spinor has rank k, then the manifold looks locally like $\mathbb{C}^k \times (\mathbb{R}^{d-2k}, J)$, as we mentioned before.

As an alternative to the use of the generalized tangent bundle E, one can just use $T \oplus T^*$ and twist the differential d by a closed three–form H such that the differential becomes $d - H\wedge$. Similar, the Courant bracket is then modified to

$$[X + \zeta, Y + \eta]_H = [X + \zeta, Y + \eta]_C + \iota_X \iota_Y H\ , \tag{21}$$

and with it the integrability condition. In terms of "integrability" of the pure spinors Φ, adding H amounts to twisting the differential condititions for integrability and for generalized Calabi-Yau. More precisely,

$$\text{"twisted" generalized Calabi-Yau} \Leftrightarrow \exists\ \Phi \text{ pure, and } H \text{ closed s.t.}$$
$$(\mathrm{d} - H\wedge)\Phi = 0$$

Alternatively, a twisted generalized Calabi-Yau can be defined by the existence of a pure spinor Φ^D on the generalized tangent bundle E such that $\mathrm{d}\Phi^D = 0$.

4. Compactifications of String Theory and Generalized Complex Geometry

As we discussed in Section 2, a consistent string theory can only be defined in a ten-dimensional spacetime. This must be linked to the four-dimensional spacetime that we observe in nature. The usual method is compactification. If some spacelike directions are compact, they play only a role at the energy scale related to the size of this compact direction and higher. This means that if these six extra dimensions are small compared to accessible energy scales, the theory appears effectively four-dimensional. The extra dimensions form an "internal" manifold whose geometry should determine the physics in four dimensions. There are two basic questions. The first one is which geometries and configurations are allowed for string theory, while the second one is what the resulting four-dimensional theory might be. In the remainder of these lectures we will focus on the first question.

4.1. *Compactifications without flux*

Viable backgrounds of string theory are those that solve the equations of motion (20) that guarantee a vanishing Weyl anomaly. Moreover, we want to find backgrounds where the ten-dimensional space splits into a four-dimensional "external" and a six-dimensional "internal" space. Moreover, the background should be a four-dimensional *vacuum*, which means that the four-dimensional space is maximally symmetric: Either it is Minkowski, de Sitter (dS) or anti-de Sitter (AdS). The background will be defined by the ten-dimensional metric, and the profile for the dilaton and fluxes. The most general ten-dimensional metric respecting Poincaré symmetry is

$$ds^2 = e^{2A(y)}\eta_{\mu\nu}\mathrm{d}x^\mu \mathrm{d}x^\nu + g_{mn}\mathrm{d}y^m \mathrm{d}y^n\,, \quad \mu = 0,1,2,3 \quad m = 1,...,6 \quad (1)$$

where A is a function of the internal coordinates called *warp factor* (we will use (1) to be the string frame metric), and $\eta_{\mu\nu}$ is the Minkowski (or AdS or dS) metric. Furthermore, Lorentz invariance dictates that the vacuum expectation values (VEVs) of all (four-dimensional) non-scalar fields must vanish in this backgrounds.

In the following we are looking for a supersymmetric solution. For such solutions, instead of solving the equations of motion directly, one can try and solve the supersymmetry conditions $< Q_\epsilon \chi > = < \delta_\epsilon \chi > = 0$, where Q is the supersymmetry generator, ϵ is the supersymmetry parameter and χ is any of the massless bosonic and fermionic fields. Supersymmetry conditions together with Bianchi identities written below in (6) determine a supersymmetric solution to the equations of motion. While the equations of motion are quadratic in the fields, supersymmetry conditions are linear, and therefore much easier to solve. One can then find for example non-supersymmetric solutions by perturbing the supersymmetric solution by a supersymmetry-breaking source. As we will see, supersymmetry enforces the internal manifold to have reduced holonomy[g], and therefore be of special type.

---

[g]The holonomy group of a connection is the group of transformations that one obtains

We are interested in vacua whose four-dimensional space has Poincare symmetry. A nonzero vacuum expectation value (VEV) for a fermionic field would break this symmetry. Therefore, the desired background should only be bosonic, i.e. it should have zero VEV for the fermionic fields. This means that the supersymmetry conditions $< \delta_\epsilon \chi >= 0$ are already solved if χ is a boson, as all of these variations are proportional to VEVs of fermionic fields. Therefore, we only have to impose those for which χ is a fermion (whose supersymmetry transformation involves bosonic fields only). As shown in (36), the fermionic fields in type II theories are two gravitinos ψ_M^A, $A = 1, 2$ and two dilatinos λ^A. Their supersymmetry transformations in the string frame read

$$\delta\psi_M = \nabla_M \epsilon + \frac{1}{4} \not{H}_M \mathcal{P} \epsilon + \frac{1}{16} e^\phi \sum_n \not{F}_n^{(10)} \Gamma_M \mathcal{P}_n \, \epsilon \,, \tag{2}$$

$$\delta\lambda = \left(\not{\partial}\phi + \frac{1}{2} \not{H} \mathcal{P} \right) \epsilon + \frac{1}{8} e^\phi \sum_n (-1)^n (5 - n) \not{F}_n^{(10)} \mathcal{P}_n \epsilon \,. \tag{3}$$

In these equations $M = 0, ..., 9$, ψ_M stands for the column vector $\psi_M = \binom{\psi_M^1}{\psi_M^2}$ containing the two Majorana-Weyl spinors of the same chirality in type IIB, and of opposite chirality in IIA, and similarly for λ and ϵ. The 2×2 matrices $\mathcal{P}$ and $\mathcal{P}_n$ are different in IIA and IIB: for IIA $\mathcal{P} = \Gamma_{11}$ and $\mathcal{P}_n = \Gamma_{11}^{(n/2)} \sigma^1$, while for IIB $\mathcal{P} = -\sigma^3$, $\mathcal{P}_n = \sigma^1$ for $\frac{n+1}{2}$ even and $\mathcal{P}_n = \mathrm{i}\sigma^2$ for $\frac{n+1}{2}$ odd. A slash means a contraction with gamma matrices in the form $\not{F}_n = \frac{1}{n!} F_{P_1 \ldots P_N} \Gamma^{P_1 \ldots P_N}$, and $H_M \equiv \frac{1}{2} H_{MNP} \Gamma^{NP}$. We are using here the democratic formulation of Ref.[37] for the RR fields, who actually considers all RR potentials ($C_1 ... C_9$ in IIA, and $C_0, C_2 ... C_{10}$ in IIB, instead of only those in Eq.(35), imposing a self-duality constraint on their field strengths to reduce the doubling of degrees of freedom. The RR field strengths are given by

$$F^{(10)} = \mathrm{d}C - H \wedge C + m \, e^B = \hat{F} - H \wedge C \tag{4}$$

where $F^{(10)}$ is the formal sum of all even (odd) fluxes in IIA (IIB), $\hat{F} = \mathrm{d}C + me^B$, and $m \equiv F_0^{(10)} = \hat{F}_0$ is the mass parameter of IIA. These RR fluxes are constrained by the Hodge duality relation

$$F_n^{(10)} = (-1)^{\mathrm{Int}[n/2]} \star F_{10-n}^{(10)} \,, \tag{5}$$

---

when parallel transporting vectors around a closed loop. The holonomy group then measures the failure to return to back to the original vector. The smaller the holonomy group, the more special the manifold. Trivial holonomy implies that the manifold is flat. On an n-dimensional Riemannian manifold, the holonomy group is a subgroup of $SO(n)$.

where $\star$ is a ten-dimensional Hodge star.

The Bianchi identities for the NS flux and the democratic RR fluxes are

$$dH = 0 \ , \quad dF^{(10)} - H \wedge F^{(10)} = 0 \ . \tag{6}$$

When no fluxes are present, demanding zero VEV for the gravitino variation (2) requires the existence of a covariantly constant spinor on the ten-dimensional manifold, i.e. $\nabla_M \epsilon = 0$. The spacetime component of this equation reads

$$\tilde{\nabla}_\mu \epsilon + \frac{e^{-A}}{2}(\gamma_\mu \gamma_5 \otimes \nabla\!\!\!/ A)\epsilon = 0 \tag{7}$$

where we have used the standard decomposition of the ten-dimensional gamma matrices ($\Gamma^\mu = \gamma^\mu \otimes \mathbf{1}, \Gamma^m = \gamma_5 \otimes \gamma^m$ for the external and internal gamma's), and written the four dimensional metric as $e^{2A}\tilde{g}_{\mu\nu}$. This yields the following integrability condition[38]

$$[\tilde{\nabla}_\mu, \tilde{\nabla}_\nu]\epsilon = -\frac{1}{2}(\nabla_m A)(\nabla^m A)\, \gamma_{\mu\nu}\, \epsilon \tag{8}$$

On the other hand,

$$[\tilde{\nabla}_\mu, \tilde{\nabla}_\nu]\epsilon = \frac{1}{4}\tilde{R}_{\mu\nu\lambda\rho}\gamma^{\lambda\rho}\epsilon = 0 \tag{9}$$

Since $\gamma_{\mu\nu}$ is invertible, the integrability condition reads $\nabla_m A \nabla^m A = 0$, i.e. the warp factor should be a constant. To analyze the internal component of the supersymmetry variation, we need to split the supersymmetry spinors into four-dimensional and six-dimensional spinors as

$$\begin{aligned} \epsilon^1_{\mathrm{IIA}} &= \xi^1_+ \otimes \eta^1_+ + \xi^1_- \otimes \eta^1_- \ , \\ \epsilon^2_{\mathrm{IIA}} &= \xi^2_+ \otimes \eta^2_- + \xi^2_- \otimes \eta^2_+ \ , \end{aligned} \tag{10}$$

for type IIA, where $\gamma_{11}\epsilon^1_{\mathrm{IIA}} = \epsilon^1_{\mathrm{IIA}}$ and $\gamma_{11}\epsilon^2_{\mathrm{IIA}} = -\epsilon^2_{\mathrm{IIA}}$, and the four and six-dimensional spinors obey $\xi^{1,2}_- = (\xi^{1,2}_+)^*$, and $\eta^A_- = (\eta^A_+)^*$. (By a slight abuse of notation we use plus and minus to indicate both four-dimensional and six-dimensional chiralities.) For type IIB both spinors have the same chirality, which we take to be positive, resulting in the decomposition

$$\epsilon^A_{\mathrm{IIB}} = \xi^A_+ \otimes \eta^A_+ + \xi^A_- \otimes \eta^A_- \ , \qquad A = 1, 2 \ . \tag{11}$$

Inserting these decompositions in the internal component of the gravitino variation, Eq.(2), we get the following condition

$$\nabla_m \eta^A_\pm = 0 \, . \tag{12}$$

The internal manifold should therefore have at least a covariantly constant spinor (at least because $\eta^1 = \eta^2$ is a perfectly valid choice), which implies that it should have reduced holonomy, $Hol(\nabla) \subseteq SU(3)$.[h] A supersymmetric string theory background in the absence of fluxes requires therefore Calabi-Yau 3-folds. When there is one covariantly constant internal spinor, the internal gravitino equation tells us that there are two four-dimensional supersymmetry parameters, ξ^1 and ξ^2. This compactification preserves therefore eight supercharges, i.e. $\mathbb{N} = 2$ in four dimensions. Internal manifolds with a further reduced holonomy group allow for more preserved supersymmetries. For example, on K3 $\times T^2$ there are two covariantly constant spinors, and a background with this as internal manifold preserves $\mathbb{N} = 4$ supersymmetry. Compactifications on T^6 preserve all the supersymmetries (32, or $\mathbb{N} = 8$ in four dimensions).

4.2. *Compactifications with flux*

Compactifications on Calabi-Yau manifolds have great mathematical sophistication but suffer from various phenomenological problems. First of all, the supersymmetry preserved in four dimensions is $\mathbb{N} = 2$, which does not allow for chiral interactions like that mediated by the electro-weak force. Supersymmetric extensions of the standard model can therefore have at most $\mathbb{N} = 1$ supersymmetry. Besides, compactifications on Calabi-Yau manifolds have a number of moduli (massless fields with no potential), like the complex structure or the Kähler deformations of the metric.[39] The magnitude of four dimensional masses and interactions depends on the VEV of these moduli, which can take any value. Furthermore, massless fields lead to a long range fifth force unobserved so far in nature. Therefore, there should be some mechanism creating a potential for these scalar fields, such that their VEV is fixed or "stabilized". The only mechanism within perturbative string theory creating a potential for the fields that we know of today is to turn on fluxes for the field strengths F_n, H_3. Due to the topological non triviality of the internal manifolds, there are sourceless solutions to the equations of motion which have nonzero VEV for some of the field strengths. String theory in non-trivial backgrounds with fluxes, known as flux compactifications, has a number of novel features that make them particularly interesting. First of all, fluxes break supersymmetry partially or completely in a stable way, leading to $\mathbb{N} = 1$ (and thus phenomenologically viable) backgrounds. Second, some or all of the moduli of traditional

[h] Under a decomposition of the spinorial **4** representation of $SO(6)$ into representations of the holonomy group, $SU(3)$ is the largest group for which there is a singlet -or in other words, SU(3) is the stability group of the **4** representation-.

compactifications get fixed, thus limiting the arbitrariness of the vacuum. Besides, fluxes allow for a non-trivial warp factor, which could be responsible for the observed hierarchy between the Planck and the electroweak scale. Non-trivial fluxes also appear in most string backgrounds with field theory duals, and they therefore play an important role in the present understanding of holography.

In this subsection, we review compactifications preserving the minimal amount of supersymmetry, i.e. $\mathbb{N} = 1$ in four dimensions, concentrating on the geometry of the six-dimensional manifold.

As we saw in the previous subsection, in the absence of fluxes, supersymmetry requires a covariantly constant spinor on the internal manifold. This condition actually splits into two parts, first the existence of such a spinor (i.e., the existence of a non-vanishing globally well defined section on the spinor bundle over T), and second the condition that it is covariantly constant. A generic spinor such as the supercurrent can be decomposed in the same way as the supersymmetry parameters, Eqs (10) and (11). The first condition implies then the existence of two four-dimensional supersymmetry parameters and thus an effective $\mathbb{N} = 2$ four-dimensional action, while the second implies that this action has an $\mathcal{N} = 2$ Minkowski vacuum. As far as the internal manifold is concerned, the fist condition is a topological requirement on the manifold, while the second one is a differential condition on the metric, or rather, on its connection. Let us first review the implications of the first condition.

A globally well defined non-vanishing spinor exists only on manifolds that have reduced structure.[40] A manifold has G-structure if the frame bundle admits a subbundle with fiber group G. A Riemannian manifold of dimension d has automatically structure group $SO(d)$. All vector, tensor and spinor representations can therefore be decomposed in representations of $SO(d)$. If the manifold has reduced structure group G, then every representation can be further decomposed in representations of G.

Let us concentrate on six dimensions, which is the case we are interested in, and the group G being $SU(3)$. We argued that supersymmetry imposes a topological plus a differential condition on the manifold. So far we have reviewed the topological condition, which amounts to the requirement that the manifold has SU(3) structure. As for the differential condition, in the case of compactifications without flux, the internal manifold is a Calabi-Yau 3-fold. A Calabi-Yau manifold has a covariantly constant spinor, which inserted in (16) yields a closed fundamental two-form J and a closed Ω. This means that the manifold is both symplectic and complex (and has additionally $c_1 = 0$). Let us now see how these differential conditions get modified in the presence of fluxes. The resulting differential conditions have a very nice unifying description in terms of generalized complex geometry.

4.3. *Flux compactifications and generalized complex geometry*

As we reviewed in the previous sections, as a result of demanding $\delta\Psi_m = \delta\lambda = 0$, supersymmetry imposes differential conditions on the internal spinor η. These differential conditions turn into differential conditions for the pure Clifford(6, 6) spinors $\Phi_\pm$, defined in (11). We quote the results of Ref. 23, skipping the technical details of the derivation. $\mathbb{N} = 1$ supersymmetry on warped Minkowski 4D vacua requires

$$\begin{aligned} e^{-2A+\phi}(\mathrm{d} + H\wedge)(e^{2A-\phi}\Phi_+) &= 0\,, \qquad (13)\\ e^{-2A+\phi}(\mathrm{d} + H\wedge)(e^{2A-\phi}\Phi_-) &= \mathrm{d}A \wedge \bar{\Phi}_- + \frac{\mathrm{i}}{16} e^{\phi+A} * F_{\mathrm{IIA}+} \end{aligned}$$

for type IIA, and

$$\begin{aligned} e^{-2A+\phi}(\mathrm{d} - H\wedge)(e^{2A-\phi}\Phi_+) &= \mathrm{d}A \wedge \bar{\Phi}_+ - \frac{\mathrm{i}}{16} e^{\phi+A} * F_{\mathrm{IIB}-}\\ e^{-2A+\phi}(\mathrm{d} - H\wedge)(e^{2A-\phi}\Phi_-) &= 0\,, \qquad (14) \end{aligned}$$

for type IIB, and the algebraic constrain

$$|\Phi_+| = |\Phi_-| = e^A \qquad (15)$$

In these equations

$$F_{\mathrm{IIA}\pm} = F_0 \pm F_2 + F_4 \pm F_6\ , \qquad F_{\mathrm{IIB}\pm} = F_1 \pm F_3 + F_5\ . \qquad (16)$$

These F are purely internal forms, that is related to the total ten–dimensional RR field strength by

$$F^{(10)} = F + \mathrm{vol}_4 \wedge \lambda(*F)\ , \qquad (17)$$

where $*$ the six–dimensional Hodge dual, we have used the self-duality property of the ten-dimensional RR fields (5) and λ is

$$\lambda(A_n) = (-1)^{\mathrm{Int}[n/2]} A_n\ . \qquad (18)$$

According to the definitions given in the previous section, Eqs. (13) and (14) tell us that all $\mathbb{N} = 1$ vacua on manifolds with SU(3)×SU(3) structure on $T \oplus T^*$ are *twisted Generalized Calabi-Yau's*, which we discussed in Section 3. We can also see from (13), (14) that RR fluxes act as an obstruction for the integrability of the second pure spinor.

Specializing to the pure SU(3) structure case, i.e. for $\Phi_\pm$ given by Eq.(15), and looking at (19), we see that the Generalized Calabi-Yau manifold is complex[i] in IIB and (twisted) symplectic in IIA. For the general SU(3)×SU(3) case, $\mathbb{N} = 1$ vacua can be realized in hybrid complex–symplectic manifolds, i.e. manifolds with k complex dimensions and $6 - 2k$ (real) symplectic ones. In particular, given the chiralities of the preserved Clifford$(6, 6)$ spinors, the rank k must be even in IIA and odd in IIB (equal respectively to 0 and 3 in the pure SU(3) case).

A final comment is that $\mathbb{N} = 2$ vacua with NS fluxes only, where shown to satisfy Eqs. (13, 14) for $\tilde{\Phi}_\pm = \Phi_\pm$, $F_{\rm IIA} = F_{\rm IIB} = A = 0$.[36]

5. Summary and outlook

In these lectures we have reviwed the construction of the massless spectrum of type II superstring theories, which includes the graviton and other massless fields coming from the so-called NSNS and RR excitations. Anomaly cancellation imposes a fixed dimension for spacetime, equal to ten. We then discussed supersymmetric solutions to the spacetime equations of motion for which the ten-dimensional spacetime is a (warped) product of four-dimensional Minkowski space times a six-dimensional (internal) compact space. The simplest solutions involve internal manifolds with special holonomy, such as T^6, $K3 \times T^2$ or CY_3. These preserve different number of supersymmetries, namely 32, 16 and 8 respectively. The geometric deformations of the internal manifold are "moduli" of the theory, meaning that there is an infinite family of T^6's parameterized by the 21 components of the metric, or an infinite family of CY_3's parameterized by $2h^{2,1}$ and $h^{1,1}$ complex structure and Kähler deformations of the metric, which are all solutions to the equations of motion. The only mechanism known within perturbative string theory that lifts these degeneracies is to introduce either closed string fluxes for the field strengths related to the NS-NS B-field and R-R C_n potentials, or open string fluxes for the D-brane gauge fields A_μ. In these lectures we concentrated on closed string fluxes, but we should note that open string ones can also stabilize moduli. We showed that within the NS-NS sector, metric and B-field combine into two "pure spinors" of the Clifford(6,6) action with opposite chirality. The one whose chirality equals that of the R-R fluxes, is closed with respect to a twisted exterior derivative. This translates into a twisted integrability of an algebraic structure to

[i] H in Eq.(14) does not "twist" the (usual) complex structure, as $(d - H\wedge)\Omega = 0$ implies in particular $d\Omega = 0$.

which the pure spinor is related. The other pure spinor is not closed, and the non-closure (or the obstruction for integrability of the associated generalized complex structure) is related to the R-R fluxes. This says that the supersymmetric solutions to the spacetime equations of motion in the presence of background fluxes preserving four-dimensional Poincare symmetry involve "twisted" Generalized Calabi-Yau threefolds. The two pure spinors describe also the supersymmetric configurations of D-branes, which should wrap generalized complex submanifolds with respect to the integrable generalized complex structure. The non-integrable pure spinor gives a stability condition for the D-brane.[41]

There are many open problems related to flux compactifications and generalized Calabi-Yau manifolds. The first obvious one is to find specific examples. Examples of Generalized Calabi-Yau manifolds with trivial structure group are nilmanifolds.[42] These serve as a very nice and easy playground for flux compactifications. However, very few of them can actually be realized as a string theory compactification.[43] It would be nice to understand better the reason for this, and find other examples of flux compactifications. Another open problem is to understand the moduli spaces of these compactifications, and whether it is possible to stabilize moduli and/or break supersymmetry by adding extra fluxes that do not change the integrability properties of the structures. Finally, it is not clear at the moment whether these manifolds are good for phenomenology. It has been shown that compactifications of the heterotic theory on Calabi-Yau manifolds can have the spectrum of the Minimal Supersymmetric Standard Model with no exotic matter.[44] In type II theories on Calabi-Yau manifolds, the Standard Model can be obtained from intersecting brane configurations on tori. Lately, compactifications of type IIB with 7-branes in their non-perturbative regime, described by F-theory, have shown to contain most of the intricate features of the Standard Model.[45] All constructions suffer from the moduli problem, though. It is very important to understand whether similar constructions can be done in generalized Calabi-Yau manifolds, but in order to get to that point and be able to address phenomenological questions, we need a much better understanding of their geometry.

References

1. M.B. Green, J. H. Schwarz and E. Witten, "Superstring Theory", Volumes I and II, Cambridge University Press (1987).
2. J. Polchinski, "String Theory", Volumes I and II, Cambridge University Press (1998).

3. C.V. Johnson, “D-branes”, Cambridge University Press (2003).
C.V. Johnson, “D-brane primer,” hep-th/0007170.
4. B. Zwiebach, “A first course in string theory”, Cambridge University Press (2004).
5. E. Witten, “New Issues In Manifolds Of SU(3) Holonomy”, Nucl. Phys. **B268**, 79 (1986).
6. B. R. Greene, K. H. Kirklin, P. J. Miron and G. G. Ross, “A Three Generation Superstring Model. 2. Symmetry Breaking And The Low-Energy Theory,” Nucl. Phys. B **292**, 606 (1987).
7. I. Antoniadis, J. R. Ellis, J. S. Hagelin and D. V. Nanopoulos, “An Improved SU(5) X U(1) Model From Four-Dimensional String,” Phys. Lett. B **208**, 209 (1988) [Addendum-ibid. B **213**, 562 (1988)].
8. G. Aldazabal, L. E. Ibañez and F. Quevedo, “A D-brane alternative to the MSSM,” JHEP **0002**, 015 (2000) [arXiv:hep-ph/0001083].
9. R. Blumenhagen, L. Goerlich, B. Kors and D. Lust, “Noncommutative compactifications of type I strings on tori with magnetic background flux,” JHEP **0010**, 006 (2000) [arXiv:hep-th/0007024].
10. L. E. Ibañez, F. Marchesano and R. Rabadan, “Getting just the standard model at intersecting branes,” JHEP **0111**, 002 (2001) [arXiv:hep-th/0105155].
11. R. Blumenhagen, B. Kors, D. Lust and T. Ott, “The standard model from stable intersecting brane world orbifolds,” Nucl. Phys. B **616**, 3 (2001) [arXiv:hep-th/0107138].
12. M. Cvetic, G. Shiu and A. M. Uranga, “Three-family supersymmetric standard like models from intersecting brane worlds,” Phys. Rev. Lett. **87**, 201801 (2001) [arXiv:hep-th/0107143].
13. D. Bailin, G. V. Kraniotis and A. Love, “Standard-like models from intersecting D4-branes,” Phys. Lett. B **530**, 202 (2002) [arXiv:hep-th/0108131]
14. R. Blumenhagen, M. Cvetic, P. Langacker and G. Shiu, “Toward realistic intersecting D-brane models,” arXiv:hep-th/0502005.
15. B. de Carlos, A. Guarino and J. M. Moreno, “Flux moduli stabilisation, Supergravity algebras and no-go theorems,” JHEP **1001**, 012 (2010) [arXiv:0907.5580 [hep-th]].
16. B. Wecht, “Lectures on Nongeometric Flux Compactifications,” Class. Quant. Grav. **24**, S773 (2007) [arXiv:0708.3984 [hep-th]].
17. S. B. Giddings, S. Kachru and J. Polchinski, “Hierarchies from fluxes in string compactifications,” Phys. Rev. D **66**, 106006 (2002) [hep-th/0105097].
18. M. Cicoli, J. P. Conlon and F. Quevedo, “General Analysis of LARGE Volume Scenarios with String Loop Moduli Stabilisation,” JHEP **0810**, 105 (2008) [arXiv:0805.1029 [hep-th]].
19. N.Hitchin, “Generalized Calabi–Yau manifolds,” [arXiv: math.DG/0209099].
20. M. Gualtieri, “Generalized Complex Geometry,” Oxford University DPhil

thesis (2004) [arXiv:math.DG/0401221].
21. F. Witt, "Special metric structures and closed forms," Oxford University DPhil thesis (2004) [arXiv:math.DG/0502443].
22. M. Graña, R. Minasian, M. Petrini and A. Tomasiello, "Supersymmetric backgrounds from generalized Calabi-Yau manifolds," JHEP **0408**, 046 (2004) [arXiv:hep-th/0406137].
23. M. Grãna, R. Minasian, M. Petrini and A. Tomasiello, "Generalized structures of N = 1 vacua," arXiv:hep-th/0505212.
24. M. Grana, "Flux compactifications in string theory: A Comprehensive review," Phys. Rept. **423**, 91 (2006) [hep-th/0509003].
M. R. Douglas and S. Kachru, "Flux compactification," Rev. Mod. Phys. **79**, 733 (2007) [hep-th/0610102].
R. Blumenhagen, B. Kors, D. Lust and S. Stieberger, "Four-dimensional String Compactifications with D-Branes, Orientifolds and Fluxes," Phys. Rept. **445**, 1 (2007) [hep-th/0610327].
H. Samtleben, "Lectures on Gauged Supergravity and Flux Compactifications," Class. Quant. Grav. **25**, 214002 (2008) [arXiv:0808.4076 [hep-th]].
P. Koerber, "Lectures on Generalized Complex Geometry for Physicists," Fortsch. Phys. **59**, 169 (2011) [arXiv:1006.1536 [hep-th]].
25. C. G. Callan, Jr., E. J. Martinec, M. J. Perry and D. Friedan, "Strings in Background Fields," Nucl. Phys. B **262**, 593 (1985).
26. A. M. Polyakov, "Quantum Geometry of Bosonic Strings," Phys. Lett. B **103**, 207 (1981).
27. D. Friedan, E. J. Martinec and S. H. Shenker, "Conformal Invariance, Supersymmetry and String Theory," Nucl. Phys. B **271**, 93 (1986).
28. A. Chodos and C. B. Thorn, "Making the Massless String Massive," Nucl. Phys. B **72**, 509 (1974).
R. C. Myers, "New Dimensions for Old Strings," Phys. Lett. B **199**, 371 (1987).
29. J. Polchinski, "Dirichlet Branes and Ramond-Ramond charges," Phys. Rev. Lett. **75**, 4724 (1995) [hep-th/9510017].
30. F. Gliozzi, J. Scherk and D. I. Olive, "Supersymmetry, Supergravity Theories and the Dual Spinor Model," Nucl. Phys. B **122**, 253 (1977).
31. A. Giveon, M. Porrati and E. Rabinovici, "Target space duality in string theory," Phys. Rept. **244**, 77 (1994) [hep-th/9401139].
32. A. A. Tseytlin, "On nonAbelian generalization of Born-Infeld action in string theory," Nucl. Phys. B **501**, 41 (1997) [hep-th/9701125].
33. D. Lust, "Intersecting brane worlds: A Path to the standard model?," Class. Quant. Grav. **21**, S1399 (2004) [hep-th/0401156].
34. U. Lindstrom, R. Minasian, A. Tomasiello and M. Zabzine, "Generalized complex manifolds and supersymmetry," Commun. Math. Phys. **257**, 235 (2005) [arXiv:hep-th/0405085].

U. Lindstrom, M. Rocek, R. von Unge and M. Zabzine, "Generalized Kaehler geometry and manifest N = (2,2) supersymmetric nonlinear sigma-models," arXiv:hep-th/0411186.

M. Zabzine, "Hamiltonian perspective on generalized complex structure," arXiv:hep-th/0502137.

35. N. J. Hitchin, "The geometry of three-forms in six and seven dimensions," arXiv:math.dg/0010054.
N. J. Hitchin, "Stable forms and special metrics," arXiv:math.dg/0107101.
36. C. Jeschek and F. Witt, "Generalised G(2)-structures and type IIB superstrings," JHEP **0503**, 053 (2005) [arXiv:hep-th/0412280].
37. E. Bergshoeff, R. Kallosh, T. Ortin, D. Roest and A. Van Proeyen, "New formulations of D = 10 supersymmetry and D8 - O8 domain walls," Class. Quant. Grav. **18**, 3359 (2001) [arXiv:hep-th/0103233].
38. P. Candelas, G. T. Horowitz, A. Strominger and E. Witten, "Vacuum Configurations For Superstrings," Nucl. Phys. B **258**, 46 (1985).
39. P. Candelas and X. de la Ossa, "Moduli Space Of Calabi-Yau Manifolds," Nucl. Phys. B **355**, 455 (1991).
40. S. Salamon, "Riemannian Geometry and Holonomy Groups", Vol. 201 of Pitman Research Notes in Mathematics, Longman, Harlow, 1989.
D. Joyce, Compact Manifolds with Special Holonomy, Oxford University Press, Oxford, 2000.
41. P. Koerber, "Stable D-branes, calibrations and generalized Calabi-Yau geometry," arXiv:hep-th/0506154.
L. Martucci and P. Smyth, "Supersymmetric D-branes and calibrations on general N = 1 backgrounds," arXiv:hep-th/0507099.
42. G. Cavalcanti and M. Gualtieri, "Generalized complex structures on nilmanifolds," J. Symplectic Geom. **2** 393 (2004) [math.DG/0404451].
43. M. Grana, R. Minasian, M. Petrini and A. Tomasiello, "A Scan for new N=1 vacua on twisted tori," JHEP **0705**, 031 (2007) [hep-th/0609124].
44. Some of the first constructions are E. Witten, "New Issues In Manifolds Of SU(3) Holonomy," Nucl. Phys. B **268**, 79 (1986).
B. R. Greene, K. H. Kirklin, P. J. Miron and G. G. Ross, "A Three Generation Superstring Model. 2. Symmetry Breaking And The Low-Energy Theory," Nucl. Phys. B **292**, 606 (1987).
I. Antoniadis, J. R. Ellis, J. S. Hagelin and D. V. Nanopoulos, "An Improved SU(5) X U(1) Model From Four-Dimensional String," Phys. Lett. B **208**, 209 (1988) [Addendum-ibid. B **213**, 562 (1988)].
For more recent heterotic string Standard Model constructions, see for instance:
V. Braun, Y. -H. He, B. A. Ovrut and T. Pantev, "The Exact MSSM spectrum from string theory," JHEP **0605**, 043 (2006) [hep-th/0512177].
L. B. Anderson, J. Gray, A. Lukas and E. Palti, "Heterotic Line Bundle

Standard Models," JHEP **1206**, 113 (2012) [arXiv:1202.1757 [hep-th]].
45. R. Donagi and M. Wijnholt, "Model Building with F-Theory," arXiv:0802.2969 [hep-th].
C. Beasley, J. J. Heckman and C. Vafa, "GUTs and Exceptional Branes in F-theory - I," JHEP **0901**, 058 (2009) [arXiv:0802.3391 [hep-th]].
C. Beasley, J. J. Heckman and C. Vafa, "GUTs and Exceptional Branes in F-theory - II: Experimental Predictions," JHEP **0901**, 059 (2009) [arXiv:0806.0102 [hep-th]].
R. Donagi and M. Wijnholt, "Breaking GUT Groups in F-Theory," arXiv:0808.2223 [hep-th].
For more references see T. Weigand, "Lectures on F-theory compactifications and model building," Class. Quant. Grav. **27**, 214004 (2010) [arXiv:1009.3497 [hep-th]].

PART B
SHORT COMMUNICATIONS

GROUPOIDS AND POISSON SIGMA MODELS WITH BOUNDARY

Alberto Cattaneo and Ivan Contreras*

Institüt für Mathematik
Universität Zürich Irchel
Winterthurerstrasse 190, CH-8057 Zürich, Switzerland
E-mail: asc@math.uzh.ch, ivan.contreras@math.uzh.ch

This note gives an overview on the construction of symplectic groupoids as reduced phase spaces of Poisson sigma models and its generalization in the infinite dimensional setting (before reduction).

Keywords: Groupoids, Poisson sigma models, manifolds with boundary.

1. Introduction

Symplectic groupoids have been studied in detail since their introduction by Coste, Dazord and Weinstein[9] and they appear naturally in Poisson and symplectic geometry, as well as in some instances of the study of topological field theories. More precisely, in,[4] it was proven that the reduced phase space of the Poisson sigma model under certain boundary conditions and assuming it is a smooth manifold, has the structure of a symplectic groupoid and it integrates the cotangent bundle of a given Poisson manifold M. This is a particular instance of the problem of integration of Lie algebroids, a generalized version of the **Lie third theorem**[12] . The general question can be stated as:

- Is there a Lie groupoid (G, M) such that its infinitesimal version corresponds to a given Lie algebroid (A, M)?

For the case where $A = T^*M$ and M is a Poisson manifold the answer is not positive in general, as there are topological obstructions encoded in what are called the monodromy groups.[10] A Poisson manifold is called integrable

*Partially supported by SNF Grant 20-131813.

if such a Lie groupoid G exists. The properties of G are of special interest in Poisson geometry, since it is possible to equip G with a symplectic structure ω compatible with the multiplication map in such a way that G is a symplectic realization for (M, Π).
For the integrable case, the symplectic groupoid integrating a given Poisson manifold (M, Π) is constructed explicitely in,[4] as the phase space modulo gauge equivalence of the Poisson Sigma model (PSM), a 2-dimensional field theory.

In a more recent perspective (see[7,8]), the study of the phase space before reduction plays a crucial role. This allows dealing with with nonintegrable Poisson structure, for which the reduced phase space is singular, on an equal footing as the integrable ones. This new approach differs from the stacky perspective of Zhu and Tseng (see[16]) and seems to be better adapted to symplectic geometry and to quantization.

In a paper in preparation,[3] we introduce a more general version of a symplectic groupoid, called *relational symplectic groupoid.* In the case at hand, it corresponds to an infinite dimensional weakly symplectic manifold equipped with structure morphisms (canonical relations, i.e. immersed Lagrangian submanifolds) compatible with the Poisson structure of M. In this work, we prove that

(i) For any Poisson manifold M (integrable or not), the relational symplectic groupoid always exists.
(ii) In the integrable case, the associated relational symplectic groupoid is equipped with a locally embedded Lagrangian submanifold.
(iii) Conjecturally, given a regular relational symplectic groupoid $\mathcal{G}$ over M (a particular type of object that admits symplectic reduction), there exists a unique Poisson structure Π on M such that the symplectic structure ω on $\mathcal{G}$ and Π are compatible. This is still work in progress.

This paper is an overview of this construction and is organized as follows. Section 2 is a brief introduction to the Poisson sigma model and its reduced phase space. Section 3 deals with the version before reduction of the phase space and the introduction of the relational symplectic groupoid. An interesting issue concerning this construction is the treatment of non integrable Poisson manifolds: even if the reduction does not exists as a smooth manifold, the relational symplectic groupoid always exists. One natural question at this point is:

- Can there be a finite dimensional relational symplectic groupoid equivalent to the infinite dimensional one for an arbitrary Poisson manifold?

The answer to this question is work in progress and it will be treated in a subsequent paper. Section 4 contains some comments on the quantized version of the relational symplectic groupoid and its possible connection with geometric and deformation quantization.

Another aspect, which will be explored later, is the connection between the relational construction and the Poisson Sigma model with branes, where the boundary conditions are understood as choices of coisotropic submanifolds of the Poisson manifold. The relational symplectic groupoid seems to admit the existence of branes and would explain in full generality the idea of dual pairs in the Poisson sigma model with boundary.[2,5]
This new program might be useful for quantization as well. Using ideas from geometric quantization, what is expected as the quantization of the relational symplectic groupoid is an algebra with a special element, which fails to be a unit, but whose action is a projector in such a way that on the image of the projector we obtain a true unital algebra. Deformation quantization of a Poisson manifold could be interpreted in this way.

Acknowledgments. *We thank Marco Zambon, David Martinez, Pavel Mnëv and the anonymous referee for useful comments and remarks.*

2. PSM and its reduced phase space

We consider the following data

(i) A compact surface Σ, possibly with boundary, called the source space.
(ii) A finite dimensional Poisson manifold (M, Π), called the target space. Recall that a bivector field $\Pi \in \Gamma(TM \wedge TM)$ is called Poisson if the the bracket $\{,\} : \mathcal{C}^\infty(M) \otimes \mathcal{C}^\infty(M) \to \mathcal{C}^\infty(M)$, defined by

$$\{f, g\} = \Pi(df, dg)$$

is a Lie bracket and it satisfies the Leibniz identity

$$\{f, gh\} = g\{f, h\} + h\{f, g\}, \forall f, g, h \in \mathcal{C}^\infty(M).$$

In local coordinates, the condition of a bivector Π to be Poisson reads as follows

$$\Pi^{sr}(x)(\partial_r)\Pi^{lk}(x) + \Pi^{kr}(x)(\partial_r)\Pi^{sl}(x) + \Pi^{lr}(x)(\partial_r)\Pi^{ks}(x) = 0, \quad (1)$$

that is, the vanishing condition for the Schouten-Nijenhuis bracket of Π.

The space of fields for this theory is denoted with Φ and corresponds to the space of vector bundle morphisms between $T\Sigma$ and T^*M. This space can be parametrized by the pair (X, η), where X is a $\mathcal{C}^{k+1}$-map from Σ to M and $\eta \in \Gamma^k(\Sigma, T^*\Sigma \otimes X^*T^*M)$, where $k \in \{0, 1, \cdots\}$ denotes the regularity type of the map.
On Φ, the following first order action is defined:

$$S(X, \eta) := \int_\Sigma \langle \eta,\, dX\rangle + \frac{1}{2}\langle \eta,\, (\Pi^\# \circ X)\eta\rangle,$$

where

- $\Pi^\#$ is the map from $T^*M \to TM$ induced from the Poisson bivector Π.
- dX and η are regarded as elements in $\Omega^1(\Sigma, X^*(TM))$ and $\Omega^1(\Sigma, X^*(T^*M))$, respectively.
- $\langle\, ,\rangle$ denotes the pairing between $\Omega^1(\Sigma, X^*(TM))$ and $\Omega^1(\Sigma, X^*(T^*M))$ induced by the natural pairing between T_xM and T_x^*M, for all $x \in M$.

The integrand, called the Lagrangian, will be denoted by $\mathcal{L}$. Associated to this action, the corresponding variational problem $\delta S = 0$ induces the following space

$$\mathrm{EL} = \{\text{Solutions of the Euler-Lagrange equations}\} \subset \Phi,$$

where, using integration by parts

$$\delta S = \int_\Sigma \frac{\delta \mathcal{L}}{\delta X}\delta X + \frac{\delta \mathcal{L}}{\delta \eta}\delta\eta + \text{boundary terms}.$$

The partial variations correspond to:

$$\frac{\delta \mathcal{L}}{\delta X} = dX + (\Pi^\# \circ X)\eta = 0 \quad (2)$$

$$\frac{\delta \mathcal{L}}{\delta \eta} = d\eta + \frac{1}{2}\langle(\partial\Pi^\# \circ X)\eta, \eta\rangle = 0. \quad (3)$$

Now, if we restrict to the boundary, the general space of boundary fields corresponds to

$$\Phi_\partial := \{\text{vector bundle morphisms between } T(\partial\Sigma) \text{ and } T^*M\}.$$

Following,[6] Φ_∂ is endowed with a symplectic form and a surjective submersion $p : \Phi \to \Phi_\partial$. We define

$$L_\Sigma := p(EL).$$

Finally, we define C_Π as the set of fields in Φ_∂ which can be completed to a field in $L_{\Sigma'}$, with $\Sigma' := \partial\Sigma \times [0, \varepsilon]$, for some ε.
It turns out that Φ_∂ can be identified with $T^*(PM)$, the cotangent bundle of the path space on M and that

$$C_\Pi := \{(X, \eta) | dX = \pi^\#(X)\eta,\ X : \partial\Sigma \to M,\ \eta \in \Gamma(T^*I \otimes X^*(T^*M))\}.$$

Furthermore the following proposition holds

Proposition 2.1. [4] *The space C_Π is a coisotropic submanifold of Φ_∂.*

In fact, the converse of this proposition also holds in the following sense. If we define $S(X, \eta)$ and C_Π in the same way as before, without assuming that Π satisfies Equation (1) it can be proven that

Proposition 2.2.[2,3] *If C_Π is a coisotropic submanifold of Φ_∂, then Π is a Poisson bivector field.*

The geometric interpretation of the Poisson sigma model will lead us to the connection between Lie algebroids and Lie groupoids in Poisson geometry. First we need some definitions.
A pair (A, ρ), where A is a vector bundle over M and ρ (called the anchor map) is a vector bundle morphism from A to TM is called a *Lie algebroid* if

(i) There is Lie bracket $[,]_A$ on $\Gamma(A)$ such that the induced map $\rho_* : \Gamma(A) \to \mathfrak{X}(M)$ is a Lie algebra homomorphism.
(ii) *Leibniz identity:*

$$[X, fY]_A = f[X, Y] + \rho_*(X)(f)Y, \forall X, Y \in \Gamma(A), f \in \mathcal{C}^\infty(M).$$

Lie algebras, Lie algebra bundles and tangent bundles appear as natural examples of Lie algebroids. For our purpose, the cotangent bundle of a Poisson manifold T^*M, where $[,]_{T^*M}$ is the Koszul bracket for 1-forms, that is defined for exact forms by

$$[df, dg] := d\{f, g\}, \forall f, g \in \mathcal{C}^\infty(M),$$

whereas for general 1-forms it is recovered by Leibniz and the anchor map given by $\Pi^{\#} : T^*M \to TM$, is a central example of Lie algebroids. To define a morphism of Lie algebroids we consider the complex $\Lambda^{\bullet}A^*$, where A^* is the dual bundle and a differential δ_A is defined by the rules

(i) for all $f \in \mathcal{C}^\infty(M)$,

$$\delta_A f := \rho^* df,$$

(ii) for all $X, Y \in \Gamma(A), \alpha \in \Gamma(A^*)$,

$$\langle \delta_A \alpha, X \wedge Y\rangle := -\langle \alpha, [X,Y]_A\rangle + \langle \delta\langle \alpha, X\rangle, Y\rangle - \langle \delta\langle \alpha, Y\rangle, X\rangle,$$

where $\langle , \rangle$ is the natural pairing between $\Gamma(A)$ and $\Gamma(A^*)$.

A vector bundle morphism $\varphi : A \to B$ is a Lie algebroid morphism if

$$\delta_A \varphi^* = \varphi^* \delta_B.$$

This condition written down in local coordinates gives rise to some PDE's the anchor maps and the structure functions for $\gamma(A)$ and $\Gamma(B)$ should satisfy. In particular, for the case of Poisson manifolds, C_Π corresponds to the space of Lie algebroid morphisms between $T(\partial\Sigma)$ and T^*M where the Lie algebroid structure on the left is given by the Lie bracket of vector fields on $T(\partial\Sigma)$ with identity anchor map and on the right is the one induced by the Poisson structure on M.

As it was mentioned before, it can be proven that this space is a coisotropic submanifold of T^*PM. Its symplectic reduction, i.e. the space of leaves of its characteristic foliation, called the reduced phase space of the PSM, when is smooth, has a particular feature, it is a symplectic groupoid over M.[4] More precisely, a groupoid is a small category with invertible morphisms. When the spaces of objects and morphisms are smooth manifolds, a Lie groupoid over M, denoted by $G \rightrightarrows M$, can be rephrased as the following data[a]

$$G \times_{(s,t)} G \xrightarrow{\ \mu\ } G \xrightarrow{\ i\ } G \underset{t}{\overset{s}{\underset{\longleftarrow\varepsilon}{\rightrightarrows}}} M$$

where s, t, ι, μ and ε denote the source, target, inverse, multiplication and unit map respectively, such that the following axioms hold (denoting $G_{(x,y)} := s^{-1}(x) \cap t^{-1}(y)$):

(A.1) $s \circ \varepsilon = t \circ \varepsilon = id_M$

(A.2) If $g \in G_{(x,y)}$ and $h \in G_{(y,z)}$ then $\mu(g,h) \in G_{(x,z)}$

[a] $G \times_{(s,t)} G$ is a smooth manifold whenever s (or t) is a surjective submersion.

(A.3) $\mu(\varepsilon \circ s \times id_G) = \mu(id_G \times \varepsilon \circ t) = id_G$
(A.4) $\mu(id_G \times i) = \varepsilon \circ t$
(A.5) $\mu(i \times id_G) = \varepsilon \circ s$
(A.6) $\mu(\mu \times id_G) = \mu(id_G \times \mu)$.
A Lie groupoid is called *symplectic* if there exists a symplectic structure ω on G such that

$$Gr_\mu := \{(a,b,c) \in G^3 | \, c = \mu(a,b)\}$$

is a lagrangian submanifold of $G \times G \times \overline{G}$, where $\overline{G}$ denotes the sign reversed symplectic strucure on G. Finally, we can state the following

Theorem 2.1. [4] *For the Poisson sigma model with source space homeomorphic to a disc, the symplectic reduction $\underline{C_\Pi}$ of C_Π (the space of leaves of the characteristic foliation), if it is smooth, is a symplectic groupoid over M.*

The smoothness of the reduced phase space has particular interest. In,[10] the necessary and sufficient conditions for integrability of Lie algebroids, i.e. whether a Lie groupoid such that its *infinitesimal version* corresponds to a given Lie algebroid exists, are stated. In,[11] these conditions have been further specialized to the Poisson case. It turns out that the reduced phase space of the PSM coincides with the space of equivalent classes of what are called $\mathcal{A}-$paths modulo $\mathcal{A}-$ homotopy,[10] with $\mathcal{A} = T^*M$.

3. The version before reduction

The main motivation for introducing the relational symplectic groupoid construction is the following. In general, the leaf space of a characteristic foliation is not a smooth finite dimensional manifold and in this particular situation, the smoothness of the space of reduced boundary fields is controlled by the integrability conditions stated in.[10] In this paper, we define a groupoid object in the extended symplectic category, where the objects are symplectic manifolds, possibly infinite dimensional, and the morphisms are immersed Lagrangian submanifolds. It is important to remark here that this extended category is not properly a category! (The composition of morphisms is not smooth in general). However, for our construction, the corresponding morphisms will be composable.

We restrict ourselves to the case when $\mathcal{C}$ is the sometimes called *Extended Symplectic Category*, denoted by Sym^{Ext} and defined as follows:

Definition 3.1. Sym^{Ext} is a category in which the objects are symplectic manifolds and the morphisms are immersed canonical relations. [b] Recall that $L : \mathcal{M} \nrightarrow \mathcal{N}$ is an immersed canonical relation between two symplectic manifolds $\mathcal{M}$ and $\mathcal{N}$ by definition if L is an immersed Lagrangian submanifold of $\bar{\mathcal{M}} \times \mathcal{N}$. [c] Sym^{Ext} carries an involution $\dagger : (\mathrm{Sym}^{Ext})^{op} \to \mathrm{Sym}^{Ext}$ that is the identity in objects and in morphisms, for $f : A \nrightarrow B$, $f^\dagger := \{(b,a) \in B \times A | (a,b) \in f\}$.

This category extends the usual symplectic category in the sense that the symplectomorphisms can be thought in terms of canonical relations.

Definition 3.2. A **relational symplectic groupoid** is a triple $(\mathcal{G}, L, I)$ where

- $\mathcal{G}$ is a weak symplectic manifold. [d]
- L is an immersed Lagrangian submanifold of $\mathcal{G}^3$.
- I is an antisymplectomorphism of $\mathcal{G}$

satisfying the following axioms

- **A.1** L is cyclically symmetric (i.e. if $(x,y,z) \in L$, then $(y,z,x) \in L$)
- **A.2** I is an involution (i.e. $I^2 = Id$).
  Notation. L is a canonical relation $\mathcal{G} \times \mathcal{G} \nrightarrow \bar{\mathcal{G}}$ and will be denoted by L_{rel}. Since the graph of I is a Lagrangian submanifold of $\mathcal{G} \times \mathcal{G}$, I is a canonical relation $\bar{\mathcal{G}} \nrightarrow \mathcal{G}$ and will be denoted by I_{rel}.
  L and I can be regarded as well as canonical relations

$$\bar{\mathcal{G}} \times \bar{\mathcal{G}} \nrightarrow \mathcal{G} \text{ and } \mathcal{G} \nrightarrow \bar{\mathcal{G}}$$

  respectively and will be denoted by $\overline{L_{rel}}$ and $\overline{I_{rel}}$. The transposition

$$\begin{aligned} T : \mathcal{G} \times \mathcal{G} &\to \mathcal{G} \times \mathcal{G} \\ (x,y) &\mapsto (y,x) \end{aligned}$$

[b]This is not exactly a category because the composition of canonical relations is not in general a smooth manifold

[c]Observe here that usually one considers embedded Lagrangian submanifolds, but we consider immersed ones.

[d]A weak symplectic manifold M has a closed 2-form ω such that the induced map $\omega^\# : T^*M \to TM$ is injective. For finite dimensional manifolds, the notion of weak symplectic and symplectic manifolds coincides.

induces canonical relations

$$T_{rel} : \mathcal{G} \times \mathcal{G} \nrightarrow \mathcal{G} \times \mathcal{G} \text{ and } \overline{T_{rel}} : \bar{\mathcal{G}} \times \bar{\mathcal{G}} \nrightarrow \bar{\mathcal{G}} \times \bar{\mathcal{G}}.$$

The identity map $Id : \mathcal{G} \to \mathcal{G}$ as a relation will be denoted by $Id_{rel} : \mathcal{G} \nrightarrow \mathcal{G}$ and by $\overline{Id_{rel}} : \overline{\mathcal{G}} \nrightarrow \overline{\mathcal{G}}$.

- **A.3** $I_{rel} \circ L_{rel} = \overline{L}_{rel} \circ \overline{T}_{rel} \circ (\overline{I_{rel}} \circ \overline{I_{rel}}) : \mathcal{G} \times \mathcal{G} \nrightarrow \mathcal{G}$.

Remark 1. Since I and T are diffeomorphisms, both sides of the equaliy correspond to immersed Lagrangian submanifolds.

Define

$$L_3 := I_{rel} \circ L_{rel} : \mathcal{G} \times \mathcal{G} \nrightarrow \mathcal{G}.$$

As a corollary of the previous axioms we get that

Corollary 3.1. $\overline{I_{rel}} \circ L_3 = \overline{L_3} \circ \overline{T_{rel}} \circ (\overline{I_{rel}} \times \overline{I_{rel}})$.

- **A.4** $L_3 \circ (L_3 \times Id) = L_3 \circ (Id \times L_3) : \mathcal{G}^3 \nrightarrow \mathcal{G}$ is an immersed Lagrangian submanifold.

The fact that the composition is Lagrangian follows from the fact that, since I is an antisymplectomorphism, its graph is Lagrangian, therefore L_3 is Lagrangian, and so $(Id \times L_3)$ and $(L_3 \times Id)$. The graph of the map I, as a relation $* \nrightarrow \mathcal{G} \times \mathcal{G}$ will be denoted by L_I.

- **A.5** $L_3 \circ L_I$ is an immersed Lagrangian submanifold of $\mathcal{G}$.

Remark 2. It can be proven that Lagrangianity in these cases is automatical if we start with a finite dimensional symplectic manifold $\mathcal{G}$. Let $L_1 := L_3 \circ L_I : * \nrightarrow \mathcal{G}$. From the definitions above we get the following

Corollary 3.2.

$$\overline{I_{rel}} \circ L_1 = \overline{L_1},$$

that is equivalent to

$$I(L_1) = \overline{L_1},$$

where L_1 is regarded as an immersed Lagrangian submanifold of $\mathcal{G}$.

Corollary 3.3.

$$L_3 \circ (L_1 \times L_1) = L_1.$$

- **A.6** $L_3 \circ (L_1 \times Id)$ is an immersed Lagrangian submanifold of $\overline{\mathcal{G}} \times \mathcal{G}$. We define

$$L_2 := L_3 \circ (L_1 \times Id) : \mathcal{G} \nrightarrow \mathcal{G}.$$

Corollary 3.4.

$$L_2 = L_3 \circ (Id \times L_1).$$

Corollary 3.5. *L_2 leaves invariant L_1, L_2 and L_3, i.e.*

$$L_2 \circ L_1 = L_1$$
$$L_2 \circ L_2 = L_2$$
$$L_2 \circ L_3 = L_3.$$

Corollary 3.6.

$$\overline{I_{rel}} \circ L_2 = \overline{L_2} \circ \overline{I_{rel}} \text{ and } L_2^\dagger = L_2.$$

The next set of axioms defines a particular type of relational symplectic groupoids, in which the relation L_2 plays the role of an equivalence relation and it allows to study the case of symplectic reductions.

Definition 3.3. A relational symplectic groupoid $(\mathcal{G}, L, I)$ is called **regular** if the following axioms are satisfied. Consider $\mathcal{G}$ as a coisotropic relation $* \nrightarrow \mathcal{G}$ denoted by $\mathcal{G}_{rel}$.

- **A.7** $L_2 \circ \mathcal{G}_{rel}$ is an immersed coisotropic relation.
Remark 3. Again in this case, the fact that this is a coisotropic relation follows automatically in the finite dimensional setting.

Corollary 3.7. *Setting $C := L_2 \circ \mathcal{G}_{rel}$ the following corollary holds.*

(i)

$$C^* = \mathcal{G}^* \circ L_2$$

(ii) *L_2 defines an equivalence relation on C.*
(iii) *This equivalence relation is the same as the one given by the characteristic foliation on C.*

- **A.8** The reduction $\underline{L_1} = L_1/L_2$ is a finite dimensional smooth manifold. We will denote $\underline{L_1}$ by M.

- **A.9** $S := \{(c,[l]) \in C \times M : \exists l \in [l], g \in \mathcal{G} | (l,c,g) \in L_3\}$ is an immersed submanifold of $\mathcal{G} \times M$.

Corollary 3.8.

$$T := \{(c,[l]) \in C \times M : \exists l \in [l], g \in \mathcal{G} | (c,l,g) \in L_3\}$$

is an immersed submanifold of $\mathcal{G} \times M$.

The following conjectures (this is part of work in progress) give rise to a connection between the symplectic structure on $\mathcal{G}$ and Poisson structures on M.

Conjecture 3.1. *Let* $(\mathcal{G}, L, I)$ *be a regular relational symplectic groupoid. Then, there exists a unique Poisson structure on* M *such that* S *is coisotropic in* $\mathcal{G} \times M$.

Conjecture 3.2. *Assume* $G := C/L_2$ *is smooth. Then* G *is a symplectic groupoid on* M *with structure maps* $s := S/L_2$, $t := T/L_2, \mu := L_{rel}/L_2, \iota = I, \varepsilon = L_1/L_2$.

Definition 3.4. A **morphism between relational symplectic groupoids** (G, L_G, I_G) and (H, L_H, I_H) is a map F from G to H satisfying the following properties:

(i) F is a Lagrangian subspace of $G \times \bar{H}$.
(ii) $F \circ I_G = I_H \circ F$.
(iii) $F^3(L_G) = L_H$.

Definition 3.5. A morphism of relational symplectic groupoids $F : G \to H$ is called an **equivalence** if the transpose canonical relation $F^\dagger$ is also a morphism.

Remark 4. For our motivational example, it can be proven that

(i) Different differentiability degrees (the $\mathcal{C}^k-$ type of the maps X and η) give raise to equivalent relational symplectic groupoids.
(ii) For regular relational symplectic groupoids, $\mathcal{G}$ and G are equivalent.

3.1. *Examples*

The following are natural examples of relational symplectic groupoids.

3.1.1. *Symplectic groupoids:*

Given a Lie symplectic groupoid G over M, we can endow it naturally with a relational symplectic structure:

$$\mathcal{G} = G.$$
$$L = \{(g_1, g_2, g_3) | (g_1, g_2) \in G \times_{(s,t)} G,\ g_3 = \mu(g_1, g_2)\}.$$
$$I = g \mapsto g^{-1},\ g \in G.$$

Remark 5. In connection to the construction in the Poisson sigma model, we can conclude that when M is integrable, the reduced space of boundary fields $\underline{C_\Pi}$ is a relational symplectic groupoid.

3.1.2. *Symplectic manifolds with a given immersed Lagrangian submanifold:*

Let (G, ω) be a symplectic manifold and $\mathcal{L}$ an immersed Lagrangian submanifold of G. We define

$$\mathcal{G} = G.$$
$$L = \mathcal{L} \times \mathcal{L} \times \mathcal{L}.$$
$$I = \{\text{identity of } G\}.$$

It is an easy check that this construction satisfies the relational axioms and furthermore

Proposition 3.1. *The previous relational symplectic groupoid is equivalent to the zero dimensional symplectic groupoid (a point with zero symplectic structure and empty relations).*

Proof: It is easy by checking that L is an equivalence from the zero manifold to $\mathcal{G}$.

3.1.3. *Powers of symplectic groupoids:*

Let us denote $G_{(1)} = G$, $G_{(2)}$ the fiber product $G \times_{(s,t)} G$, $G_{(3)} = G \times_{(s,t)} (G \times_{(s,t)} G)$ and so on . It can be proven the following

Lemma 3.1.[3] *Let $G \rightrightarrows M$ be a symplectic groupoid.*

(i) $G_{(n)}$ is a coisotropic submanifold of G^n.

(ii) The reduced spaces $\underline{G_{(n)}}$ are symplectomorphic to G. Furthermore, there exists a natural symplectic groupoid structure on $\underline{G_{(n)}}$ coming from the quotient, isomorphic to the groupoid structure on G.

We have natural canonical relations $P_n : \underline{G_{(n)}} \to G^n$ defined as:

$$P := \{(x, \alpha, \beta) | x \in \underline{G_{(n)}},\ [\alpha] = [\beta] = x\},$$

satisfying the following relations:

$$P^\dagger \circ P = Gr(Id_G),\ P \circ P^\dagger = \{(g, h) \in G^n | [g] = [h]\}.$$

It can be checked that

Proposition 3.2. *$G_{(i)}$ is equivalent to $G_{(j)}, \forall i,\ j \geq 1$ and the equivalence is given by $P_i \circ P_j^\dagger$.*

3.1.4. *The cotangent bundle of the path space of a Poisson manifold.*

This is the motivational example for the construction of relational symplectic groupoids. In this case, the coisotropic submanifold C_Π is equipped with an equivalence relation, called T^*M- homotopy,[10] and denoted by $\sim$. More precisely, to points of C_Π are $\sim$- equivalent if they belong to the same leaf of the characteristic foliation of C_Π. We get the following relational symplectic groupoid (where L is the restriction to the boundary of the solutions of the Euler-Lagrange equations in the bulk)

$\mathcal{G} = T^*(PM)$.
$L = \{(X_1, \eta_1), (X_2, \eta_2), (X_3, \eta_3) \in C_\Pi^3 | (X_1 * X_2, \eta_1 * \eta_2) \sim (X_3 * \eta_3)\}$.
$I = (X, \eta) \mapsto (\phi^* X, \phi^* \eta)\}$.

Here $*$ denotes path concatenation and

$$\begin{aligned} \phi : [0,1] &\to [0,1] \\ t &\mapsto 1 - t \end{aligned}$$

Theorem 3.1.[3] *The relational symplectic groupoid $\mathcal{G}$ defined above is regular.*

The improvement of Theorem 2.1 in terms of the relational symplectic groupoids can be summarized as follows. L_1 can be understood as the space of T^*M- paths that are T^*M- homotopy equivalent to the trivial T^*M- paths and

$$\overline{L_1} := \cup_{x_0 \in M} T^*_{\overline{(X,\eta)}} PM \cap L_1,$$

where $(\overline{X,\eta}) = \{(X,\eta)|X \equiv X_0, \eta \in \ker \Pi^{\#}\}$, we can prove the following

Theorem 3.2.[3] *If the Poisson manifold M is integrable, then there exists a tubular neigborhood of the zero section of T^*PM, denoted by $N(\Gamma_0(T^*PM))$ such that $\overline{L_1} \cap N(\Gamma_0(T^*PM))$ is an embedded submanifold of T^*PM.*

Theorem 3.3.[3] *If M is integrable, then $L_1 \cap N(\Gamma_0(T^*PM))$, $L_2 \cap N(\Gamma_0(T^*PM))^2$ and $L_3 \cap N(\Gamma_0(T^*PM))^3$ are embedded Lagrangian submanifolds.*

4. Quantization

The structure of relational symplectic groupoid may be reformulated in the category *Hilb* of Hilbert spaces. We define a *preunital Frobenius algebra* as the following data

- A Hilbert space H, equipped with an inner product $\langle \, , \rangle$ and an associative map
$$m : H \otimes H \to H.$$
- Defining
$$\langle a, b \rangle_H := \langle \overline{a}, b \rangle,$$
where $\bar{}$ denotes complex conjugation, the following axioms holds:
  (i) *Cyclicity or Frobenius condition:*
$$\langle m(a,b), c \rangle_H = \langle a, m(b,c) \rangle_H$$
  (ii) *Projectability:* Choosing an orthonormal basis $\{e_i\}$ of H and assuming that
$$e := \sum_i m(e_i, \bar{e_i})$$
  is a well defined element in H, the operator
$$\begin{aligned} P : H &\to H \\ a &\mapsto m(a,e) \end{aligned}$$
  is an orthogonal projection.

Remark 6. Under the assumptions above, $(H, m, \langle\,,\rangle_H)$ is not an unital algebra, however, the image of the operator P, called the *reduced algebra*, is unital.

The relational symplectic groupoid may be seen as the dequantization of this structure. The hard problem consists in going the other way around, namely, in quantizing a relational symplectic groupoid.

In the finite dimensional examples, methods of geometric quantization might be available, the problem being that of finding an appropriate polarization compatible with the structures. This question, in the case of a symplectic groupoid, has been addressed by Weinstein[17] and Hawkins.[13] The relational structure might allow more flexibility.

In the infinite-dimensional case, notably the example in 3.1.4., perturbative functional integral techniques might be available. The reduced algebra should give back a deformation quantization of the underlying Poisson manifold.

Finally notice that quantization might require weakening a bit the notion of preunital Frobenius algebra, allowing for example nonassociative products. However, one expects that the reduced algebra should always be associative.

References

1. A.S. Cattaneo, *On the integration of Poisson manifolds, Lie algebroids, and coisotropic submanifolds*, Letters in Mathematical Physics 67: 33-48, 2004.
2. A.S. Cattaneo, *Coisotropic submanifolds and dual pairs*, preprint.
3. A.S. Cattaneo and I. Contreras, *Relational symplectic groupoids and Poisson sigma models with boundary*, in preparation.
4. A. S. Cattaneo and G. Felder, *Poisson sigma models and symplectic groupoids*, in *Quantization of Singular Symplectic Quotients*, (ed. N.P. Landsman, M. Pflaum, M. Schlichenmeier), Progress in Mathematics 198 (Birkhäuser), 61-93, 2001.
5. A. S. Cattaneo and G.Felder, *Coisotropic submanifolds in Poisson geometry and branes in the Poisson sigma model*, Lett. Math. Phys. 69. 157-175. 2004.
6. A.S. Cattaneo, P. Mñev and N. Reshetikin, *Classical and Quantum Lagrangian Field Theories with Boundary*, Proceedings of the Corfu Summer Institute 2011 School and Workshops on Elementary Particle Physics and Gravity,Corfu, Greece, 2011.
7. A. S. Cattaneo, P. Mnëv and N. Reshetikhin, *Classical BV theories on manifolds with boundaries.* math-ph/1201.0290, 2012.
8. A. S. Cattaneo, P. Mnëv and N. Reshetikhin, *Classical and Quantum La-*

grangian Field Theories with Boundary, preprint.

9. A. Coste, P. Dazord, A. Weinstein, *Groupoïdes symplectiques*, Publ. Dept. Math. Univ. Claude-Bernard Lyon I, 1987.
10. M. Crainic and R. L. Fernandes. *Integrability of Lie brackets.* Ann. of Math.(2) 157, 575–620, 2003.
11. M. Crainic and R. L. Fernandes. *Integrability of Poisson brackets*, J. Differential Geometry 66, 71-137, 2004.
12. J.J. Duistermaat and J.A.C. Kolk, *Lie Groups* Universitext, Springer, 1999.
13. E. Hawkins, *A groupoid approach to quantization*, J. Symplectic Geom. Vol. 6,1, 61-125, 2008.
14. I. Moerdjik and J.Mrcun. *Intoduction to Foliations and Lie Groupoids*, Cambridge studies in advanced mathematics, 91, 2003.
15. P. Severa, *Some title containing the words homotopy and symplectic, e.g. this one*, preprint math.SG-0105080.
16. H.H Tseng, C. Zhu, *Integrating Lie algebroids via stacks*, Compositio Mathematica, Volume 142 Issue 01, pp 251-270, 2006.
17. A. Weinstein, *Symplectic groupoids, geometric quantization, and irrational rotation algebras*, in Symplectic geometry, groupoids, and integrable systems (Berkeley, CA, 1989), 281-290, Springer, New York, 1991.

A SURVEY ON ORBIFOLD STRING TOPOLOGY

Andrés Angel

Departamento de Matematicas
Universidad de los Andes
Bogotá, Colombia
E-mail: ja.angel908@uniandes.edu.co

The aim of this short communication is to review some classical results on string topology of manifolds and discuss recent extensions of this theory to orbifolds. In particular, we review the relation between the loop homology of the classiying space of the orbifold and the Hochschild cohomology of dg-ring naturally associated to the orbifold.

Keywords: String Topology; Orbifolds

Introduction

String topology is the study of the topological properties of to the space of smooth loops LM on a closed and oriented manifold M. In particular algebraic structures defined on the homology of the loop space. The area started with the paper[5] by Chas and Sullivan where they defined an intersection product in the homology of the free loop space

$$H_p(LM) \otimes H_q(LM) \to H_{p+q-n}(LM),$$

having total degree $-n$.

Orbifolds, originally known as V-manifolds by I. Satake,[18] and named this way by W. Thurston,[19] are a generalizations of manifolds, locally they look like the quotient of euclidean space by the action of a finite group. They appear naturally in many areas such as moduli problems, noncommutative geometry and topology. It is natural to ask, *What is the string topology of an orbifold?* In this survey article we give different answers in the case where the orbifold is a global quotient.

The character of this paper is expository and is organized as follows. In section one we review some of the classical theory for manifolds, and state the results of Chas-Sullivan[5] and Cohen-Jones[7] on the string topology of manifolds. In section two we discuss the necessary background on orbifolds. In section three, we review the construction of[14] of the string topology of quotient orbifolds over the rationals and the construction of[3] of the string topology of quotient orbifolds over the integers. In section four we specialize to the case of $[*/G]$ and give geometric descriptions of the string topology product on the cohomology $H^*(LBG)$.

Acknowledgements

The author would like the organizers of the 2011 summer school "Geometrical, Algebraic and Topological Methods in Quantum Field Theory" for the opportunity to present some of the results mentioned in this work.

1. String topology of manifolds

In this section we present the basic results on the classical theory of string topology of manifolds. We give the definition of the string topology product using intersection theory on loop spaces, and the description at the level of spectra of this product.

1.1. *Intersection theory*

From the point of view of algebraic topology, the most fundamental property of closed oriented manifold of finite dimension n is Poincare duality, which asserts that the bilinear pairing giving by cup product and evaluating on the fundamental class is a non-degenerate pairing

$$H^{n-p}(M) \times H^p(M) \to \mathbf{k}$$
$$(\alpha, \beta) \to \langle \alpha \cup \beta, [M] \rangle$$

where homology and cohomology are taken with values in a field $\mathbf{k}$. By the universal coefficients theorem $H_p(M) \cong (H^p(M))^*$ and we have that the Poincare duality pairing identifies homology with cohomology

$$P.D : H_p(M) \cong (H^p(M))^* \cong H^{n-p}(M).$$

Since the cohomology of space is a ring, we have that the homology $H_{*+n}(M)$ (shifted by the dimension) of a closed oriented manifold has a ring structure $\star$ (grading preserving because of the shift).

$$\begin{array}{ccc} H_p(M) \otimes H_q(M) & \overset{\star}{\dashrightarrow} & H_{p+q-n}(M) \\ \downarrow{\scriptstyle P.D \otimes P.D} & & \uparrow{\scriptstyle P.D^{-1}} \\ H^{n-p}(M) \otimes H^{n-q}(M) & \xrightarrow{\sqcup} & H^{2n-p-q}(M) \end{array}$$

This ring structure is called *the intersection ring of the manifold.* This ring structure can be interpreted in terms of intersection theory because when two homology classes $\sigma \in H_p(M), \tau \in H_q(M)$ can be represented by submanifolds [a] $P^p \subseteq M$ and $Q^q \subseteq M$ respectively, then the product is represented by the tranverse intersection $P \cap Q$.

The intersection ring can be realized at the level of spaces by a Thom-Pontrjagin collapse map. Take the diagonal embedding $\Delta : M \to M \times M$ which has a tubular neighborhood ν_Δ homeomorphic to the total space of the normal bundle, which is TM. Consider the projection map that collapses the complement of the tubular neighborhood to a point

$$\tau_\Delta : M \times M \to (M \times M)/\nu_\Delta^c$$

$(M \times M)/\nu_\Delta^c$ is homeomorphic to the Thom space M^{TM} of the normal bundle (the one point compactification). Thus we have a Thom-Pontrjagin collapse map

$$\tau_\Delta : M \times M \to M^{TM}$$

which induces on homology a map that composed with the Thom isomorphism and the Kunneth map

$$H_p(M) \otimes H_q(M) \longrightarrow H_{p+q}(M \times M) \xrightarrow{\tau_{\Delta *}} H_{p+q}(M^{TM}) \overset{Thom}{\longrightarrow} H_{p+q-n}(M)$$

realizes the intersection product.

They key fact of what we did is that if a map $e : P \to Q$ is an embedding with finite dimensional normal bundle, then we have a Thom-Pontrjagin collapse map $Q \to P^{\nu(e)}$ from Q to the Thom space of the normal bundle $\nu(e) \to P$. If the normal bundle is orientable after applying the Thom isomorphism we obtain a wrong way or umkehr map $e_! : H_*(Q) \to H_{*-\dim \nu(e)}(P)$.

[a] We say that a homology class $\sigma \in H_p(M)$ can be represented by submanifold if there exists a p-dimensional closed oriented submanifold P, such that σ is the pushforward of the fundamental class of P under the inclusion map.

1.2. *String topology of manifolds*

Let LM be the space of piecewise smooth maps from S^1 to M, the free loop space. Since LM is infinite dimensional and we do not have Poincare duality we do not expect to have a non-trivial ring structure on the homology. But still, Chas and Sullivan defined a "loop product" in the homology of the free loop space of a closed oriented n-dimensional manifold

$$\circ : H_p(LM) \otimes H_q(LM) \to H_{p+q-n}(LM).$$

Let $Map(8, M)$ be the space of piecewise smooth maps from the figure 8 to M. This space fits into a pullback square

$$\begin{array}{ccc} Map(8,M) = LM \times_M LM & \longrightarrow & LM \times LM \\ \downarrow & & \downarrow{\scriptstyle ev\times ev} \\ M & \xrightarrow{\quad\Delta\quad} & M \times M \end{array}$$

Since the evaluation at zero $ev : LM \to M$ can be considered as a fiber bundle of infinite dimensional manifolds, the map $Map(8, M) \to LM \times LM$ is an embedding of finite codimension, its normal bundle is given by the inverse image of the normal bundle to $M \to M \times M$. Thus we have a Thom-Pontrjagin collapse map $\tau_\Delta : LM \times LM \to Map(8, M)^{ev^*(TM)}$.

There is a concatenation map $Map(8, M) \to LM$ that extends the usual Pontrjagin product $\Omega M \times \Omega M \to \Omega M$. Composing the Kunneth map with the map induced by the Thom-Pontrjagin collapse map, together with the Thom isomorphism and the map induced by the concatenation map gives the Chas-Sullivan loop product.

$$\begin{array}{llll} H_*(LM)\otimes H_*(LM) \to & H_*(LM \times LM) & & \\ & \downarrow{\scriptstyle \tau_{\Delta *}} & & \\ & H_*(Map(8,M)^{ev^*(TM)}) & \overset{Thom}{\to} & H_{*-n}(Map(8,M)) \\ & & & \downarrow \\ & & & H_{*-n}(LM) \end{array}$$

1.3. *String Topology spectra*

Cohen and Jones[7] realized homotopically this operation, by constructing a ring spectrum

$$LM^{-TM} \wedge LM^{-TM} \to LM^{-TM},$$

where LM^{-TM} is the Thom spectrum of a virtual bundle over the loop space.

The Poincare duality identification is realized at the level of stable homotopy theory by the composition of the Thom isomorphism for stable normal bundle $-TM \to M$ (a virtual bundle of dimension $-n$)

$$H_p(M^{-TM}) \cong H_{p+n}(M)$$

and the Atiyah duality $M^{-TM} \simeq F(M_+, S)$, that identifies M^{-TM} as the Spanier-Whitehead dual of M_+[b]. Therefore $H_*(M^{-TM}) \cong H_*(F(M_+, S)) \cong H^{-*}(M)$.

The classical construction of the Thom spectrum M^{-TM} is as follows, given an embedding $e : M \hookrightarrow \mathbb{R}^k$, let $\nu(e) \to M$ the normal bundle of the embedding and denote by $M^{\nu(e)}$ its Thom space. Define the spectrum

$$M^{-TM} := \Sigma^{-k} M^{\nu(e)}$$

as the k-th de-suspension of the Thom space $M^{\nu(e)}$.

The diagonal embedding $\Delta : M \to M \times M$ has a Thom-Pontrjagin collapse map $M \times M \to M^{TM}$ that can be twisted by the bundle $-2TM$ to obtain a map spectra

$$\Delta^* : M^{-TM} \wedge M^{-TM} = (M \times M)^{-2TM} \to M^{TM-2TM} = M^{-TM}$$

that is associative up to homotopy.

Theorem 1.1 (Atiyah Duality). *There exists a ring spectrum structure on the Thom spectrum M^{-TM} which induced (after applying the Thom isomorphism) the intersection product on homology.*

In a similar fashion, we can twist the concatenation map $Map(8, M) \to LM$ by the virtual bundle $ev^*(-TM)$ to obtain a map of spectra $Map(8, M)^{ev^*(-TM)} \to LM^{ev^*(-TM)}$, and the Thom-Pontrjagin collapse map $\tau_\Delta : LM \times LM \to Map(8, M)^{ev^*(TM)}$ by $ev^*(-2TM)$ to get a map of spectra

$$LM^{ev^*(-TM)} \wedge LM^{ev^*(-TM)} \to Map(8, M)^{ev^*(-TM)}.$$

Composing these maps we get a map of spectra

$$\begin{array}{rl} LM^{ev^*(-TM)} \wedge LM^{ev^*(-TM)} = (LM \times LM)^{ev*(-2TM)} & \\ \downarrow & \\ Map(8, M)^{ev^*(-TM)} & \to LM^{ev^*(-TM)} \end{array}$$

[b]For a space X, the disjoint union of X with a point is denoted by X_+. For a space X the collection of unbased maps from X to a spectrum Y is denoted by $F(X_+, Y)$.

that is associative up to homotopy.

Theorem 1.2 ([7]). *There exists a ring spectrum structure on the Thom spectrum $LM^{ev^*(-TM)}$ which induces the Chas-Sullivan product on homology.*

In the same paper Cohen and Jones give a cosimplicial model of LM^{-TM} which is used to prove that for a simply connected manifold the string topology is related with Hochschild cohomology of the cochains of the manifold.

Recall that for an associative k-algebra A and a A-bimodule M, we have the Hochschild homology and cohomology of A with coefficients in M are

$$HH_*(A, M) \cong \mathrm{Tor}^*_{A \otimes A^{op}}(A, M)$$

$$HH^*(A, M) \cong \mathrm{Ext}^*_{A \otimes A^{op}}(A, M)$$

and these functors can be extended to the category of differential graded algebras. This extension can be done by using the bar construction for differential graded algebras to obtain complexes

$$A \overset{L}{\otimes}_{A^e} M = B(A) \overset{L}{\otimes}_{A^e} M$$

$$\mathcal{RH}om_{A^e}(A, M) = Hom_{A^e}(B(A), M)$$

that calculate the Hochschild homolgogy and cohomology of the dga A with coefficients in the dg-bimodule M.

$$HH_*(A, M) = H^*(A \overset{L}{\otimes}_{A^e} M)$$

$$HH^*(A, M) = H_*(\mathcal{RH}om_{A^e}(A, M))$$

The cosimplicial model of LM^{-TM} is used to prove,

Theorem 1.3 ([7]). *For M a simply connected, oriented, closed manifold, there is a quasi-isomorphism*

$$C_*(LM^{-TM}) \xrightarrow{\simeq} \mathcal{RH}om_{C^*(M)^e}(C^*(M), C^*(M))$$

that induces an isomorphism of rings

$$H_{*+n}(LM) \cong HH^*(C^*(M), C^*(M)).$$

2. Orbifolds

Orbifolds were first introduced by I. Satake[18] in the fifties as generalizations of smooth manifolds that allow mild singularities. In this section we follow the classical perspective on orbifolds by defining orbifold charts and atlases akin to the way manifolds are defined.

2.1. *Charts*

Let X be a paracompact Hausdorff topological space. An *n-dimensional orbifold chart* on X is a triple $(\overline{U}, G, U)$, where $\overline{U}$ is a connected manifold, G is a finite group acting on $\overline{U}$, and U is an open subset of X, homeomorphic to $\overline{U}/G$.

As with manifolds, an *orbifold atlas* on X is a family of charts $(\overline{U}_\alpha, G_\alpha, U_\alpha)$, that is compatible and covers X. An *orbifold structure* on X is just an equivalence class of orbifold atlases, where two atlases are equivalent if there is a zig-zag of common refinements.

Example 2.1. A manifold M with an action of a finite group G gives rise to an orbifold structure that we will denoted by $[M/G]$.

Example 2.2. For a finite group G, the orbifold $[*/G]$ is a zero dimensional orbifold, the underlying space has only one point.

Given a point $x \in X$, take a chart $(\overline{U}, G, U)$ around x, let $\overline{x} \in \overline{U}$ be a lift of x. Then we define the *isotropy group* of x to be,

$$G_x = \{g \in G \mid g\overline{x} = \overline{x}\}$$

This group is well defined up to isomorphism. A point is called *non-singular* if G_x is trivial, and *singular* otherwise.

Example 2.3. Consider the action of S^1 on $\mathbb{C}^2 - \{0\}$ given by $\lambda(z_1, z_2) = (\lambda^2 z, \lambda z)$, it is a an action with finite stabilizers, and therefore the underlying space, which topologically is just a sphere, has an orbifold structure coming from this action, it has one singular point with isotropy group $\mathbb{Z}_2$, but it can be seen that is not the quotient of a manifold by a finite group (even though it is a quotient by a compact Lie group).

To an orbifold $\mathcal{X}$ we can associate a topological space $B\mathcal{X}$ called the *classifying space of the orbifold*, see,[161] For example for a global quotient $\mathcal{X} = [M/G]$ with G a finite group acting on M, $B\mathcal{X}$ is homotopy equivalent to the homotopy orbit space $EG \times_G M$, where $EG \to BG$ is the universal principal G-bundle.

To an orbifold $\mathcal{X}$ we have associated two topological spaces, the underlying space $|\mathcal{X}|$ and the classifying space $B\mathcal{X}$; there is a map $B\mathcal{X} \to |\mathcal{X}|$, but in general this map is far from being a homotopy equivalence.

3. String topology of orbifolds

3.1. *Rational string topology*

In the case of global quotient orbifolds of the form $[M/G]$ for G finite group, Lupercio and Uribe[13] constructed the loop groupoid $[P_G M/G]$ as the natural model for the free loop space of the orbifold, whose homotopic quotient turns out to be homotopy equivalent to the space of free loops of the homotopy quotient $M \times_G EG$, i.e.

$$P_G M \times_G EG \simeq L(M \times_G EG).$$

Lupercio, Xicoténcatl and Uribe[14] showed that the rational homology of the free loop space of an oriented orbifold

$$H_*(\mathcal{L}(M \times_G EG); \mathbb{Q}) \cong H_*(P_G M; \mathbb{Q})^G$$

could also be endowed with a ring structure; the authors called this structure with the name **orbifold string topology**.

Let X be a topological space and G a discrete group acting on X. Denote by $[X/G]$ the topological groupoid whose objects are X and whose morphisms are $X \times G$ with $s(x,g) = x$ and $t(x,g) = xg$.

The loops on $[X/G]$ can be understood as the groupoid whose objects are the functors from the groupoid $[\mathbf{R}/\mathbf{Z}]$ to $[X/G]$ and whose morphisms are given by natural transformations. More explicitly, we have

Definition 3.1. The loop groupoid for $[X/G]$ is the groupoid $[P_G X/G]$ whose space of objects is

$$P_G X := \bigsqcup_{g \in G} P_g X \times \{g\} \quad \text{with} \quad P_g X := \{f : [0,1] \to X | f(0)g = f(1)\}$$

and which are endowed with the G-action

$$\begin{aligned} G \times P_G X &\to P_G X \\ ((f,k),g) &\mapsto (fg, g^{-1}kg). \end{aligned}$$

Now, for $t \in \mathbb{R}$ let $e_t : P_gM \to M$ be the evaluation of a map at the time t: $e_t(f) := f(t)$. Consider the space of composable maps

$$P_gM :\, _1\times_0\, P_hM = \{(\phi, \psi) \in P_gM \times P_hM | \phi(1) = \psi(0)\}$$

and note that they fit into the pullback square

$$\begin{array}{ccc} P_gM :\, _1\times_0\, P_hM & \longrightarrow & P_gM \times P_hM \\ \big\downarrow{\scriptstyle e_0 \circ \pi_1} & & \big\downarrow{\scriptstyle e_0 \times e_0} \\ M & \xrightarrow{\quad \Delta_g \quad} & M \times M \end{array}$$

where the downward arrow $e_0 \circ \pi_1$ is the evaluation at zero of the first map.

The normal bundle of the upper horizontal map becomes isomorphic to the pullback under $e_0 \circ \pi_1$ of the normal bundle of the map Δ_g and therefore we can construct the Thom-Pontrjagin collapse maps $P_gM \times P_hM \to P_gM :\, _1\times_0\, P_hM^{\pi_1^* e_0^* N_g}$, and obtain a map of spectra $P_gM^{-e_0^*TM} \wedge P_hM^{-e_0^*TM} \to P_gM :\, _1\times_0\, P_hM^{-\pi_1^* e_0^* TM}$. The concatenation of paths in $P_gM :\, _1\times_0 P_hM$ produces a map $P_gM :\, _1\times_0 P_hM \to P_{gh}M$ which induces the map of spectra $P_gM :\, _1\times_0\, P_hM^{-\pi_1^* e_0^* TM} \to P_{gh}M^{-e_0^*TM}$ that defines the map $\mu_{g,h} : P_gM^{-e_0^*TM} \wedge P_hM^{-e_0^*TM} \to P_{gh}M^{-e_0^*TM}$. Define

$$P_GM^{-TM} := \bigsqcup_{g \in G} P_gM^{-TM}$$

which by assembling the maps $\mu_{g,h}$ define a map that is associative up to homotopy.

$$\mu : P_GM^{-TM} \wedge P_GM^{-TM} \to P_GM^{-TM}.$$

Theorem 3.1 ([14]). *P_GM^{-TM} is a ring spectrum with a G-action in the homotopy category, taking homology, we get a G-module ring $H_*(P_GM^{-TM}; \mathbb{Z})$, the induced ring structure on the invariant set*

$$H_*(P_GM^{-TM}; \mathbb{Q})^G \cong H_{*+n}(L(M \times_G EG); \mathbb{Q})$$

is an invariant of the orbifold $[M/G]$.

This ring was called in[14] the **orbifold string topology ring**.

3.2. *Hochschild cohomology of global quotient orbifolds*

In the case that M is simply connected, Angel, Backelin and Uribe in[3] showed that the orbifold string topology ring of a global quotient can be recovered as the Hochschild cohomology ring of the dga $C^*(M;\mathbb{Q})\#G$, i.e. there is an isomorphism of rings

$$HH^*(C^*(M;\mathbb{Q})\#G, C^*(M;\mathbb{Q})\#G) \cong H_*(P_G M^{-TM};\mathbb{Q})^G$$

where $C^*(M)\#G$ is the convolution differential graded algebra,

$$C^*(M)\#G := C^*(M) \otimes \mathbb{Q}G \qquad (\alpha, g)\cdot(\beta, h) := (\alpha \cdot g^*\beta, gh).$$

This isomorphism is obtained by carefully decomposing the Hochschild cohomology ring into smaller parts, which leads to the ring isomorphism

$$HH^*(C^*(M;\mathbb{Q})\#G, C^*(M;\mathbb{Q})\#G) \cong \mathrm{Ext}^*_{\mathbb{Q}G}(\mathbb{Q}, C_*(P_G M^{-TM};\mathbb{Q})).$$

The decomposition works also over the integers and starts by taking any space X and giving an explicit cofibrant replacement for $C^*(X)\#G$ as a $(C^*(X)\#G)^e$-module, together with an explicit diagonal map. It is shown in[3] that if we consider $C^*(X)\#G$ as a G-module-dg-ring by the action $g\cdot(x\otimes h) \mapsto g(x)\otimes ghg^{-1}$, then we have that

Theorem 3.2. *For any space X there are isomorphisms of graded groups*

$$HH_*((C^*(X)\#G, C^*(X)\#G) \cong \mathrm{Tor}^*_{\mathbb{Z}G}(\mathbb{Z}, C^*(X) \overset{L}{\otimes}_{C^*(X)^e} C^*(X)\#G)$$

$$HH^*(C^*(X)\#G, C^*(X)\#G) \cong \mathrm{Ext}^*_{\mathbb{Z}G}(\mathbb{Z}, \mathcal{RH}om_{C^*(X)^e}(C^*(X), C^*(X)\#G))$$

defined in an appropriate way such that the second isomorphism becomes one of graded rings.

Using a cosimplicial description for the space $P_G X$, in[3] it is proved that

Theorem 3.3. *If X is simply connected space and G is finite, then there is an equivariant quasi-isomorphism*

$$C^*(X) \overset{L}{\otimes}_{C^*(X)^e} C^*(X)\#G \xrightarrow{\simeq} C^*(P_G X)$$

that induces the equivariant isomorphism

$$\mathrm{Tor}_{C^*(X)^e}(C^*(X), C^*(X)\#G) \cong H^*(P_G X).$$

Whenever M is a closed oriented manifold of dimension l, Atiyah duality gives a homomorphism of graded groups

$$C^{-*}(M) \to C_*(M^{-TM})$$

that induces an isomorphism in homology. By using the multiplicative properties of Atiyah duality and a cosimplicial model of $P_G M$ in[3] it is proved that

Theorem 3.4. *If M is a simply connected closed oriented manifold and G is finite, then there is an equivariant quasi-isomorphism*

$$C_*(P_G M^{-TM}) \overset{\simeq}{\to} \mathcal{RH}om_{C^*(M)^e}(C^*(M), C^*(M)\#G)$$

Using the theorem 3.2 and the previous theorem we obtain that

$$HH^*(C^*(M;\mathbb{Q})\#G, C^*(M;\mathbb{Q})\#G) \cong \mathrm{Ext}^*_{\mathbb{Q}G}(\mathbb{Q}, C_*(P_G M^{-TM};\mathbb{Q}))$$

and, since rationally taking invariants is exact,

$$HH^*(C^*(M;\mathbb{Q})\#G, C^*(M;\mathbb{Q})\#G) \cong H_*(P_G M^{-TM};\mathbb{Q})^G.$$

Therefore, when M is simply connected, it is natural to consider the ring

$$HH^*(C^*(M)\#G, C^*(M)\#G)$$

as the **orbifold string topology ring with integer coefficients.**

3.3. *Orbifold string topology pro-spectrum*

We would like to have a topological construction associated to the Hochschild cohomology of $C^*(M)\#G$. Since

$$\mathrm{Ext}^*_{\mathbb{Z}G}(\mathbb{Z}, C_*(P_G M^{-TM})) \cong H^*\left(\left(C^*(EG)\otimes C_*(P_G M^{-TM})\right)^G\right)$$

has a mixture of chain with cochain complexes, we need to introduce some sort of Poincare duality for EG.

Let $EG_1 \subset \cdots \subset EG_n \subset EG_{n+1} \subset \cdots EG$ be a finite dimensional closed manifold approximation of the universal G-principal bundle $EG \to BG$. Consider the maps

$$P_G M \times_G EG_n \overset{e_0}{\to} M \times_G EG_n$$

where by abuse of notation we denote the evaluation at the time t of a pair $(f,\lambda) \in P_G M \times_G EG_n$ also by e_t. Let us call $M_n = M \times_G EG_n$.

Take the Thom spectra formed by pulling back the negative of the tangent bundle of M_n under e_0

$$(P_G M \times_G EG_n)^{-e_0^* TM_n}.$$

The inclusions $EG_n \subset EG_{n+1} \subset \cdots EG$ induce an inverse system of maps

$$(P_G M \times_G EG_n)^{-e_0^* T(M_n)} \leftarrow (P_G M \times_G EG_{n+1})^{-e_0^* T(M_{n+1})} \leftarrow \cdots$$

that defines a pro-ring spectrum called $\mathbb{L}(M \times_G EG)^{-T(M\times_G EG)}$. The homology of this pro-spectrum

$$H_*^{pro}(\mathbb{L}(M\times_G EG)^{-T(M\times_G EG)}) := \lim_{\leftarrow n} H_*((P_G M\times_G EG_n)^{-e_0^* T(M\times_G EG_n)})$$

is the orbifold string topology ring of the orbifold $[M/G]$ that in the simply connected case recovers the Hochschild cohomology of $C^*(M)\#G$.

Theorem 3.5 ([3]). *If M is simply connected and G is finite*

$$HH^*(C^*(M)\#G, C^*(M)\#G) \cong H_*^{\text{pro}}\left(\mathbb{L}(M \times_G EG)^{-T(M\times_G EG)}\right).$$

4. String topology of point orbifolds

When $M = *$ the orbifold string topology of $[*/G]$ is precisely the pro-ring spectrum LBG^{-TBG} which has been studied in,[10][11] and[12] for general compact Lie groups G. In the case of a finite group G, the loop groupoid $\mathsf{L}[*/G]$ is the groupoid $[G^{ad}/G]$ that is the translation groupoid associated to the action of G on itself by conjugation, since G is a discrete group we have an equivalence of groupoids

$$[G^{ad}/G] \cong \bigsqcup_{(g)} [*/C(g)]$$

where the union runs over the conjugacy classes of elements in G and $C(g)$ is the centralizer of g in G. Rationally, since al centralizers are finite, we have $H_*(P_G M^{-TM}; \mathbb{Q})^G \cong \mathbb{Q}G$ concentraded in degree zero and where the ring product described in section 3.1 is just the product of the group ring.

Approximate EG by $EG_1 \subset \cdots \subset EG_n \subset EG_{n+1} \subset \cdots EG$ finite dimensional closed manifolds. By definition $BG_n := EG_n/G$ and the pro-ring spectrum LBG^{-TBG} is therefore the inverse system formed by the spaces $\left(G^{ad} \times_G EG_n\right)^{-TBG_n}$.

The space $G^{ad} \times_G EG_n$ decomposes into a disjoint union of spaces $\bigsqcup_{(g)} EG_n/C(g)$ and the tangent space $T\left(G^{ad} \times_G EG_n\right)$ is isomorphic to $\bigsqcup_{(g)} T\left(EG_n/C(g)\right)$. Since any model of EG is also a model of EH for any subgroup $H \subseteq G$ we can take $EG_1 \subset \cdots \subset EG_n \subset EG_{n+1} \subset \cdots$ to be an approximation of each $EC(g)$. Therefore we have that the pro-ring spectrum LBG^{-TBG} is formed by the spaces

$$\bigsqcup_{(g)} BC(g)_n^{-TBC(g)_n}$$

The homology of LBG^{-TBG} is therefore the inverse limit of $\bigoplus_{(g)} H_*\left((BC(g)_n)^{-TBC(g)_n}\right)$, which is by Atiyah Duality the inverse system of $\bigoplus_{(g)} H^*(BC(g)_n)$, that is just $\bigoplus_{(g)} H^*(BC(g))$. To get the ring structure we have to understand the map induced by the correspondance:

$$EG_n/C(g) \times EG_n/C(h) \xleftarrow{\ \Delta\ } EG_n/(C(g)\cap C(h)) \longrightarrow EG_n/C(gh)$$

First the Thom-Pontrjagin construction applied to Δ

$$H_*(EG_n/C(g) \times EG_n/C(h)) \to H_{*-k_n}(EG_n/(C(g) \cap C(h)))$$

with $k_n = \dim(EG_n)$, is Poincaré dual to the pull-back map in cohomology. Second, the natural map in homology

$$H_*(EG_n/(C(g) \cap C(h))) \to H_*(EG_n/C(gh))$$

is Poincaré dual to the push-foward map in cohomology

$$H^*(EG_n/(C(g) \cap C(h))) \to H^*(EG_n/C(gh))$$

that defines the induction map

$$H^*(B(C(g) \cap C(h))) \to H^*(BC(gh)).$$

Therefore the ring structure in $H_*^{\mathrm{pro}}(\mathbb{L}BG^{-TBG})$ is obtained by taking classes in $H^*(BC(g))$ and $H^*(BC(h))$, pulling them back to $H^*(B(C(g) \cap C(h)))$ and then pushing them forward to $H^*(BC(gh))$. This procedure is what is known as the pull-push formalism. For a finite abelian group, this is just $H^*(BG) \otimes \mathbb{Z}G$.

For a finite group G theorem 3.5 gives an isormorphism of rings

$$HH^*(\mathbb{Z}G, \mathbb{Z}G) \cong H_*^{\mathrm{pro}}(LBG^{-TBG}).$$

The prospectrum LBG^{-TBG} is Spanier-Whitehead dual (in the sense of Christensen and Isaksen[8]) to a spectrum LBG^{-ad}. When G is finite this is just the suspension spectrum of the space LBG and therefore we have an isomorphism of groups

$$H_*^{pro}(LBG^{-TBG}) \cong H^*(LBG).$$

There is a product on the cohomology of LBG that can be described by a coproduct at the level of spectra defined using a transfer map of the multiplication of the group on the suspension spectrum of LBG. This coproduct is described as follows:

For any topological group G, the space of free loops on the classifying space is homotopically equivalent to the Borel construction of the action of G on itself by conjugation $LBG \cong EG \times_G G$.

Furthermore when G is finite the multiplication map $\mu : G \times G \to G$ is a (trivial) covering space and defines a stable transfer map $m_! : G_+ \to G \wedge G_+$. This transfer map is equivariant and induces a map on the suspension spectrum of the Borel construction $EG \times_G G \to EG \times_G (G \times G)$.

Composing the map $EG \times_G G \to EG \times_G (G \times G)$ with the natural map $EG \times_G (G \times G) \to (EG \times_G G) \times (EG \times_G G)$ we obtain a map $EG \times_G G \to$

$(EG \times_G G) \times (EG \times_G G)$, that induces on cohomology a product which makes the isomorphism

$$H_*^{pro}(LBG^{-TBG}) \cong H^*(LBG)$$

an isomorphism of rings.

In,[6] Chateur and Menichi proved that when taken with coefficients in a field the homology of LBG is a co-unital non unital homological conformal field theory, and the cohomology $H^*(LBG)$ is a Batalin-Vilkovisky algebra structure. In particular they give another way to describe the product on $H^*(LBG)$. Consider the correspondence

$$LBG \leftarrow Map(\Theta, BG) \rightarrow LBG \times LBG$$

where Θ is the topological space obtained by glueing two circles along a common interval and the left map restricts to the outer circle, and the right map restricts to the upper and lower circle respectively.

The map $Map(\Theta, BG) \rightarrow LBG$ is a fibration with fiber ΩBG, which is homotopically equivalent to G, that is finite. For fibrations with homotopically finite fiber Dwyer[9] has defined a transfer map on homology and cohomology. Using this transfer map we obtain a map $H^*(Map(\Theta, BG)) \rightarrow H^*(LBG)$.

Composing the map $H^*(Map(\Theta, BG)) \rightarrow H^*(LBG)$ with the natural map $H^*(LBG) \otimes H^*(LBG) \rightarrow H^*(Map(\Theta, BG)$ we obtain a product

$$H^*(LBG) \otimes H^*(LBG) \rightarrow H^*(LBG)$$

that in[2] it is shown to define a ring structure isomorphic to the orbifold string topology of $[*/G]$.

References

1. A. Adem,J. Leida ,Y. Ruan. *Orbifolds and Stringy Topology*, Cambridge Press. (2007).
2. A. Angel, L. Menichi. *Homology coproduct of classifying spaces* In progress.
3. A. Angel, E Backelin and B. Uribe. *Hochschild cohomology and string topology of global quotient orbifolds* . Accepted for publication on the Journal of Topology. arXiv:1004.4427
4. Kai Behrend, Gregory Ginot, Behrang Noohi, and Ping Xu, String topology for loop stacks, C. R. Math. Acad. Sci. Paris 344 (2007), no. 4
5. M. Chas and D. Sullivan, *String topology*, Annals of Math., to appear, arXiv: mathGT/9911159.
6. Chataur, D., Menichi L. *String topology of classifying spaces*, arXiv:0801.0174.

7. R.L Cohen and J.D.S. Jones, *A homotopy theoretic realization of string topology, Math. Annalen*, vol. **324**, 773-798 (2002).
8. J. D. Christensen and D. C. Isaksen, *Duality and pro-spectra* , Algebr. Geom. Topol. **4** (2004).
9. W. G. Dwyer, *Transfer maps for brations*, Math. Proc. Cambridge Philos. Soc. 120 (1996), no. 2, 221-235.
10. Gruher K., Salvatore P. *Generalized string topology operations.*, Proc. London Math. Soc. **96** (2008), 78-106
11. Gruher K. *A duality between string topology and the fusion product in equivariant K-theory.*, Math. Res. Lett. **14** (2007), 303-313
12. Gruher K. Westerland C. *String topology prospectra and Hochschild cohomology.* Journal of Topology. *4* (2008), 837-856.
13. Lupercio E., Uribe B., *Loop groupoids, gerbes, and twisted sectors on orbifolds*, Orbifolds in mathematics and physics (Madison, 2001), 163–184, Contemp. Math., **310**, AMS.
14. Lupercio E., Uribe B., Xicotecatl M. A. *Orbifold string topology*, Geom. Topol. **12** (2008) , 2203-2247.
15. L. Menichi. *Batalin-Vilkovisky algebras and cyclic cohomology of Hopf algebras*, K-Theory 32(2004), no. 3.
16. I. Moerdijk. *Orbifolds as groupoids: an introduction*, Orbifolds in mathematics and physics (Madison, WI, 2001), Contemp. Math., **310**, Amer. Math. Soc., Providence, RI, (2002), pp. 205-222.
17. B. Noohi, *Mapping stacks of topological stacks*, J. Reine Angew. Math. 646 (2010), 117-133.
18. I. Satake. *On a generalization of the notion of manifold*, Proc. Nat. Acad. Sc., **42**, (1956), 359-363.
19. W. Thurston. *The geometry and topology of 3-manifolds*, Lectures Notes. Princeton, (1978).

GROTHENDIECK RING CLASS OF BANANA AND FLOWER GRAPHS

Pedro Morales-Almazán

Department of Mathematics
Baylor University
Waco, TX 76798-7328, USA
E-mail: Pedro_Morales@baylor.edu

We define a special type of hypersurface varieties inside $\mathbb{P}_k^{n-1}$ arising from connected planar graphs and then find their equivalence classes inside the Grothendieck ring of projective varieties. Then we find a characterization for graphs in order to define irreducible hypersurfaces in general.

Keywords: Grothendieck ring, Banana graphs, Flower graphs.

1. Introduction

This results from a short communication session during the summer school Geometric, Algebraic and Topological Methods in Quantum Field Theory held in Villa de Leyva, Colombia, during the summer of 2011.

The first three sections are mainly based on.[1] In the first section we introduce the concept of the Grothendieck ring of varieties together with some important results in this ring such as the inclusion-exclusion principle. In the second section we define a graph polynomial and a graph hypersurface and find some special properties about such polynomials. The third section is mainly concerned with the Grothendieck class of some special type of graphs, namely star, flower, polygon and banana graphs. In section four we explore graph hypersurfaces for graphs in general and find a necessary and sufficient condition for a graph to produce an irreducible graph hypersurface.

2. Grothendieck Ring of Varieties

The Grothendieck ring of varieties can be thought as a generalization of the Euler characteristic,[4] It can be thought of as the quotient of the free abelian group generated by isomorphism classes $[X]$ of k-varieties, modulo inclusion-exclusion type relations, together with the product of two classes given by the class of the product, [1,4].

We start off with a characteristic zero field k and the the category of algebraic varieties $\mathcal{V}_k$ defined over it. Then we look at the abelian group $K_0(\mathcal{V}_k)$ of isomorphism classes $[X]$ of varieties X over the field k with the relations

$$[X] - [Y] = [X \setminus Y], \tag{1}$$

where $Y \subset X$ is a closed subvariety of X. Notice that setting $X = Y$ gives that the Grothendieck class of the empty set is zero,

$$[\emptyset] = 0\,. \tag{2}$$

$K_0(\mathcal{V}_k)$ is called the *Grothendieck group of varieties* which then can be turned into a ring by defining the product of two isomorphism classes by

$$[X] \cdot [Y] = [X \times Y]. \tag{3}$$

This can be also thought of as the quotient of the free abelian group generated by the symbols $[X]$ by the relation (1) and with the product (3),[4]

The use of the Grothendieck ring of varieties is very useful when looking for invariants. We call an *additive invariant* a map $\chi : \mathcal{V}_k \to R$, with values on a commutative ring R, that satisfies

(i) $\chi(X) = \chi(Y)$ if X and Y are isomorphic
(ii) $\chi(X \setminus Y) = \chi(X) - \chi(Y)$, for $Y \subset X$ closed
(iii) $\chi(X \times Y) = \chi(X)\chi(Y)$

Thus, an additive invariant is the same as a ring homomorphism χ from the Grothendieck ring of varieties to the ring R,.[1] Some interesting additive invariants are the Euler characteristic, as mentioned before, and the Hodge polynomial,[1,4]

For the following discussion, let $\mathbb{L} = [\mathbb{A}^1_k]$ be the equivalence class in the Grothendieck ring of varieties of the one dimensional affine space $\mathbb{A}^1_k$. From the product (3) of $K_0(\mathcal{V}_k)$ we have that the Grothendieck class of the affine spaces $\mathbb{A}^n_k$ are given in terms of $\mathbb{L}$ as

$$[\mathbb{A}^n_k] = \mathbb{L}^n\,. \tag{4}$$

Also, let $\mathbb{T}$ be the class of the multiplicative group $k^\times$, which is just $\mathbb{A}^1_k$ without a point. Thus we have that

$$\mathbb{T} = [\mathbb{A}^1_k] - [\mathbb{A}^0_k] = \mathbb{L} - 1, \tag{5}$$

where 1 is the class of a point in $K_0(\mathcal{V}_k)$.

$\mathbb{L}$ and $\mathbb{T}$ are useful to find the Grothendieck class of a variety together with the inclusion-exclusion principle

Theorem 2.1 (Inclusion-Exclusion Principle). *Let X, Y be varieties over k, then*

$$[X \cup Y] = [X] + [Y] - [X \cap Y] \tag{6}$$

Proof.

Since we have that $X \setminus (X \cap Y) = (X \cup Y) \setminus Y$, then

$$[X] - [X \cap Y] = [X \setminus (X \cap Y)] = [(X \cup Y) \setminus Y] = [X \cup Y] - [Y], \tag{7}$$

and hence the result follows. □

From this we find as a corollary the classes the projective spaces,

Corollary 2.1. *We have that*

$$[\mathbb{P}^n_k] = \frac{\mathbb{L}^{n+1} - 1}{\mathbb{L} - 1}, \tag{8}$$

where the fraction is taken as a short hand notation for the corresponding summation.

Proof. *Since $\mathbb{P}^{n+1}_k \setminus \mathbb{A}^{n+1}_k \simeq \mathbb{P}^n_k$ for all $n \in \mathbb{N}$, by induction and the previous result, we have that*

$$[\mathbb{P}^n_k] = 1 + \mathbb{L} + \mathbb{L}^2 + \cdots + \mathbb{L}^n, \tag{9}$$

from which the result follows. □

3. Graphs

3.1. *Graph Hypersufaces*

In this section, let Γ be a connected planar graph with n edges and label them with the variables $t_1, t_2, \ldots, t_n$ of the polynomial ring $Z[t_1, t_2, \ldots, t_n]$. With this, define the graph polynomial associated to Γ by

$$\Psi_\Gamma(t) = \sum_{T \subset \Gamma} \prod_{e \notin E(T)} t_e, \tag{1}$$

where T runs through all spanning trees of Γ and $E(T)$ is the set of all edges of T.

For a graph with v vertices, the pigeon hole principle gives that all spanning trees have $v-1$ edges, as otherwise there would be a cycle. Hence the polynomial $\Psi_\Gamma(t)$ is a homogeneous polynomial of degree $n-v+1$ and one can define the graph hypersurface associated to the graph Γ by

$$X_\Gamma = \{t = (t_1 : t_2 : \cdots : t_n) \in \mathbb{P}_k^{n-1} | \Psi_\Gamma(t) = 0\}. \tag{2}$$

Since $\Psi_\Gamma(t)$ is homogeneous, X_Γ is well defined as a projective variety.

3.2. *Dual Graph*

Given a planar connected graph Γ, define the dual graph $\Gamma^\vee$ by the following:

(i) embed Γ in $\mathbb{S}^2$
(ii) for each region defined by Γ on $\mathbb{S}^2$ assign a vertex of $\Gamma^\vee$
(iii) if two regions share an edge, join the corresponding vertices on $\Gamma^\vee$

This definition depends on the particular embedding used. Different embeddings of the same graph Γ might lead to different dual graphs $\Gamma^\vee$, but the resulting graph polynomials $\Psi_{\Gamma^\vee}(t)$ are the same up to relabeling. This fact follows from the relation between the graph polynomial of Γ and the graph polynomial of the dual $\Gamma^\vee$ via the Cremona transformation described below,[1,2] This makes the graph hypersurface of the dual graph $X_{\Gamma^\vee}$ a well defined object.

3.3. *Cremona Transformation*

The Cremona transformation on $\mathbb{P}_k^{n-1}$ is given by

$$\mathcal{C} : \mathbb{P}_k^{n-1} \to \mathbb{P}_k^{n-1} \tag{3}$$

$$(t_1 : t_2 : \cdots : t_n) \mapsto \left(\frac{1}{t_1} : \frac{1}{t_2} : \cdots : \frac{1}{t_n}\right) \tag{4}$$

which is well defined outside the coordinate hyperplanes

$$\Sigma_n = \left\{(t_1 : t_2 : \cdots : t_n) \in \mathbb{P}_k^{n-1} | \prod_{i=1}^{n} t_i = 0\right\}. \tag{5}$$

This Cremona transform is useful to relate the graph hypersurfaces of a graph Γ and its dual $\Gamma^\vee$. For this, consider the graph of the Cremona transform in $\mathbb{P}_k^{n-1} \times \mathbb{P}_k^{n-1}$ and define $\mathcal{G}(\mathcal{C})$ to be its closure. Then we have the two projections π_1 and π_2 of $\mathcal{G}(\mathcal{C})$ into $\mathbb{P}_k^{n-1} \times \mathbb{P}_k^{n-1}$ related by

$$\mathcal{C} \circ \pi_1 = \pi_2 \,. \tag{6}$$

$\mathcal{G}(\mathcal{C})$ is then a closed subvariety of $\mathbb{P}_k^{n-1} \times \mathbb{P}_k^{n-1}$ with equations (see,[1] Lemma 1.1)

$$t_1 t_{n+1} = t_2 t_{n+2} = \cdots = t_n t_{2n} \,, \tag{7}$$

where we used the coordinates $(t_1 : t_2 : \cdots : t_{2n})$ for $\mathbb{P}_k^{n-1} \times \mathbb{P}_k^{n-1}$.

Now, we have a relation between the graph polynomials of a graph Γ and its dual $\Gamma^\vee$ given by

Proposition 3.1. *Let Γ be a graph with n edges. Then the graph polynomials of Γ and its dual $\Gamma^\vee$ are related by*

$$\Psi_\Gamma(t) = \left(\prod_{i=1}^{n} t_i\right) \Psi_{\Gamma^\vee}\left(\frac{1}{t}\right) , \tag{8}$$

and hence the corresponding graph hypersurfaces of Γ and $\Gamma^\vee$ are related via the Cremona transformation by

$$\mathcal{C}\left(X_\Gamma \cap \left(\mathbb{P}_k^{n-1} \setminus \Sigma_n\right)\right) = X_{\Gamma^\vee} \cap \left(\mathbb{P}_k^{n-1} \setminus \Sigma_n\right) \tag{9}$$

Proof. *The proof follows from the definition of the graph polynomial and the combinatorial properties between Γ and $\Gamma^\vee$ (see,[1] Lemma 1.3).* □

This result also gives an isomorphism between the graph hypersurfaces of Γ and $\Gamma^\vee$ away from the coordinate hyperplanes which can be summarized in the following

Corollary 3.1. *Given a graph Γ, we have that the graph hypersurface of its dual is*

$$X_{\Gamma^\vee} = \pi_2(\pi_1^{-1}(X_\Gamma)) \,, \tag{10}$$

where π_i are the projections given on (6). Also, we have that the Cremona transform gives a (biregular) isomorphism

$$\mathcal{C} : X_\Gamma \setminus \Sigma_n \to X_{\Gamma^\vee} \setminus \Sigma_n \,, \tag{11}$$

and the projection $\pi_2 : \mathcal{G}(\mathcal{C}) \to \mathbb{P}_k^{n-1}$ restricts to an isomorphism

$$\pi_2 : \pi_1^{-1}\left(X_\Gamma \setminus \Sigma_n\right) \to X_{\Gamma^\vee} \setminus \Sigma_n \,. \tag{12}$$

Proof. *The result follows from Proposition 3.1 and the geometric properties of $\mathcal{G}(C)$ (see,[1] Lemma 1.2, Lemma 1.3, Corollary 1.4).* □

4. Banana and Flower graph hypersurfaces

4.1. *Flower graphs*

Let us start with the simplest of the graphs that define a graph hypersurface. Let Γ be a star graph with n edges. Then

$$\Psi_\Gamma(t) = 1\,, \tag{1}$$

and hence

$$X_\Gamma = \emptyset\,. \tag{2}$$

On the other hand, the dual graph $\Gamma^\vee$ is a flower graph consisting with only one vertex and n loops. Therefore the graph polynomial associated to $\Gamma^\vee$ is given by

$$\Psi_{\Gamma^\vee}(t) = t_1 t_2 \cdots t_n\,, \tag{3}$$

which shows that the graph hypersurface is

$$X_{\Gamma^\vee} = \{t \in \mathbb{P}_k^{n-1} | \Psi_{\Gamma^\vee}(t) = 0\} = \Sigma_n\,. \tag{4}$$

Therefore, by means of Corollary 3.1, we have an isomorphism between the graph hypersurfaces of the star graph and the flower graph away from the coordinate hyperplanes, which is indeed the case,

$$X_\Gamma \setminus \Sigma_n = \emptyset = X_{\Gamma^\vee} \setminus \Sigma_n\,. \tag{5}$$

From (2) we have that the Grothendieck class of X_Γ is

$$[X_\Gamma] = 0\,. \tag{6}$$

Now, for $X_{\Gamma^\vee}$, recall that from (5), we have that the complement of Σ_n consists of all tuples where all the elements are different from zero, which is a copy of $(k^\times)^{n-1}$ and gives the class of $\mathbb{T}^{n-1}$.

Proposition 4.1. *The Grothendieck class of Σ_n is given by*

$$[X_{\Gamma^\vee}] = [\Sigma_n] = \frac{(1+\mathbb{T})^n - 1 - \mathbb{T}^n}{\mathbb{T}} = \sum_{i=1}^{n-1} \binom{n}{i} \mathbb{T}^{n-1-i}\,. \tag{7}$$

Proof. *Since $X_{\Gamma^\vee} = \Sigma_n = \mathbb{P}_k^{n-1} \setminus (k^\times)^{n-1}$, the result follows from the definition of Grothendieck class, the relations given in (8) and (5) of section 2.* □

4.2. *Banana graphs*

The next interesting type of graphs that give rise to graph hypersurfaces are the polygons. Let Γ be a polygon with n edges, i.e. a graph with n edges, n vertices in which each vertex has degree 2. Here, the graph polynomial is then given by

$$\Psi_\Gamma(t) = t_1 + t_2 + \cdots + t_n \,, \tag{8}$$

and hence the graph hypersurface is the hyperplane

$$X_\Gamma = \{t \in \mathbb{P}_k^{n-1} | t_1 + t_2 + \cdots + t_n = 0\} =: \mathcal{L} \,. \tag{9}$$

Notice that choosing any $n-1$ points for $t_1, t_2, \ldots, t_{n-1}$ in $\mathcal{L}$ gives a unique value for t_n, and hence we can identify $\mathcal{L}$ with $\mathbb{P}_k^{n-2}$. Thus we find the Groethendieck class of $\mathcal{L}$ to be given by

$$[\mathcal{L}] = [X_\Gamma] = [\mathbb{P}_k^{n-2}] = \frac{\mathbb{L}^{n-1} - 1}{\mathbb{L} - 1} = \frac{(\mathbb{T}+1)^{n-1} - 1}{\mathbb{T}} \,. \tag{10}$$

The dual graph for Γ is a banana graph, which consists of 2 vertices of degree n each, and n edges joining them. The corresponding graph polynomial is then given by the $(n-1)$–th elementary symmetrical polynomial in t_i,

$$\Psi_{\Gamma^\vee}(t) = t_2 t_3 \cdots t_n + t_1 t_3 \cdots t_n + \cdots + t_1 t_2 \cdots \hat{t_i} \cdots t_n + \cdots + t_1 t_2 \cdots t_{n-1} \,, \tag{11}$$

where $\hat{t_i}$ means that the variable t_i is omitted from the product. In order to find its Grothendieck class, we first use the isomorphism between $X_\Gamma \setminus \Sigma_n$ and $X_{\Gamma^\vee} \setminus \Sigma_n$ to have

$$[X_\Gamma \setminus \Sigma_n] = [X_{\Gamma^\vee} \setminus \Sigma_n] \,. \tag{12}$$

From this, we can find the class of $X_\Gamma \setminus \Sigma$ by studying $\mathcal{L} \cap \Sigma_n$. First, we need a result that tells us how to find the Grothendieck class of a hyperplane section of a given class in the Grothendieck ring.

Lemma 4.1. *Let C be a class in the Grothendieck ring that can be written as a function of the torus class $\mathbb{T}$ by means of a polynomial expression $C = g(\mathbb{T})$. Then the transformation*

$$\mathcal{H} : g(\mathbb{T}) \mapsto \frac{g(\mathbb{T}) - g(-1)}{\mathbb{T}+1} \tag{13}$$

gives an operation on the set of classes in the Grothendieck ring that are polynomial functions of the torus class $\mathbb{T}$ that can be interpreted as taking a hyperplane section.

Proof. *Notice that for* $g(\mathbb{T}) = [\mathbb{P}^n] = \frac{(\mathbb{T}-1)^n - 1}{\mathbb{T}}$, *equation (13) gives*

$$\frac{g(\mathbb{T} - g(-1))}{\mathbb{T}+1} = \frac{(\mathbb{T}-1)^{n-1} - 1}{\mathbb{T}} = [\mathbb{P}^{n-1}], \tag{14}$$

which effectively is the same as taking a hyperplane section. Since $\mathcal{H}$ *is linear in* g, *it works for any* g. □

Since $[\Sigma_n] = \frac{(\mathbb{T}+1)^n - 1 - \mathbb{T}^n}{\mathbb{T}} = g(\mathbb{T})$, using the previous result gives that

$$[\mathcal{L} \cap \Sigma_n] = \frac{g(\mathbb{T}) - g(-1)}{\mathbb{T}+1} = \frac{(1+\mathbb{T})^{n-1} - 1}{\mathbb{T}} - \frac{\mathbb{T}^{n-1} - (-1)^{n-1}}{\mathbb{T}+1}, \tag{15}$$

and with this, we find that

$$[X_{\Gamma^\vee} \setminus \Sigma_n] = [\mathcal{L} \setminus \Sigma_n] = [\mathcal{L}] - [\mathcal{L} \cap \Sigma_n] = \frac{\mathbb{T}^{n-1} - (-1)^{n-1}}{\mathbb{T}+1}. \tag{16}$$

On the other hand, consider the variety $\mathcal{S}_n$ generated by the ideal

$$\mathcal{I} = (t_2 t_3 \cdots t_n, t_1 t_3 \cdots t_n, \ldots, t_1 t_2 \cdots \hat{t_i} \cdots t_n, \ldots, t_1 t_2 \cdots t_{n-1}). \tag{17}$$

This is the singularity subvariety of the divisor of singular normal crossings Σ_n given by the union of coordinate hyperplanes (see[1] Section 1.3).

We can find the Grothendieck class of $\mathcal{S}_n$ by

Lemma 4.2. *The class of* $\mathcal{S}_n$ *is given by*

$$[\mathcal{S}_n] = [\Sigma_n] - n\mathbb{T}^{n-2} = \sum_{i=2}^{n-1} \binom{n}{i} \mathbb{T}^{n-1-i}. \tag{18}$$

Proof. *Each coordinate hyperplane* $\mathbb{P}_k^{n-2}$ *in* Σ_n *intersects the others along the union of its coordinate hyperplanes* Σ_{n-1}. *Thus, to obtain the class of* $\mathcal{S}_n$ *from the class of* Σ_n, *we just need to subtract the class of* n *complements of* Σ_{n-1} *in the* n *components of* Σ_n, *from which the result follows.* □

Finally, this results lead to the Grothendieck class of the graph hypersurface of the banana graph by

Theorem 4.3. *The class of the graph hypersurface of the banana graph is given by*

$$[X_{\Gamma^\vee}] = \frac{(\mathbb{T}+1)^n - 1}{\mathbb{T}} - \frac{\mathbb{T}^n - (-1)^n}{\mathbb{T}} - n\mathbb{T}^{n-2}. \tag{19}$$

Proof. *Writing* $[X_{\Gamma^\vee}]$ *as*

$$[X_{\Gamma^\vee}] = [X_{\Gamma^\vee} \setminus \Sigma] + [X_{\Gamma^\vee} \cap \Sigma] = [X_{\Gamma^\vee} \setminus \Sigma] + [\mathcal{S}_n]\,, \tag{20}$$

and using the previous results for $[X_{\Gamma^\vee} \setminus \Sigma]$ *and* $[\mathcal{S}_n]$ *we find the desired result.* □

5. Irreducible Graph Hypersurfaces

To start, notice that if a graph Γ produces a reducible graph hypersurface X_Γ so will its dual graph $\Gamma^\vee$.

Theorem 5.1. X_Γ *is irreducible if and only if* $X_{\Gamma^\vee}$ *is irreducible.*

Proof. *This result follows directly from Proposition 3.1.* □

Even though every planar connected graph gives rise to a graph hypersurface, not every graph produces an irreducible graph hypersurface. To see this, consider a planar connected graph Γ and following[3] define a *separation* to be a decomposition of the graph into two subgraphs which share only one vertex. This common vertex is called a *separating vertex.*

With this, we can give a relation between separating vertices and irreducibility of the graph hypersufaces by the following result

Theorem 5.2. *If the graph* Γ *has a separating vertex such that each of the two components contain a cycle, then the corresponding graph hypersurface* X_Γ *is reducible.*

Proof. *Let* a *be a separating vertex in* Γ *with the properties required. Consider the two associated components of the graph separately. Since the components do not share edges, the choice of edges to remove in order to achieve a spanning tree on each component is independent of one another. Thus every spanning tree of* Γ *can be obtained by spanning trees of the two components joint through* a*. Then the graph polynomial for* Γ *is the product of the graph polynomials of each component. Since each component contains a cycle, the corresponding graph polynomials are non-constant, and hence* X_Γ *is reducible.* □

The concept of independence in the graph is the key for reducibility. With this discussion, we can refine the previous result by means of defining separability on graphs. A connected graph that has no separating vertices is called *non-separable*, otherwise it is called *separable*. The next result, due to Whitney, will provide the key ingredient to characterize reducibility on graph hypersurfaces by means of graph theoretical properties of Γ.

Lemma 5.3. *A connected graph is non-separable if and only if any two of its edges lie on a common cycle.*

Proof. *The proof can be found in Theorem 5.2 in.*[3] *The result follows from the definition of separating vertex and properties of cycles.* □

With this result, we can characterize reducibility of graph hypersurfaces.

Theorem 5.4. *X_Γ is reducible if and only if Γ is separable and each component contains a cycle.*

Proof. *By Theorem 5.2 we have that if Γ is separable, and hence has a separating vertex, the corresponding graph polynomial $\Psi_\Gamma(t)$ is reducible.*

Now, suppose that Γ is such that $\Psi_\Gamma(t)$ is reducible. Thus

$$\Psi_\Gamma(t) = p(t)q(t) \tag{1}$$

for $p(t)$ and $q(t)$ non constant polynomials. Now, let $t_i = x$ be variable and fix all other t_j, $j \neq i$. Therefore $\Psi_\Gamma(x)$ is a linear polynomial, and hence either $p(x)$ or $q(x)$ is a constant for any choice of t_j. Without loss of generality, suppose that $p(x)$ is a constant and that $q(x)$ is a linear polynomial. Now regard $p(x)$ and $q(x)$ as polynomials in x whose coefficients depend on t_j. Since the coefficient of x in $q(x)$ is non-zero for our initial choice of t_j, by continuity on t_j it remains non-zero in a neighborhood of these initial values. Hence the coefficient of x in $p(x)$ is zero for a neighborhood of t_j, and since $p(t)$ is a polynomial, $p(x)$ must be constant for all t_j.

Therefore, we have that $p(t)$ and $q(t)$ separate the variables $t_1, t_2, \ldots, t_n$, that is, if t_i is present in $p(t)$, then it is not in $q(t)$. Hence we have that $p(t)$ and $q(t)$ are homogeneous polynomials where the exponent of each t_i is at most 1.

Since $p(t)$ and $q(t)$ separate the variables t_i associated to the edges of Γ, the choice of the edges appearing in $p(t)$ is independent of the choice of the edges appearing in $q(t)$ in order to break cycles and obtain a spanning tree *of* Γ. *Hence the edges appearing in* $p(t)$ *do not lie in any cycle in which are any of the edges that appear in* $q(t)$.

Let

$$P = \{t_i | \text{ such that } t_i \text{ appears in } p(t)\} \tag{2}$$

and

$$Q = \{t_i | \text{ such that } t_i \text{ appears in } q(t)\}\,. \tag{3}$$

Since $P \cap Q = \emptyset$, $P \cup Q$ *are all the variables appearing in* Ψ_Γ, *and no cycle in* Γ *contains both elements of* P *and* Q, *we have that, by Lemma 5.3,* Γ *is separable.* □

References

1. Aluffi, P., Marcolli, M., *Feymann motives of banana graphs*, Comm. in Number Theory and Physics (2009) 1-57.
2. S. Bloch, *Motives associated to graphs*, Japan J. Math., Vol.2 (2007) 165196.
3. Bondy, A., Murty, U.S.R., *Graph Theory*, Springer, (2008).
4. Sahasrabudhe, N., *Grothendieck Ring of Varieties*, Master Thesis, ALGANT Erasmus Mundus, Bordeaux, 2008.

ON THE GEOMETRY UNDERLYING A REAL LIE ALGEBRA REPRESENTATION

Rodrigo Vargas Le-Bert*

Instituto de Matemática y Física
Universidad de Talca
Casilla 747, Talca, Chile
E-mail: rvargas@inst-mat.utalca.cl

Let G be a real Lie group with Lie algebra $\tilde{m}g$. Given a unitary representation π of G, one obtains by differentiation a representation $\mathbb{D}\pi$ of $\tilde{m}g$ by unbounded, skew-adjoint operators. Representations of $\tilde{m}g$ admitting such a description are called *integrable,* and any cyclic vector ν enables one to see them as the action of $\tilde{m}g$ by derivations on the algebra of representative functions $g \mapsto \langle \xi, \pi(g)\eta \rangle$, which are naturally defined on the homogeneous space $M = G/H$, where H is the stabilizer of ν. In other words, integrable and cyclic representations of a real Lie algebra can always be seen as realizations of that algebra by vector fields on a homogeneous manifold. Here we show how to use the coproduct of the universal enveloping algebra $\mathcal{E}(\tilde{m}g)$ to generalize this to representations which are not necessarily integrable. This provides a first step towards a geometric approach to integrability questions regarding Lie algebra representations.

2010 MSC: 16W25, 17B15.

Keywords: Lie algebras, representations, integrability.

1. Introduction

Let G be the simply connected, real Lie group with Lie algebra $\tilde{m}g$. Recall that the *group algebra* $\mathbb{C}[G]$ is the involutive algebra consisting of finite, complex linear combinations of elements of G with product and involution obtained by linearly extending the maps $g \otimes h \mapsto gh$ and $g \mapsto g^{-1}$, respectively. We can also define a complex conjugation on $\mathbb{C}[G]$ by anti-linearly extending $\overline{g} = g$. The infinitesimal version of $\mathbb{C}[G]$ is the *universal enveloping algebra* $\mathcal{E} = \mathcal{E}(\tilde{m}g)$, defined as the quotient of the complex tensor

*Supported by Fondecyt Postdoctoral Grant N°3110045.

algebra $\mathcal{T}(\tilde{mg}) = \left(\bigoplus_{n\in\mathbb{N}} \tilde{mg}^{\otimes n}\right) \otimes \mathbb{C}$ by the ideal generated by the elements

$$XY - YX - [X, Y], \quad X, Y \in \tilde{mg}.$$

It comes equipped with the involution and complex conjugation obtained by linear extension from $(X_1 \cdots X_n)^\dagger = (-1)^n X_n \cdots X_1$, and anti-linear extension from $\overline{X_1 \cdots X_n} = X_1 \cdots X_n$, respectively. The subalgebra of real elements with respect to this conjugation will be denoted by $\mathcal{E}_\mathbb{R} = \mathcal{E}_\mathbb{R}(\tilde{mg})$.

1.1. *Integrable representations*

Consider a representation π of G by unitary operators on a Hilbert space $\mathcal{H}$. Given $g \in G$ and $\xi \in \mathcal{H}$, it is sometimes convenient to write $g\xi$ instead of $\pi(g)\xi$, and consequently we can also think of $\mathcal{H}$ as a $\mathbb{C}[G]$-module. In what follows we will use interchangeably the languages of representation and module theory.

We want to understand $\mathcal{H}$ as a submodule of $C(G)$, which is naturally a $\mathbb{C}[G]$-module via

$$gf(x) = f(g^{-1}x), \quad f \in C(M).$$

In order to do so, suppose that there is a distinguished cyclic vector $\nu \in \mathcal{H}$, i.e. $\mathcal{H} = \overline{\mathbb{C}[G]\nu}$, and an anti-unitary operator $J : \mathcal{H} \to \mathcal{H}$ whose action commutes with that of G. Then, to each $\xi \in \mathcal{H}$, associate the representative function

$$e_\xi : g \in G \mapsto \langle J\xi, g\nu \rangle \in \mathbb{C}.$$

Since $ge_\xi = e_{g\xi}$, we see that $\xi \mapsto e_\xi$ exposes $\mathcal{H}$ as a submodule of $C(G)$.

Remark 1.1. The anti-unitary J is needed in order for $\xi \mapsto e_\xi$ to be linear. Its use can be easily avoided in several ways, such as for instance working with the *anti-linear* $\mathbb{C}[G]$-module associated to π, or considering the anti-holomorphic "representative functions" $g \mapsto \langle g\nu, \xi \rangle$. Instead of using any such non-standard convention, we have decided to restrict ourselves to modules admitting an anti-unitary J as above.

Consider, then, the subalgebra of $C(G)$ generated by $\{: e_\xi \mid \xi \in \mathcal{H} : \}$. It does not separate points of G, because each e_ξ is constant along the cosets gH, where $H \subseteq G$ is the stabilizer of ν. Thus, it is essentially an algebra of functions on $M = G/H$, and this homogeneous space is, in that sense, the geometry underlying the representation π.

Now, let $\mathcal{H}^\infty = \mathcal{H}^\infty(\pi) \subseteq \mathcal{H}$ be the subspace of *smooth vectors* for π, i.e. those $\xi \in \mathcal{H}$ such that the map $g \in G \mapsto \pi(g)\xi \in \mathcal{H}$ is smooth.[a] Then, given $X \in \tilde{m}g$ and $\xi \in \mathcal{H}^\infty$, we can define

$$\mathbb{D}\pi(X)\xi = X\xi = \lim_{t\to 0} \frac{1}{t}\big(\exp(tX)\xi - \xi\big),$$

where $\exp : \tilde{m}g \to G$ is the exponential map. This way, the unitary representation π of G gives rise to a representation $\mathbb{D}\pi$ of $\tilde{m}g$ by unbounded, skew-adjoint operators, thus naturally turning $\mathcal{H}^\infty$ into an $\mathcal{E}$-module. We will also say, under this circumstances, that $\mathcal{H}$ is an *unbounded* $\mathcal{E}$-module. As above, we can expose $\mathcal{H}^\infty$ as a submodule of $C^\infty(M)$, by identifying it with $\{\, e_\xi \mid \xi \in \mathcal{H}^\infty \,\}$. Now, the action of $\tilde{m}g$ is by derivation along the vector fields which generate the flows $gH \in M \mapsto \epsilon^{tX} gH \in M$.

It will be helpful to be more precise regarding the notion of representation by unbounded operators, which we do in the next subsection. For the time being, however, we anticipate the following definition.

Definition 1.1. A representation of $\tilde{m}g$ is *integrable* if it is of the form $X \mapsto \mathbb{D}\pi(X)|_{\mathcal{F}}$, where π is a unitary representation of G, and $\mathcal{F} \subseteq \mathcal{H}^\infty(\pi)$ is "telling enough," in the sense that the closure of $\mathbb{D}\pi(X)|_{\mathcal{F}}$ coincides with the closure of $\mathbb{D}\pi(X)$, for each $X \in \tilde{m}g$.

Remark 1.2. Recall that an unbounded operator is *closed* if its graph is closed, and is *closable* if the closure of its graph is the graph of another operator, called its *closure.* An operator is closable if, and only if, its adjoint is densely defined; thus, the operators $\mathbb{D}\pi(X)$ and $\mathbb{D}\pi(X)|_{\mathcal{F}}$ above are closable.

1.2. *Unbounded operator algebras*

For a complete exposition of the material in this and the next subsection see.[7] Here we follow.[8]

Let $\mathcal{F}$ be a complex vector space sitting densely inside a Hilbert space $\mathcal{H}$.

[a] $\mathcal{H}^\infty$ is also called the *Gårding space* of π. As Harish-Chandra pointed out,[2] the space of *analytic vectors* (see also[6,7]) is more adequate in studying representations of G. By contrast, when studying non-integrable representations of $\tilde{m}g$ the right domain is $\mathcal{H}^\infty$ (representations having a dense subspace of analytic vectors are automatically integrable,[3] even if $\tilde{m}g$ is infinite dimensional[5]).

Definition 1.2. We write $\mathcal{L}^\dagger(\mathcal{F})$ for the algebra of linear operators $A : \mathcal{F} \to \mathcal{F}$ such that $\mathcal{F} \subseteq \mathrm{dom}(A^*)$. It comes equipped with the involution $A^\dagger = A^*|_\mathcal{F}$.

Definition 1.3. We write $\mathcal{C}^*(\mathcal{F})$ for the set of closed operators A on $\mathcal{H}$ which satisfy $A|_\mathcal{F} \in \mathcal{L}^\dagger(\mathcal{F})$.

Analytically speaking, closed operators behave better than arbitrary ones, but this is not so from an algebraic viewpoint. Indeed, given two closed, densely defined operators A and B on $\mathcal{H}$, we define—whenever it makes sense—their *strong sum* by

$$A + B = \text{the closure of } A|_\mathcal{D} + B|_\mathcal{D}, \quad \mathcal{D} = \mathrm{dom}\, A \cap \mathrm{dom}\, B.$$

Analogously, we define their *strong product* by

$$AB = \text{the closure of } AB|_\mathcal{D}, \quad \mathcal{D} = B^{-1}\, \mathrm{dom}\, A.$$

As it turns out, strong operations are always well-defined for elements of $\mathcal{C}^*(\mathcal{F})$, but the axioms of associativity and distributivity might well not hold.

Definition 1.4. An *unbounded representation* of a complex involutive algebra $\mathcal{A}$ is an involutive algebra morphism $\pi : \mathcal{A} \to \mathcal{L}^\dagger(\mathcal{F})$, where $\mathcal{F}$ sits densely in a Hilbert space $\mathcal{H}$. Under this circumstances, we also say that $\mathcal{H}$ is an unbounded $\mathcal{A}$-module. Finally, we say that π is *cyclic* if there exists a so-called *cyclic vector* $\nu \in \mathcal{F}$ such that $\mathcal{F} = \mathcal{A}\nu$.

Recall that there is a natural order defined on any involutive algebra $\mathcal{A}$: the one with positive cone generated by the elements of the form $A^\dagger A$. This, in turn, induces an order on the algebraic dual of $\mathcal{A}$, which will be written $\mathcal{A}^\dagger$: an $\omega \in \mathcal{A}^\dagger$ is positive if, and only if, it is positive on positive elements. We say that such an ω is a *state* if, besides being positive, it is normalized by $\omega(1) = 1$. We have introduced all this notions just to recall that there is a correspondence between cyclic representations and states. Given a representation $\pi : \mathcal{A} \to \mathcal{L}^\dagger(\mathcal{F})$, one obtains a state $\omega \in \mathcal{A}^\dagger$ from any vector $\nu \in \mathcal{F}$ with $\|\nu\| = 1$ by the formula

$$\omega(A) = \langle \nu, A\nu \rangle, \quad A \in \mathcal{A}.$$

The GNS construction allows for a recovery of both $\pi|_{\mathcal{A}\nu}$ (modulo unitary conjugation) and ν (modulo a phase factor) from ω, thus establishing the aforementioned correspondence. We proceed to describe it briefly.

Let ω be a state of $\mathcal{A}$ and consider the set $\mathcal{I} = \{ A \in \mathcal{A} \mid \omega(A^\dagger A) = 0 \}$. Using the Cauchy-Schwarz inequality it is easily seen that $\mathcal{I}$ is a left ideal. The quotient $\mathcal{F} = \mathcal{A}/\mathcal{I}$ is densely embedded in its Hilbert space completion $\mathcal{H}$ with respect to the scalar product

$$\langle [A], [B] \rangle = \omega(A^\dagger B), \quad A, B \in \mathcal{A},$$

where $[\cdot]$ denotes the equivalence class in $\mathcal{F}$ of its argument. The GNS representation π is simply given by the canonical left $\mathcal{A}$-module structure of $\mathcal{F}$. It admits the cyclic vector $\nu = [1] \in \mathcal{F}$.

1.3. *Non-integrable representations*

As we saw above, integrable $\mathcal{E}$-modules are, essentially, spaces of smooth functions on homogeneous manifolds. Our objective in this paper is to work out the analogous geometric point of view for non-integrable representations. Thus, it is pertinent to discuss briefly the phenomenon of non-integrability.

From an operator algebraic point of view, integrability is just good behaviour, in the following sense. Consider a representation $\pi : \mathcal{E} \to \mathcal{L}^\dagger(\mathcal{F})$. Operators in $\mathcal{L}^\dagger(\mathcal{F})$ are closable, because their adjoints are densely defined. Denote the closure of (the image under π of) $E \in \mathcal{E}$ by $cl(E)$. Then, integrability of π is equivalent to:

(i) $cl(E + F) = cl(E) + cl(F)$,
(ii) $cl(EF) = cl(E)cl(F)$,
(iii) $cl(E^\dagger) = cl(E)^*$,

where on the right hand sides strong operations are meant (the last property actually implies the first two). Hence, integrability implies that $cl(1 + E^\dagger E)$ is invertible, for all $E \in \mathcal{E}$. Let us mention, by the way, that this relationship between integrability and invertibility of "uniformly positive" elements is deep: it can be shown,[8] whenever

$$\mathcal{S} = \text{the multiplicative set generated by } \{ : 1 + E^\dagger E \mid E \in \mathcal{E} : \}$$

satisfies the Ore condition, that integrable representations are, modulo trivial domain modifications, in bijection with representations of the localization $\mathcal{E}\mathcal{S}^{-1}$.

An important question is whether a given representation admits an integrable extension. One can extend the algebra, the module, or both. Thus, the integrable extension problem in its full generality consists in completing the following diagram:

$$\begin{array}{ccc} \tilde{\mathcal{E}} & \dashrightarrow & \mathcal{C}^*(\tilde{\mathcal{F}}) \\ \downarrow & & \downarrow \\ \mathcal{E} & \longrightarrow & \mathcal{L}^\dagger(\mathcal{F}), \end{array}$$

with $\tilde{\mathcal{E}} \supseteq \mathcal{E}$ and $\tilde{\mathcal{F}} \supseteq \mathcal{F}$. The image of $\tilde{\mathcal{E}}$ under the upper arrow must be a subset of $\mathcal{C}^*(\tilde{\mathcal{F}})$ which is an algebra with respect to the strong operations; all arrows must be involutive algebra morphisms; and the diagram must commute. We do not necessarily require that $\tilde{\mathcal{F}} \subseteq \mathcal{H}$. Again under the hypothesis that the set $\mathcal{S}$ above satisfies the Ore condition, cyclic representations always admit an integrable extension with $\tilde{\mathcal{E}} = \mathcal{E}$, see.[8]

2. The construction of M

2.1. *Algebraic structure of* $\mathcal{E}^\dagger$

In this subsection we set up the needed algebraic preliminaries regarding the dual $\mathcal{E}^\dagger$ of the enveloping algebra $\mathcal{E} = \mathcal{E}(\tilde{m}g)$. A full exposition is found in.[1] We will also make use of some coalgebra notions, for which we mention.[4]

The universal enveloping algebra $\mathcal{E}$ is not only an algebra: it comes naturally endowed with a coproduct, too. Recall that coproducts are dual to products, so that a coproduct is a linear function $\Delta : \mathcal{E} \to \mathcal{E} \otimes \mathcal{E}$ which is coassociative, meaning that the following diagram commutes:

$$\begin{array}{ccc} \mathcal{E} & \xrightarrow{\Delta} & \mathcal{E} \otimes \mathcal{E} \\ \downarrow{\scriptstyle \Delta} & & \downarrow{\scriptstyle \mathrm{Id} \otimes \Delta} \\ \mathcal{E} \otimes \mathcal{E} & \xrightarrow{\Delta \otimes \mathrm{Id}} & \mathcal{E} \otimes \mathcal{E} \otimes \mathcal{E}. \end{array}$$

In order to define the coproduct of $\mathcal{E}$, we define first a coproduct Δ on the tensor algebra $\mathcal{T}(\tilde{m}g)$, by extending the linear map

$$X \in \tilde{m}g \mapsto 1 \otimes X + X \otimes 1 \in \mathcal{T}(\tilde{m}g) \otimes \mathcal{T}(\tilde{m}g)$$

to all of $\mathcal{T}(\tilde{m}g)$ as an algebra morphism (we are appealing here to the universal property of tensor algebras). Now, the ideal $\mathcal{I} \subseteq \mathcal{T}(\tilde{m}g)$ generated by the elements $XY - YX - [X, Y]$, with $X, Y \in \tilde{m}g$, is also a *coideal*, i.e.

$$\Delta\mathcal{I} \subseteq \mathcal{I} \otimes \mathcal{T}(\tilde{m}g) + \mathcal{T}(\tilde{m}g) \otimes \mathcal{I}.$$

Thus, it can be readily seen that Δ passes to the quotient $\mathcal{E} = \mathcal{T}(\tilde{m}g)/\mathcal{I}$.

It will be helpful to have an explicit formula for Δ. Let us choose a basis $(X_1, \ldots, X_n)$ of $\tilde{m}g$ and use the multi-index notation $X^\alpha = X_1^{\alpha_1} \cdots X_n^{\alpha_n}$. Using the fact that Δ is an algebra morphism together with the binomial formula, we see that

$$\Delta X^\alpha = \sum_{\beta \leq \alpha} \binom{\alpha}{\beta} X^\beta \otimes X^{\alpha - \beta},$$

where $\binom{\alpha}{\beta} = \frac{\alpha!}{\beta!(\alpha-\beta)!} = \binom{\alpha_1}{\beta_1} \cdots \binom{\alpha_d}{\beta_d}$.

For us, the importance of the coproduct resides in the fact that it enables one to turn the algebraic dual $\mathcal{E}^\dagger$ into an algebra, with product given by

$$ab(E) = a \otimes b(\Delta E), \quad a, b \in \mathcal{E}^\dagger, \ E \in \mathcal{E}.$$

Recall that Δ is cocommutative, meaning that the following diagram commutes:

$$\begin{array}{ccc} & \overset{\Delta}{\nearrow} & \mathcal{E} \otimes \mathcal{E} \\ \mathcal{E} & & \downarrow \tau \\ & \underset{\Delta}{\searrow} & \mathcal{E} \otimes \mathcal{E}, \end{array}$$

where τ is obtained by linear extension from $E \otimes F \mapsto F \otimes E$, with $E, F \in \mathcal{E}$. This implies that $\mathcal{E}^\dagger$ is a commutative complex algebra. It is naturally an $\mathcal{E}$-module, too, with

$$Ea(F) = a(\overline{E}^\dagger F), \quad E, F \in \mathcal{E}, \ a \in \mathcal{E}^\dagger. \tag{1}$$

The main properties of $\mathcal{E}^\dagger$ are given in the next two propositions.

Prop 2.1. The action (1) defines a morphism $\tilde{m}g \to der(\mathcal{E}^\dagger)$, where $der(\mathcal{E}^\dagger)$ stands for the Lie algebra of linear maps $\delta : \mathcal{E}^\dagger \to \mathcal{E}^\dagger$ such that

$$(\forall a, b \in \mathcal{E}^\dagger) \ \delta(ab) = \delta(a)b + a\delta(b).$$

Proof. Indeed, let $X \in \tilde{m}g$ and $a, b \in \mathcal{E}^\dagger$. One has that

$$X(ab) = X((a \otimes b) \circ \Delta) = (a \otimes b) \circ \Delta \circ X^\dagger(\cdot).$$

On the other hand, given $E \in \mathcal{E}$,

$$\begin{aligned} \Delta(X^\dagger E) &= \Delta(X^\dagger)\Delta(E) = (X^\dagger \otimes 1 + 1 \otimes X^\dagger)\Delta(E) \\ &= \big((X^\dagger(\cdot) \otimes \mathrm{Id} + \mathrm{Id} \otimes X^\dagger(\cdot)) \circ \Delta\big)(E). \end{aligned}$$

Thus, replacing above,

$$\begin{aligned} X(ab) &= (a \otimes b) \circ \big(X^\dagger(\cdot) \otimes \mathrm{Id} + \mathrm{Id} \otimes X^\dagger(\cdot)\big) \circ \Delta \\ &= (Xa \otimes b + a \otimes Xb) \circ \Delta = (Xa)b + a(Xb). \end{aligned}$$ □

Prop 2.2. Let $X_1, \ldots, X_n$ be a basis of $\tilde{m}g$. One has that $\mathcal{E}^\dagger \cong \mathbb{C}[\![x_1, \ldots, x_n]\!]$, with algebra isomorphism given by

$$a \in \mathcal{E}^\dagger \mapsto f_a \in \mathbb{C}[\![x_1, \ldots x_n]\!], \quad f_a(x) = \sum_{\alpha \in \mathbb{N}^n} \frac{x^\alpha}{\alpha!} a(X^\alpha).$$

Proof. Indeed, given $a, b \in \mathcal{E}^\dagger$,

$$\begin{aligned} f_{ab}(x) &= \sum_\alpha \frac{x^\alpha}{\alpha!} ab(X^\alpha) = \sum_\alpha \frac{x^\alpha}{\alpha!}(a \otimes b)(\Delta X^\alpha) \\ &= \sum_\alpha \frac{x^\alpha}{\alpha!}(a \otimes b)\left(\sum_{\beta \le \alpha} \binom{\alpha}{\beta} X^\beta \otimes X^{\alpha-\beta}\right) \\ &= \sum_\beta \sum_\gamma \frac{x^\beta x^\gamma}{\beta!\gamma!} a(X^\beta) b(X^\gamma) = f_a(x) f_b(x). \end{aligned}$$ □

Remark 2.1. Observe that

$$ab(1) = a \otimes b(\Delta 1) = a(1)b(1).$$

Thus, evaluation at $1 \in \mathcal{E}$ is a character of $\mathcal{E}^\dagger$, which we will denote by χ_0, and elements of $\mathcal{E}^\dagger$ are, essentially, formal Taylor expansions around χ_0.

The fact that $\mathcal{E}^\dagger$ is a formal power series algebra suggests endowing it with an involution, corresponding to complex conjugation of the coefficients. Explicitly, this is given by

$$a^\dagger(E + ⨍\, F) = \overline{a(E)} + ⨍\, \overline{a(F)}, \quad E, F \in \mathcal{E}_\mathbb{R}.$$

With respect to this conjugation, χ_0 becomes a *real* character:

$$\chi_0(a^\dagger) = a^\dagger(1) = \overline{a(1)} = \overline{\chi_0(a)}.$$

Prop 2.3. For all $a \in \mathcal{E}^\dagger$ and $E \in \mathcal{E}$, $(Ea)^\dagger = \overline{E}a^\dagger$.

Proof. For $E, F \in \mathcal{E}_\mathbb{R}$, one has $(Ea)^\dagger(F) = \overline{a(E^\dagger F)} = Ea^\dagger(F)$, and the general case follows immediately. □

2.2. *M as an affine scheme*

Let $\pi : \mathcal{E} \to \mathcal{L}^\dagger(\mathcal{F})$ be a representation of $\mathcal{E} = \mathcal{E}(\tilde{m}g)$ admitting a cyclic vector $\nu \in \mathcal{F}$, where $\tilde{m}g$ is a real Lie algebra and $\mathcal{F}$ is a dense subspace of a Hilbert space $\mathcal{H}$. We will suppose that there exists an anti-unitary $J : \mathcal{H} \to \mathcal{H}$ commuting with π and such that $J\nu = \nu$, even though this assumption is superfluous, see Remark 1.1. Recall that we look for some naturally defined space M together with a realization of $\tilde{m}g$ by vector fields on it, such that the corresponding coordinate algebra $\mathcal{A}$, which is an $\mathcal{E}$-module, contains $\mathcal{F}$ as a submodule. In this subsection we realize $\mathcal{A}$ as a subalgebra of $\mathcal{E}^\dagger$.

Consider the linear map

$$\xi \in \mathcal{F} \mapsto e_\xi \in \mathcal{E}^\dagger, \quad e_\xi(E) = \langle J\xi, E\nu \rangle,$$

and let $\mathcal{A}$ be the involutive subalgebra of $\mathcal{E}^\dagger$ generated by $\{\, e_\xi \mid \xi \in \mathcal{F} \,\}$. We define

$$M = \left\{ : \chi \in \mathcal{A}^\dagger \;\middle|\; \chi(ab) = \chi(a)\chi(b),\ \chi(a^\dagger) = \overline{\chi(a)} : \right\}.$$

Note that $\chi_0 \in M$.

Remark 2.2. The map $\xi \mapsto e_\xi$ is well defined all over $\mathcal{H}$. However, we regard the functions e_ξ with $\xi \in \mathcal{H} \setminus \mathcal{F}$ as not belonging to $\mathcal{A}$, for they are not smooth in the sense that, given $E \in \mathcal{E}$, Ee_ξ might not be of the form e_η, for any $\eta \in \mathcal{H}$.

Remark 2.3. The existence of pathological characters should be expected, for $\mathcal{A}$ is not necessarily finitely generated.

Prop 2.4. One has that $Ee_\xi = e_{E\xi}$, for all $\xi \in \mathcal{F}$ and $E \in \mathcal{E}$. In particular, $\mathcal{A}$ is a submodule of $\mathcal{E}^\dagger$.

Proof. Indeed, given $F \in \mathcal{E}$,

$$Ee_\xi(F) = \langle J\xi, \overline{E}^\dagger F\nu \rangle = \langle \overline{E}J\xi, F\nu \rangle = \langle JE\xi, F\nu \rangle = e_{E\xi}(F). \qquad \square$$

By Proposition 2.3, Proposition 2.4 and cyclicity, $\mathcal{A}$ is the involutive subalgebra of $\mathcal{E}^\dagger$ generated by the orbit of the state

$$\omega \in \mathcal{E}^\dagger, \quad \omega(E) = e_\nu(E) = \langle \nu, E\nu \rangle$$

under the action of $\mathcal{E}$. Thus, the fundamental object is $\omega \in \mathcal{E}^\dagger$, and the representation π can be recovered from it by GNS construction.

The affine scheme M can be quite far from being a differential manifold, but it is at least *reduced* (has no nilpotents) and *irreducible* (has exactly one minimal prime ideal), as follows from the fact that $\mathcal{E}^\dagger \cong \mathbb{C}[\![x_1, \ldots, x_n]\!]$ is an integral domain.

Consider the case

$$\tilde{mg} = \mathbb{R}X, \quad \mathcal{F} = \mathcal{H} = \mathbb{C}^2, \quad X = \begin{pmatrix} 0 & 1 \\ -1 & 0 \end{pmatrix}, \quad \nu = \begin{pmatrix} 1 \\ 0 \end{pmatrix}.$$

Recall that the elements of $\mathcal{A}$ can be seen as formal power series around χ_0, where $\chi_0(a) = a(1)$. Here, we have only the derivation X, which will be, say, with respect to the coordinate x. The power series corresponding to ω is computed as follows:

$$\begin{aligned} \omega|_{x=0} &\cong \chi_0(\omega) = \omega(1) = \langle \nu, \nu \rangle = 1, \\ \partial_x \omega|_{x=0} &\cong \chi_0(X\omega) = \langle \nu, X^\dagger \nu \rangle = \left\langle \begin{pmatrix} 1 \\ 0 \end{pmatrix}, \begin{pmatrix} 0 & -1 \\ 1 & 0 \end{pmatrix} \begin{pmatrix} 1 \\ 0 \end{pmatrix} \right\rangle = 0, \\ \partial_x^2 \omega|_{x=0} &\cong \chi_0(X^2\omega) = \langle \nu, (X^\dagger)^2 \nu \rangle = \left\langle \begin{pmatrix} 1 \\ 0 \end{pmatrix}, \begin{pmatrix} -1 & 0 \\ 0 & -1 \end{pmatrix} \begin{pmatrix} 1 \\ 0 \end{pmatrix} \right\rangle = -1, \end{aligned}$$

and so on. Thus, $\omega \cong \cos(x)$, as one might have anticipated, and $\mathcal{A}$ is the algebra of trigonometric polynomials. Writing u for ω and v for $X\omega$, we see that $\mathcal{A} \cong \mathbb{C}[u, v]/(u^2 + v^2 - 1)$, and M is a circle.

In the previous example the algebra $\mathcal{A}$ was finitely generated, but non-integrable representations are necessarily infinite dimensional. Thus, an important problem is to find an adequate topology for it. Then, assuming a suitable topology has been defined, an obvious question arises: will the space of continuous characters be a smooth manifold? If not, a study of its possible singularities should shed some light on the structure of non-integrable representations of real Lie algebras.

References

1. J. Dixmier, *Enveloping algebras*, Graduate Studies in Mathematics, 1996.
2. M. Harish-Chandra, *Representations of a semisimple Lie group on a Banach space. I*, Transactions of the American Mathematical Society (1953), 185–243.
3. Z. Magyar, *Continuous linear representations*, vol. 168, North Holland, 1992.
4. S. Montgomery, *Hopf algebras and their actions on rings*, vol. 82, American Mathematical Society, 1993.
5. K.H. Neeb, *On analytic vectors for unitary representations of infinite dimensional Lie groups*, Annales de l'Institut Fourier **61** (2011), no. 5, 1839–1874.
6. E. Nelson, *Analytic vectors*, Annals of Mathematics **70** (1959), no. 2, 572–615.
7. K. Schmüdgen, *Unbounded operator algebras and representation theory*, Birkhäuser, 1990.

8. R. Vargas Le-Bert, *Integrable representations of involutive algebras and Ore localization.* Algebras and Representation Theory **15** (2012), no. 6, pp. 1147–1161.